Human Ageing

A Unique Experience

Implications for the Disease Concept

Human Ageing

A Unique Experience

Implications for the Disease Concept

Michael A Singer

Queen's University, Canada

World Scientific

NEW JERSEY • LONDON • SINGAPORE • BEIJING • SHANGHAI • HONG KONG • TAIPEI • CHENNAI

Published by

World Scientific Publishing Co. Pte. Ltd.
5 Toh Tuck Link, Singapore 596224
USA office: 27 Warren Street, Suite 401-402, Hackensack, NJ 07601
UK office: 57 Shelton Street, Covent Garden, London WC2H 9HE

Library of Congress Cataloging-in-Publication Data
Singer, Michael A. (Michael Allan), 1940– author.
Human ageing : a unique experience : implications for the disease concept / Michael A. Singer.
p. ; cm.
Includes bibliographical references and index.
ISBN 978-9814619127 (hardcover : alk. paper) -- ISBN 978-9814619134 (electronic)
I. Title.
[DNLM: 1. Aging--physiology. 2. Environment. 3. Gene Expression Regulation.
4. Socioeconomic Factors. WT 104]
QP86
612.6'7--dc23
2014025860

British Library Cataloguing-in-Publication Data
A catalogue record for this book is available from the British Library.

Typeset by Stallion Press
Email: enquiries@stallionpress.com

Printed in Singapore

Abstract

The ageing process can be understood at the level of both the species and the individual. Although much has been written about ageing at the species level, ageing at the level of the individual human has not received the same focus. Each species has evolved a characteristic maximum lifespan and ageing pattern with these phenotypic traits set by species-specific developmental programs. Within these broad species-set boundaries for maximum lifespan and ageing trajectory, there is considerable variation across individuals. Each individual will have their own developmental programs based upon the unique genome and epigenome of that individual. Developmental programs are plastic and can be modified by environmental cues acting at critical time windows of the life cycle such as the gestational and early postnatal periods. Individual longevity and ageing patterns are also shaped by experiences occurring in childhood and adult life. Each individual responds to objective experiences in a specific manner such that the subjective experiences of each individual are unique. The influence of experiences on phenotypic traits is generally mediated by epigenetic reprogramming of gene expression patterns.

Several themes are developed in this book. One theme is that the ageing pattern of each individual is unique. A second theme is that it is impossible to dissect the ageing process from the living state itself; there are no specific ageing or longevity genes. A third theme is that the ageing process is molded by the social and cultural context in which the individual lives. Culture is an inheritance system unique to our species and gene–culture co-evolution is the major driver of human evolution. A framework for conceptualizing the ageing process is developed and the disease construct is examined using this framework as a lens. The disease construct dominates the way our medical care system has been organized and the ways in which physicians deliver medical care. The disease construct has also been a major determinant of how we educate future physicians.

Contents

Introduction

I have titled this book, *Human Ageing: A Unique Experience*. I have chosen this title for very specific reasons. Although much has been written about ageing at the species level, ageing at the level of the individual animal has not received the same focus. The observation that there is extensive genomic and epigenomic variation across individuals and that environmental cues can reprogram epigenetic marks indicate that each individual human will have a unique ageing phenotype in terms of longevity and ageing trajectory. Given that epigenetic marks are considered reversible, aspects of an individual's ageing phenotype could be potentially modified. The uniqueness of each individual's ageing pattern is the primary theme explored in this monograph and hence the title.

Ageing and longevity can be conceptualized at both the species and individual level (Singer, 2013). Species-specific longevity and, by inference, ageing are developmentally programmed (Singer, 2013). Across taxonomic groups and species within a given group, there are vast differences in maximum lifespan (Deweerdt, 2012). In mammals for example, the range of maximum lifespan encompasses four years for the mouse to over 200 years for the bowhead whale. Among primates, humans are developmentally programmed to live longer than other primates (e.g., the lifespan of a chimpanzee is about 60 years and that of humans about 100 years). Within the rodent family itself, the lifespan of four years for the mouse is completely eclipsed by the 30-year lifespan of the naked mole rat (Deweerdt, 2012). Although the link between lifespan and ageing pattern is complex, animals that are very long lived tend to show negligible senescence. One observation based upon the study of natural animal models displaying an extensive variation in both lifespan and ageing pattern is that no single mechanism accounts for this variation (Singer, 2013). The biology of senescence is highly complex and involves perturbations in multiple intersecting signaling and metabolic networks. Hence, it is hard to imagine that some type of "master switch" will be responsible for the extreme variations in species-specific longevity (Deweerdt, 2012). The classical evolutionary theory of ageing posits that as a consequence of the decline of natural selection with age, the onset of senescence corresponds to the time of reproductive maturity (Rebke *et al.*, 2010). However, there are many exceptions to this construct. The common tern is a long-lived bird that reaches reproductive maturity at about the age of three. However, in this species, individuals improve in reproductive performance until the age of 14 to 15 before a decline occurs (Rebke *et al.*, 2010). The albatross is another long-lived bird which reaches reproductive maturity between the ages of five and nine. One particular albatross named "Wisdom" has been able to successfully breed at age 62, her current age (U.S. Geological Survey, 2013).

Although maximum lifespan is developmentally set for a given species, within that species there will be considerable variation in both lifespan and ageing phenotype across individuals. Ageing at the species and individual level are mechanistically intertwined and in essence inseparable. A good example illustrating the mix of species and individual components of the ageing process is the domestic dog (Kraus *et al.*, 2013; Galis *et al.*, 2007). Across mammalian species, there is a strong positive correlation between body size and lifespan. However, if one examines the data relating lifespan and size in the species (domestic) dog, the opposite relationship holds. Across breeds within this species, there is a negative correlation between size and lifespan; i.e., larger dogs die earlier than smaller dogs (Kraus *et al.*, 2013). There appear to be several reasons underlying this observation. Larger dogs show an increased rate of ageing and a higher death rate from cancer than smaller dogs. However, in contrast, within breeds, larger dogs live longer on average than smaller dogs (Galis *et al.*, 2007). The situation in dogs is further complicated by the extreme artificial selection that has occurred within this species giving rise to roughly 400 breeds known today. Many of these artificially selected breeds have maladaptive traits that would have been eliminated in nature, but persist under the artificial selection provided by human care (Galis *et al.*, 2007). Regardless, the dog as a species has a programmed maximum lifespan, but different breeds within the species and individuals within each breed show quite variable lifespans and ageing trajectories.

The objective of this review is to develop several themes. One of these themes is that it is impossible to dissect the ageing process from the living state itself. Animals are exposed to external circumstances and agents that cause damage to biological systems. Damage to biological systems can also arise as a byproduct of the biological processes themselves. Genomic structural variants (mutations) are an ongoing feature of animals that undergo DNA replication and reproduction. Various DNA repair mechanisms are in place to minimize mutational damage as a result of errors that occur during reproduction. However, repair mechanisms are imperfect and damage cannot always be completely repaired. In addition, damage resulting from biological processes may be a consequence of trade-offs that are made during the course of evolution. A second important theme is that the pattern of the ageing process is strongly determined by environmental cues and how the individual animal responds to those cues. Each animal is unique by virtue of its own genome and epigenome and each animal's experiences (responses to environmental cues) are also unique. Experiences (environmental cues) modify gene expression patterns through epigenetic mechanisms, and the resultant mix of epigenetic marks will be specific for each animal. Hence, within broad species-determined boundaries, the ageing pattern will vary considerably across individuals and be unique for each individual. Finally, a third theme is that the ageing process will be molded by the social and cultural context in which the animal lives. For humans, this social context consists of a complex urban environment containing a high density of people with relatively ready access to food, shelter, and medical care. Culture is a system of inheritance that is unique to the human species and gene–culture co-evolution is the main driver of human evolution.

It is important to remember that human structural and functional biology has been shaped by evolutionary processes. To fully understand human biological characteristics, one must dissect out their evolutionary roots, which means studying these biological features across many animal species. Evolution is a "tinkerer" and uses the material at hand to create something worthwhile (Flicek, 2013). This is an important point. Evolution is a conservative process that builds upon successful body designs; evolution is generally not characterized by sudden giant leaps. Within this context, the human represents a "tinkered" form built upon the designs present in our ancestral lineage. In addition, across the diversity of organisms, the "problem-solving" capacity of evolution is incredible. For example, multicellular animals and multicellular plants need to supply all of their cells with fuel and water. For animals, the solution was the evolution of a vascular system with a pump to circulate isotonic blood plasma that prevented cell rupture through the osmotic inflow of water. Plants took a different route to solve both the problem of osmoregulation and the circulation of fluid. Plant cells are rigid due to an exoskeleton and evolved a circulatory system designed to transport nearly pure water driven by the force of surface tension acting on the cohesion of water molecules to the conduit walls — the cohesion–tension theory (Tyree, 2003).

The concept of evolution as a "tinkerer" is an important one and deserves a digression within this introductory section. One example of evolution as a "tinkerer" was described by Cotney *et al.* (2013), who examined the molecular and genetic mechanisms that have driven human-specific morphologic features.

Cotney *et al.* (2013) looked at embryonic limbs in the human, rhesus monkey, and mouse at multiple stages of development. Compared with other mammals and primates, humans have several specific limb characteristics including the relative lengthening of the thumb and multiple adaptations of the foot to support upright walking. They measured the chromatin modification, acetylation of lysine 27 on histone 3 (H3K27ac). This chromatin mark identifies active promoters and enhancers (gene regulatory regions) and correlates with the level of nearby gene expression. H3K27ac regions within 1 kb upstream of a transcription start site were defined as promoters, whereas intergenic or intronic H3K27ac regions were defined as enhancers. The level of chromatin H3K27ac was measured in the embryonic limb of each species at comparable time points from the bud stage through to the onset of digit separation. The differences between the extents of H3K27ac marks in the three species at different embryonic time points were assessed; these differences in the extent of H3K27ac marking reflect differences in the numbers of active promoters and enhancers between the three species at that stage of limb development. Increases in H3K27ac signaling between the human compared to the monkey and mouse were considered gains in promoters and enhancers by the human lineage. At the earliest developmental time point examined, the human embryonic limb showed only a 2% increase in numbers of active promoters and a 7% increase in numbers of active enhancers compared to orthologous promoters and enhancers in the rhesus monkey and mouse. At the latest time point examined, the difference had grown to a 16% increase in the numbers of both promoters and enhancers. Genes associated with these gained promoters and enhancers in the human lineage were enriched for functions related to proliferation and growth, bone morphogenesis, and connective tissue. Cotney *et al.* (2013) examined the evolutionary origin of these human lineage gains in H3K27ac-marked promoters and enhancers. The gained human enhancers were evolutionarily young, dating back to the common ancestor of the opossum and placental mammals. In contrast, the orthologous enhancers across all three species that did not show a specific increase in the human lineage were significantly older. The same held for gained human promoter sequences; they had a more recent evolutionary origin than the H3K27ac-labeled promoters common to all three species. Human lineage gains (increase in numbers of promoters and enhancers) could have arisen from modification of an existing ancestral regulatory element, co-option of a regulatory element active in another tissue, or *de novo* formation. For gained human lineage promoters, about 91% arose by modification of ancestral elements, whereas for gained human lineage enhancers, modification of existing ancestral enhancers accounted for 62% of the increase. In general, gained human promoters and enhancers did not display a human-specific increased nucleotide substitution rate or enrichment for particular transcription factor motifs or repetitive elements. The human lineage gains at promoters and enhancers were associated with significant increases in the expression of 302 genes in the human compared to the orthologous genes in the mouse.

The observations of Cotney *et al.* (2013) indicate that the developmental processes resulting in unique human morphologies such as limb architecture are orchestrated by regulatory activities that have arisen primarily from modification of ancestral regulatory regions. Interestingly, in previous publications (e.g., Pollard *et al.* 2006; Amadio and Walsh, 2006), so-called human accelerated genomic regions (HARs) were described. These are regions of non-coding DNA that show an accelerated rate of nucleotide substitution in the human lineage since divergence from the common ancestor with the chimpanzee. Many of these HARs are involved in transcriptional regulation and neurodevelopment. Pollard *et al.* (2006) described one HAR, denoted as HAR1 which contained a non-coding gene *HAR1F*. This gene was strongly expressed in the embryonic neocortex of the human early in development. Pollard *et al.* (2006) speculated that since about one quarter of identified HARs were in the vicinity of a neurodevelopment gene such as *HAR1F*, they may be responsible for the evolved unique features of the human brain. Cotney *et al.* (2013) point out that the HARs described by Pollard and others are essentially DNA sequences with human specific regulatory activity that can be identified by H3K27ac marking. According to Cotney *et al.* (2013), these HARs are actually human-specific enhancers. Therefore, in essence, the evidence is growing that morphological features "unique" to the human have primarily evolved through the activity of human lineage-specific genomic regulatory elements (enhancers and promoters). These elements have arisen largely from modifications to existing ancestral regulatory regions rather than being due to the emergence of *de novo* genomic sequences.

This book has been organized such as to develop the themes referred to in the previous paragraphs. Ageing is presented as a journey, but not a linear one. In fact, the pattern of this journey can be characterized as consisting of concentric circles. The center of these concentric circles has been set as the well-described premature ageing disorder — Hutchinson–Gilford Progeria syndrome (HGPS). This choice was somewhat arbitrary on my part, but I suspect that no matter what ageing disorder I used as a center point, the radiation out from this point would lead to the same web of biological processes that occurs with HGPS as the center point.

HGPS is the result of a mutation in the gene encoding nuclear lamins A/C. From this central point, the path of the journey radiates out concentrically to encompass a web of biological processes that connect the domain of lamin biology with other biological domains and processes. This web of interacting biological processes comprises the innermost circle. A larger concentric circle "contains" experiences that are inputted into that individual. Through epigenetic mechanisms, these experiences can alter gene expression patterns in that individual and hence modulate the ageing process. An even larger concentric circle encompasses the environmental–social–cultural landscape in which that individual lives. This landscape plays a very special role in human evolution through the process of gene–culture co-evolution, a process unique to the human species. As discussed in subsequent sections, gene–culture co-evolution is the major driver of human evolution. The sum total of all of these concentric circles "defines" the ageing process. Although for each species this journey has a set of broad boundaries, within these broad species-specific boundaries the journey is unique for each individual member of that species.

In **Chapter 1**, HGPS, which is the starting point of the journey, is described. This syndrome exemplifies a premature ageing state and the underpinnings of this syndrome are abnormalities in lamin biology. From this starting point, the various links that connect lamin biology to other biological processes such as alternative splicing, telomere maintenance, mTOR signaling, and molecular clocks are outlined. This web of multiple threads weaves a fabric (a complex road map) which is an integral part of both the living state and the ageing process. The biological processes that maintain cellular and tissue functions are, by virtue of various necessary trade-offs and functional errors, also responsible for the ageing of these cells and tissues. Since all of us are born with unique genomes, this interwoven fabric of biological processes will not be identical across individuals. **Chapter 1** explores the first theme, in which the biological processes that maintain the living state are the same set of processes that are associated with ageing.

In **Chapter 2**, the effects of environmental cues or experiences on ageing are explored. Experiences can be decomposed into sensory components that are inputted through associated sensory pathways. The activation of central neurons upregulates the transcription of immediate early genes and through epigenetic mechanisms influences processes such as adult neurogenesis and synaptic (structural and strength) plasticity. This sequence of steps links objective experiences to cognitive functions such as learning and memory. The nature of this sequence of steps differs across individuals so that two individuals can have the same objective experience but very different subjective experiences. Differing subjective experiences help to explain individual variations in the ageing process.

Nutrition is a critical environmental cue, and changes in the type of nutrition and/or the life cycle stage of the animal at which nutritional changes occur can have profound phenotypic effects. These phenotypic effects include metabolic programming as well as resetting of the function and morphology of organs such as those comprising the cardiovascular system.

In **Chapter 3**, the process of domestication is discussed. The behavioral phenotypes characterizing domesticated animals are developmentally determined. In addition, domesticated animals, particularly the dog, experience human-type health issues and hence, domesticated animals represent a valuable ageing model. The social–cultural context in which humans reside is an important factor molding the ageing process. Cultural evolution has been an important driver of human evolution through the process of gene–culture co-evolution. This process is unique to humans. This social–cultural context is described in **Chapter 4**. In **Chapter 5**, the various themes and ideas explored in **Chapters 1 to 4** are woven together to provide a conceptual framework for the ageing process. Finally, in **Chapter 6**, the implications of ageing as a process unique to each individual, for our understanding of what constitutes a disease, are discussed. The question as to whether current molecular biology renders the disease concept outmoded is also discussed in this chapter.

Human ageing can be considered on multiple levels: in the context of mammals in general, at the level of a specific mammalian species, and at the level of the individual human. The evidence indicates that no "single" biological process is preeminent in accounting for the ageing phenotype. Every individual has a unique genome and epigenome, and an individual's interaction with environmental cues within their social context will result in epigenetic reprogramming specific for that individual. Hence, the ageing phenotype both in terms of its trajectory and underlying mechanisms will differ between individuals.

Even though this review is quite focused, the complexities of the biological processes discussed as well as the many unknowns concerning these processes are quite apparent. However, as pointed out by Ball (2013), many scientists instead of embracing these unknowns as a challenge, sometimes present our current understanding of the relationship between genomic and epigenomic mechanisms and phenotypic expression in an overly simplistic and somewhat sanitized manner. Perhaps this oversimplifying approach is most evident in the area of ageing in which scientists tend to promote unitary mechanisms such as oxidative stress, telomere dysfunctioning, or "chronic inflammation" as primarily responsible for senescence.

One additional note: the fields of genomics, proteomics, transcriptomics, and related disciplines are rapidly expanding. It is likely that many of the current interpretations of experimental observations reviewed in this book will also change. However, the trend line is one of increasing complexity. That is to say, future research in these disciplines will no doubt uncover layers of complexity in the molecular biology of the living state that are not appreciated today. I believe that the discovery of new layers of complexity will actually add support to the themes developed in this book and corroborate the proposal that chronic degenerative diseases are variants of the ageing process.

CHAPTER 1

The Biology of Ageing

Lamins: Hutchinson–Gilford Progeria Syndrome

Recently, alterations in the functioning of lamin proteins have been implicated in contributing to the ageing phenotype. This involvement of lamins, which are important components of the nuclear envelope, in ageing was revealed as a result of a better understanding of the genetic underpinnings of the rare syndrome known as Hutchinson–Gilford Progeria (HGPS). Research into the genetic mechanisms underlying HPGS has uncovered a web of connections between lamins, telomeres, alternative splicing, gene expression patterns, and the ageing process. Although HGPS is rare, understanding the underlying molecular biology can translate into important insights into the ageing process in general.

Genetics

HPGS is an autosomal dominant genetic disorder expressed as an accelerated ageing phenotype. Individuals with HPGS generally die at about 13 years of age of cardiovascular complications (Olive *et al.*, 2010). The genetic basis of HPGS was discovered in 2003: sporadic *de novo* mutations in the *LMNA* gene with the most common variant being a C to T nucleotide substitution at position 1824 of the coding sequence in exon 11 (D'Apice *et al.*, 2004; Gordon *et al.*, 2012).

LMNA encodes three different proteins: lamin A, lamin C, and progerin. Alternative splicing gives rise to lamins A and C. Lamin A and lamin C together with lamin B form a dynamic meshwork located just inside the nuclear membrane (Cao *et al.*, 2011). The C>T substitution mutation at position 1824 activates a cryptic 5′ splice site (5′ss) within exon 11 of *LMNA* and this activation leads to transcription of a lamin A mRNA that contains a 150-nucleotide deletion (^150 LMNA mRNA). This mRNA is translated into the protein, progerin, which lacks 50 amino acids near the C terminus. D'Apice *et al.* (2004) studied three nuclear families with an affected (HGPS) offspring due to a C>T substitution in the *LMNA* gene, at position 1824 exon 11. In all three families, the *de novo* mutation was transmitted by the father to the offspring.

The post-translational handlings of the lamin A protein and progerin differ as a result of this amino acid deletion. Immature lamin A (prelamin A) bears a Ca1a2X motif at the carboxy terminus where C is cysteine, a1 and a2 are aliphatic amino acid residues, and X is usually methionine. Prelamin has a farnesyl group added to the cysteine residue followed by cleavage of the last three amino acid residues (a1a2X). Next, the carboxy terminus undergoes methylation carried out by isoprenyl-cysteine carboxyl methytransferase (ICMT). Farnesylation and carboxy methylation function to tether the protein to the inner surface of the nuclear membrane (Gordon *et al.*, 2012; Lopez-Mejia *et al.*, 2011). The protein is subsequently released from this tether by a zinc metalloprotease (Zmpste 24) which removes the 15 amino acids at the C-terminus including the farnesyl and carboxymethyl groups. On the other hand, the truncated progerin protein, due to the deletion of 50 internal amino acids, resists cleavage by the enzyme Zmpste 24 and hence remains permanently farnesylated and carboxymethylated, resulting in abnormal anchoring to the nuclear membrane throughout the cell cycle (Lopez-Mejia *et al.*, 2011). Unlike the A-type lamins, B-type lamins remain farnesylated.

Rusinol and Sinensky (2006) examined the steps involved in the post-translational modification of (pre)lamin A. The farnesylation of the cysteine residue is directed by the Ca1a2X motif at the carboxy terminal. The presence of a methionine residue at the X position in human prelamin A determines the preference for farnesylation over geranylgeranylgeranylation. Methylation of the carboxy terminus can only occur after cleavage of the last three amino acid residues (a1a2X). Two enzymes can accomplish this step, one of them being Zmpste24. The second cleavage step can only be accomplished by Zmpste24, and this step does not occur in progerin since the 50-amino acid deletion contains the cleavage site. According to Rusinol and Sinensky (2006), lamin A is the only farnesylated mammalian protein known to undergo a second upstream proteolytic cleavage step. Protein farnesylation and carboxymethylation add hydrophobic groups which facilitate protein–protein and protein–membrane associations. Presumably, the initial farnesylation and carboxymethylation of lamin A are necessary for membrane attachment, but the reasons for the subsequent removal of these two hydrophobic moieties are unclear. The farnesyl and carboxymethyl groups are not necessary for the assembly of (mature) lamin A into the lamina since (mature) lamin A is disassembled and reassembled into the nuclear lamina during the course of mitosis.

Recently, the three-dimensional crystal structure of Zmpste24 has been determined (Michaelis and Hrycyna, 2013). The biological reasons underlying why cells go to the "expense" of farnesylating and carboxymethylating lamin A only to subsequently remove these groups remain a mystery. The zinc metalloprotease Zmpste24 is a bulky transmembrane protein complex containing a large hollow barrel-shaped chamber which encloses the active zinc catalytic site. The structure is consistent with the following reaction model. The C-terminus of lamin A enters the chamber and the a1a2X tripetide is cleaved. Lamin A is then carboxymethylated while in the chamber. The speculation is that the lamin A remains within the chamber and subsequently undergoes the second cleavage with the release of the last 15 amino acids at the C-terminus including the farnesyl and carboxymethyl groups. The proposal is that the large chamber functions to sequester the farnesylated 15-amino acid lamin tail where it is further degraded. Such a mechanism would be necessary if the farnesylated tail itself had "toxic"properties (Michaelis and Hrycyna, 2013).

The production of progerin protein is not restricted to individuals suffering from HGPS. Scaffidi and Misteli (2006) examined for nuclear defects and accumulation of the progerin protein in normal human skin fibroblasts. Cells from individuals with HGPS display a dysmorphic nuclear shape, downregulation of several nuclear proteins (heterochromatin protein HP1 and LAP2 group of lamin A-associated proteins), and reduced lysine methylation on histone H3. Dermal fibroblasts from healthy old individuals (81 to 96 years of age) showed similar nuclear abnormalities. Scaffidi and Misteli (2006) demonstrated very small quantities of ^150 lamin A mRNA and progerin protein in dermal fibroblasts and liver and heart tissues from healthy individuals of all ages. In dermal fibroblasts, the relative amount of ^150 LMNA mRNA was about 50-fold less than that occurring in dermal fibroblasts from individuals with HGPS. These observations indicate that the cryptic 5′ss in exon 11 is also utilized in healthy individuals of all ages. Although small quantities of progerin protein were found in dermal fibroblasts of healthy individuals, the effects of this progerin protein varied with the age of the individual. In cells from young donors, lamin A and lamin C were diffusely distributed throughout the nucleoplasm, whereas in cells from old individuals, lamin A and lamin C accumulated at the nuclear rim (a distribution similar to that seen in fibroblasts from HGPS individuals). These nuclear abnormalities in cells from old donors could be reversed by inhibiting utilization of the cryptic 5′ss using a targeted morpholino oligonucleotide, thus confirming that the nuclear changes were due to the production of the progerin protein.

Alternative splicing

In order to better appreciate the specifics of alternative splicing that pertain to the HGPS phenotype, it is useful to first discuss the process of alternative splicing within a broader context. Although vertebrate species display very diverse phenotypic characteristics, they share similar sets of protein coding genes. Evolved species-specific gene regulatory mechanisms are thought to be part of the answer to this apparent paradox. Another biological process that can account for phenotypic differences without invoking changes in protein-coding genes is that of alternative splicing. This process allows a single gene to code for multiple proteins; particular exons may be included or excluded from the transcribed pre-mRNA to form alternatively spliced variants of the final mRNA. Proteins translated from alternatively spliced mRNAs will have different amino acid sequences and will often have different biological functions. The production of alternatively spliced mRNAs involves the action of *trans*-acting regulatory proteins binding to *cis*-acting sequences on the pre-mRNA itself. The complex structure containing the *trans*-acting proteins is known as the spliceosome and the binding of this complex to *cis*-acting sites determines the specific locations where cleavage occurs and hence, which exons are included in or excluded from the final mRNAs.

Barbosa-Morais *et al.* (2012) have surveyed the process of alternative splicing across multiple organ tissues derived from different vertebrate species: human, macaque, mouse, opossum, platypus, chicken, lizard, and frog. The frequency of alternative splicing (percent of surveyed internal exons that were spliced) was much higher in primate organs compared to the organs of the other species. Alternative splicing profiles across organs segregated according to species, whereas gene expression (transcript) levels segregated according to organs rather than species. Hence, alternative splicing patterns reflect phylogenetic differences between species better than gene expression levels. The species-specific patterns observed with alternative slicing were predominantly due to changes in *cis*-regulatory sequences rather than in *trans*-acting factors. These observations indicate that the process of alternative splicing makes a significant contribution to the genetic underpinnings of phenotypic diversity across species. Among various vertebrate species, the contribution of alternative splicing to phenotypic diversity varies; primates demonstrated the highest frequencies of alternative splicing compared to other species surveyed.

Patterns of alternative splicing have also been examined specifically in human populations (Gonzalez-Porta *et al.*, 2012; Lappalainen *et al.*, 2013). Gonzalez-Porta *et al.* (2012) used RNA sequencing data from lymphoblastoid cell lines from two HapMap populations: Nigerian and Caucasian. These data gave a measure of each gene's transcriptional output which encompasses the sum total expression of all the spliced variants coded by that gene. The analysis was restricted to protein-coding genes. Variation in the abundance of alternative spliced transcripts of a given gene (across individuals in the study population) could be due to variation in transcriptional output or variation in splicing ratios. The model developed by the investigators allowed them to assess the relative contributions of these two components. They constructed a ratio (range of values from zero to one) that estimated the relative contributions of these two components. A value of one indicated that the variation in abundance of transcripts was due only to variation in transcription output. A value of zero indicated that the variation in abundance of transcripts was due only to the variation in splice ratios. Within both populations (Nigerian and Caucasian), there were many genes that showed very little variation in splice ratio or a constant splice ratio. In addition, approximately 10% of the genes assayed exhibited population specific splicing ratios. On average, within both populations, about 60% of variation in the abundance of transcripts was due to variation in expression. Most of the remaining 40% of variation was due to variation in splice ratios. Variation in splicing without variation in expression appeared to be very uncommon, while the converse appeared to be quite common.

Further in 2013, Lappalainen *et al.* measured alternative splicing patterns based on RNA sequencing data of lymphoblastoid cell lines from five populations: four from Europe and one from Nigeria. They used the methodology of Gonzalez-Porta *et al.* (2012) to separate the contributions of variation in splice ratios and variation in transcriptional output to the variation in the abundance of alternative spliced transcripts for a given gene. In their study, Gonzalez-Porta *et al.* found that the distribution of variation in splice ratios and transcriptional output to variation in total transcript abundance within each population was quite asymmetric with a skew in the direction of genes in which most of the variation in total transcript abundance was due to variation in transcriptional output. There were very few genes in which most of the variation in transcript abundance was the result of variation in splice ratios, but there was a significant number of genes which essentially showed no variation in splice ratios, and in this case, variation in transcript abundance was completely due to variation in gene transcriptional output. In the study by Lappalainen *et al.* (2013), within each population, variation in splice ratios significantly contributed to variation in the transcript abundance .The distribution pattern was similar for all of the populations and showed a skew in the direction of genes in which most of the variation in transcript abundance was due to variation in splice ratios. This observation is somewhat different than that reported by Gonzalez-Porta and co-workers for their two study populations, one of which was Nigerian. Lappalainen and co-workers also found that about 3% of the variation in transcript abundance was population-specific. In addition, they identified a small subset of genes that showed differences in variation of transcriptional output and/or variation of splice ratios between population pairs, i.e., population-specific differences. In particular, when they compared Nigerian and European population pairs, 75–90% of the assayed genes showed different splice ratio variation, whereas when European population pairs were compared, only 40–75% of assayed genes showed different splice ratio variation. This observation indicates that between African and European populations, variation in splice ratios is the dominant component contributing to variation in transcript abundance, whereas between European populations, variation in transcript output generally contributes more importantly to the variation in transcript abundance.

The observation that the patterns of pre-mRNA alternative splicing differ between vertebrate species and between certain human populations underscores the role of alternative splicing in the regulation of gene expression. However, as noted by Gonzalez–Porta *et al.* (2012), splicing is governed by generic conserved sequence motifs (splice sites), which are under strong purifying (negative) selection.

Within this broad context, the key to understanding the biological processes underlying the HGPS phenotype starts with defining why an SNP (single-nucleotide polymorphism) at position 1824 of the *LMNA* gene activates a cryptic 5′ss.

Alternative splicing of pre-mRNA is carried out by a structure known as the spliceosome (Chen and Manley, 2009). The first step in the splicing process involves splice-site recognition. Splice-site recognition is achieved through the binding of small ribonucleoprotein particles (snRNPs) to specific locations on the pre-mRNA. The intron is defined by an upstream 5′ss and a downstream 3′ss. A branch point is located just upstream of the 3′ss. U1 snRNP binds to the 5′ss, while other snRNPs (U2 snRNP, U2 auxiliary factor) bind downstream to the branch point and the 3′ss. The complex formed by these snRNPs (plus other auxiliary proteins) bound to the 3′ss, 5′ss, and branch point is known as the spliceosome. The complex subsequently undergoes conformational changes and remodeling to become catalytically active. Two transesterification steps (phosphoryl transfer reactions) take place with the cleavage of the intron at the 5′ss and 3′ss. The spliceosome uses magnesium ions at its active site as catalytic cofactors, and recent evidence indicates that the ligands coordinating the catalytic metal cofactors are the snRNAs and not proteins (Strobel, 2013). The protein components of the spliceosome appear to be mainly involved in regulation, in determining splice-site specificity, and acting as a scaffold to promote the proper conformation of the snRNAs (Strobel, 2013). Which exons are included and which exons are removed during the splicing process will be determined by the regulation of splice-site recognition and selection. This process involves a number of *trans*-acting protein regulators acting on *cis*-regulatory elements in the pre-mRNA (Chen and Manley, 2009).

The *cis*-regulatory elements consist of exonic splicing enhancers (ESEs) and silencers (ESSs) and intronic splicing enhancers (ISEs) and silencers (ISSs). Enhancer elements are generally regulated by members of the SR (serine/arginine-rich) protein family, while silencer elements are generally regulated by hnRNPs (heterogeneous nuclear RNPs). SR proteins bind to an ESE and by so doing recruit snRNPs to the 5′ss and 3′ss. Other proteins such as T-cell restricted intracellular antigen 1 (TIA-1) bind to an ISE and recruit snRNPs to the nearby upstream 5′ss. There are a number of mechanisms by which proteins such as hnRNPs can inhibit splice-site recognition and selection (Chen and Manley, 2009). One mechanism involves the binding of specific hnRNPs to the *cis*-regulatory elements ISSs and ESSs. This binding prevents the recruitment of components of the spliceosome to the splice site. There is also evidence that the enzyme RNA polymerase II (RNAP II) is involved in the regulation of alternative splicing. One possibility is that RNAP II and different transcription factors can influence the recruitment of various splicing factors to the pre-mRNA molecule. Alternative splicing is also both tissue-specific (Chen and Manley, 2009) and species-specific (Papasaikas and Valcarcel, 2012). In summary, the spliceosome recognizes the sequence boundaries (splice sites) between exons and introns. Regulation of splice-site selection involves the interplay between *cis*-regulating elements (enhancers and silencers) in introns and exons and *trans*-acting proteins that recognize these elements to promote or inhibit spliceosome assembly on particular splice sites (Tejedor and Valcarcel, 2010). If the selected 5′ss and 3′ss straddle an exon, then that exon will be removed along with its associated introns. In the case of the *LMNA* gene, splicing at the 5′ss of the "regular" exon 11 removes only intron 11, while splicing at the progerin 5′ss, which is located within exon 11, removes 150 nucleotides from the exon as well as intron 11.

Lopez-Mejia *et al.* (2011) have done a detailed examination of the differences in usage between the lamin A 5′ss and the progerin 5′ss (cryptic 5′ss). The lamin A 5′ss is located at the boundary between exon 11 and intron 11, and was found in a highly accessible single-strand region. The progerin 5′ss is located within exon 11 between positions 1804 to 1851, and was "buried" in the stem of a compact two-dimensional structure. This sequestration of the progerin 5′ss by the surrounding RNA structure would significantly reduce its usage compared to the more readily accessible lamin A 5′ss. Lopez-Mejia *et al.* (2011) observed that the mutation at position 1824 characteristic of HGPS induced an opening of the secondary RNA structure around the progerin 5′ss, making this cryptic 5′ss more accessible. Two SR proteins, SRSF1 and SRSF6, were able to bind to the progerin 5′ss. SRSF6 acted to repress the usage of the progerin 5′ss, while SRSF1 mainly acts to increase utilization of the lamin 5′ss.The data indicated that the progerin 5′ss is a bona fide but naturally weak 5′ss and not a *de novo* creation of the HGPS mutation. The HGPS mutation increases the selection of the progerin 5′ss by opening up the enclosing secondary RNA structure and thus allowing access of snRNPs to the splice site. In addition, differential selection between the progerin 5′ss and the lamin A 5′ss is at least partially regulated by the SR proteins SRSF1 and SRSF6 which have opposing effects on *LMNA* gene splicing.

Telomere connection

Cao *et al.* (2011) examined the relationship between telomere dysfunction and utilization of the progerin 5′ss. Dermal fibroblasts from normal donor individuals (10 to 92 years of age) showed distinct but low levels of utilization of the progerin 5′ss; the extent of utilization was independent of donor age but did increase with the number of cell passages. In addition, fibroblasts that utilized the progerin 5′ss had on average shorter telomeres than fibroblasts which did not display activity of this cryptic 5′ss. Fibroblasts with engineered ectopic expression of telomerase (enzyme complex that elongates telomeres) showed much less progerin production (measured as progerin mRNA) than parent cells that did not express telomerase. In addition, fibroblasts that had been genetically modified to have telomeres with their capping proteins dislodged, had increased progerin production compared to control cells. These observations support an association between shortened or disrupted telomeres and the extent to which dermal fibroblasts utilize the progerin 5′ss. Cao *et al.* (2011) also observed that fibroblasts with shortened telomeres displayed not only increased utilization of the cryptic progerin 5′ss, but also a broader set of changes in alternative splicing and gene expression. A common set of 82 genes were found that showed changes in alternative splicing when all the comparisons between fibroblasts with short versus long telomeres were pooled. These genes were primarily involved in cytoskeleton organization and function. Cao and co-workers postulated that shortened or uncapped telomeres acted as an upstream signal to change patterns of alternative splicing, including activation of the progerin 5′ss in dermal fibroblasts.

Two earlier reports are consistent with the observations of Cao *et al.* (2011). Jagatheesan *et al.* (1999), using both HeLa cells and a rat fibroblast line, demonstrated that antibody-labeled nuclear lamin A co-localized with two components of the spliceosome: SC-35, a member of the SR family and U5-16 kD, an snRNP. This observation implies that changes in the pattern of alternative splicing can be induced by lamin A itself in addition to shortened telomeres. Csoka *et al.*, (2004) examined gene expression profiles in early-passage fibroblasts from individuals with HGPS. They noted a set of 361 genes that were differentially expressed in these fibroblasts; 193 were upregulated while 168 were downregulated. The largest functional category of genes consisted of transcription factors. On the basis of the putative functions of the defined genes, Csoka *et al.* (2004) considered HGPS to be a systemic mesodermal dystrophy with associated early ageing of the cardiovascular system and connective tissues. These investigators were unable to clarify the mechanisms underlying the specificity of the gene expression pattern in HGPS, but several possible mechanisms have been described. First, considering the observations of Cao *et al.* (2011) and Jagatheesan *et al.* (1999), telomere- and lamin A-induced regulation of alternative splicing is one possible mechanism. A second mechanism is based on the observation that lamin A can bind to a number of transcription regulators (Vlcek and Foisner, 2007).

Decker *et al.* (2009) also examined telomere length in fibroblasts and hematopoietic cells from individuals with HGPS. In two patients, mean fibroblast telomere length was considerably shorter than those in the control samples, whereas for one patient, mean fibroblast telomere length was about the same as those of controls. For the two individuals with HGPS and short telomeres, there was considerable variation of telomere length between chromosomes. Also, signal-free chromosome ends, i.e., telomeres, too short to be detected by the technique used, were observed in the fibroblasts of all three individuals with HGPS but not in control fibroblasts. Hematopoietic cells (granulocytes, T lymphocytes) do not express the lamin gene or at much lower levels than fibroblasts. Mean telomere length was compared in hematopoietic cells (granulocytes, T lymphocytes, B lymphocytes) between four patients with HGPS and control individuals. Three out of the four HGPS subjects showed telomeres of similar length to those of controls. These observations imply that the expression of lamin A or progerin (mutant lamin A) is a general requirement for the generation of short telomeres in HGPS subjects.

Huang *et al.* (2008) measured the effects of wild-type and mutant lamin A on replicative senescence and telomere shortening in isogenic lines of human diploid fibroblasts. Transgenic fibroblasts containing either the wild-type lamin A gene or various mutant forms of the gene, including the mutation characteristic of HGPS, were created. The resultant fibroblast lines expressed wild-type lamin A or mutant forms of lamin A, equivalent to four copies of the lamin A gene. Hence, these fibroblast lines were overexpressing wild-type or mutant forms of lamin A. Six percent of the fibroblasts overexpressing wild-type lamin A showed misshapen nuclei compared to 2% of control fibroblasts. In fibroblasts containing the gene with the HGPS mutation, about one quarter showed misshapen nuclei. Fibroblasts overexpressing wild-type lamin A or mutant forms showed replicative senescence compared to parental control fibroblasts. Control fibroblasts had a replicative lifespan of at least 45 cumulative population doublings (CPDs), whereas for fibroblasts overexpressing wild-type lamin A or the HGPS mutant form of lamin A, the replicative lifespans were about 22 and 20 CPDs, respectively. In addition, fibroblasts overexpressing wild-type lamin A or mutant forms had accelerated rates of telomere shortening (loss) with successive cell divisions compared to parental control strains. Interestingly, the relative telomere length at which replicative senescence occurred was similar for control fibroblasts and fibroblasts overexpressing wild-type lamin A. Fibroblasts expressing mutant forms of lamin A showed replicative senescence at considerably shorter telomere lengths than control cells. In mammalian cells, telomeres are complexed within specific chromatin domains attached to the nuclear matrix. Huang *et al.* (2008) speculated that the overexpression of wild-type lamin A or the presence of mutant forms of lamin A could result in disruptions in chromatin–nuclear lamina interactions that normally protect telomere DNA. In such a situation, accelerated telomere shortening could occur with successive cell divisions and associated DNA replications.

Benson *et al.* (2010) looked at the effects of progerin on telomere function. Lamins have a structural role in supporting the nuclear envelope, but also subserve a number of other nuclear functions such as DNA synthesis and repair, RNA transcription, and maintenance of chromatin structure. When HGPS fibroblasts near the end of their replicative lifespan were transfected with a gene encoding TERT, the catalytic subunit (reverse transcriptase enzyme) of telomerase, they were able to replicate continuously. Hence, ectopic TERT expression extends the lifespan of HGPS fibroblasts even though these HGPS fibroblasts containing ectopic TERT continue to make the progerin protein. DNA damage triggers a response which involves the recruitment of specific proteins to the damaged site. HGPS fibroblasts show signs of DNA damage, but the extent of this DNA damage can be reduced by the ectopic expression of TERT. Hence, in HGPS fibroblasts, the expression of TERT can reduce the extent of foci of DNA damage as well as extend the replicative lifespan of the fibroblasts. The DNA damage observed in HGPS fibroblasts was localized to telomeric DNA as confirmed by the following experiment. Normal fibroblasts were transfected with a gene encoding for progerin and these transgenic cells developed evidence of DNA–telomere damage within 5 days. The DNA damage was detected using a stain for a protein that is recruited to double-strand DNA breaks, and the DNA damage was localized to telomeres by demonstrating that the damage marker co-localized with a specific telomere marker. Progerin-induced telomeric damage occurs time-wise prior to the actual loss of telomere length. Although the mechanism by which progerin induces telomere damage is unknown, a likely possibility is the disruption of normal chromatin structure at telomeres by the mutant lamin, progerin.

The experimental observations of *Cao et al.* (2011), Decker *et al.* (2009), Huang *et al.* (2008), and Benson *et al.* (2010) describe a complex interaction between lamins (wild-type and mutant), telomeres, and cellular senescence. HPGS fibroblasts have structurally altered (uncapped) and shortened telomeres. The telomere damage and loss induced by progerin disrupts normal chromatin–nuclear lamina interactions. Telomere damage and loss can also occur in fibroblasts which overexpress the normal lamin A protein. The occurrence of damaged and shortened telomeres is associated with extensive changes in alternative gene-splicing patterns. One of these changes in alternative splicing involves activation of the cryptic 5′ss in the lamin A gene. Cao *et al.* (2011) speculated that the signaling links between telomere damage and length and the spliceosome is mediated by components of the telomere-capping shelterin complex. The nuclear structural abnormalities due to the presence of mutant lamins plus the damage to and loss of telomeres are responsible for the early replicative senescence of HGPS fibroblasts. Since progerin is produced, although to a limited extent, in normal tissues, these observations have relevance to normal cellular senescence. The data indicate that wild-type lamins and progerin are important regulators of telomere integrity and length and indirectly a regulator of gene–spliceosome functioning.

Telomere biology

The biological roles of telomeres were initially considered to be restricted to DNA protection, i.e., to protect chromosome ends from being mistaken for double-strand breaks and eliciting a DNA damage response (DDR) and to prevent loss of important DNA coding sequences during cell division (Monaghan, 2014). Now, it is appreciated that telomeres have more widespread functions involving the regulation of gene transcription patterns.

Telomere dysfunction has been implicated in a number of biomedical disorders including an association with chronic stress (Armanios and Blackburn, 2012; Blackburn and Epel, 2012). The mechanisms by which a chronic stressful experience is translated into a shortening of telomeres are not understood. Mitchell *et al.* (2014) studied the relationships between social disadvantage (a form of chronic stress), genetic variants, and telomere length in a group of 40 African-American male children nine years of age. Social environment was measured using an index derived from four parameters: economic conditions, parenting quality, family structure and stability, and the presence of maternal depression. Half of the children in the sample were raised in what were categorized as very disadvantaged social environments, while the other half were raised in advantaged social environments. Saliva was used as the source of DNA and total telomere length per diploid genome was measured. This value was then divided by 92 to arrive at the average telomere length per chromosome.

Mean telomere length was about 7% shorter in the group of children exposed to harsh social environments compared to the half exposed to nurturing social environments. However, social environmental parameters did not all have the same influence. For example, exposure to multiple transitions within the family structure was associated with a 40% reduction in telomere length. The investigators explored the effects of genomic variants on the association between social environment and telomere length. The genomic variants chosen were alleles related to two serotonergic pathway genes (*5-HTT*, serotonin transporter gene; *TPH2*, tryptophan hydroxylase gene 2) and alleles related to four dopaminergic pathway genes (*DAT1*, dopamine transporter gene; *DRD2* and *DRD4*, dopamine receptor genes; *COMT*, catechol-o-methyltransferase gene). A sensitivity score was created for each pathway by combing the number of "risk" alleles present; "risk" alleles were defined as alleles associated with decreased transcriptional activity. The combination of a harsh social environment with a high sensitivity score was associated with the presence of, on average, the shortest telomeres, whereas an advantageous social environment plus a high sensitivity score correlated with, on average, the longest telomeres. These relationships appeared to be more robust for serotonergic genes compared to dopaminergic genes. The observations of Mitchell and coworkers describe, at least in children, complex interactions between social environment, genomic variants, and telomere length.

Telomeres are specialized DNA–protein structures that function to protect the ends of linear chromosomes from being recognized and processed as DNA double-strand breaks. If chromosome ends are recognized as a break and are fused together, the result will be a deleterious gross genomic rearrangement. Telomeric sequences do not contain protein-encoding genes (Bekaert *et al.*, 2004). In humans and other vertebrates, the telomere at the 3′ end of the DNA strand has a tract of tandem repeats of the G-rich six-nucleotide unit sequence TTAGGG. The 5′ end of the other DNA strand has a complementary C-rich tract. At each end of the chromosome, the 3′ G-rich telomere tract extends beyond the 5′ C-rich telomere strand. This extension is known as the G-overhang, and depending upon the species, it ranges from 16 to 200 nucleotides. The G-overhang acts as a primer for telomere elongation mediated by telomerase and serves as a binding site for DNA- binding proteins such as shelterin and the CST complex. The 3′ overhang can also form protective secondary DNA structures such as T-loops. A component of the shelterin complex is a protein known as telomere repeat binding factor 2 (TRF2). TRF2 plays a crucial role in suppressing the activation of the DDR (Okamoto *et al.*, 2013). TRF2 is probably involved in T-loop formation, a structural arrangement that protects the chromosome ends from being accessed by the DNA damage machinery. TRF2 can also disrupt the DDR signaling pathway, preventing 53BP1 localization at the telomere.

Recently, the role of the transcription factor p53 in regulating telomere length was reported by Simeonova *et al.* (2013). The p53 protein contains a core DNA-binding domain as well as a second DNA-binding site in its C-terminus. This C-terminal domain (CTD) is not dependent on the DNA sequence. Simeonova *et al.* (2013) constructed mutations in the mouse gene encoding p53 which deleted the last 31 amino acid residues including the CTD. Mouse embryonic fibroblasts (MEFs) containing the mutation showed increased p53 transcriptional activity as measured by increased mRNA levels of p53 target genes. Mice homozygous for the truncated p53 showed skin hyperpigmentation, bone marrow failure, pulmonary fibrosis, and premature death — features associated with short telomeres in the human. MEFs homozygous for the mutation displayed a telomere length distribution shifted in the direction of shorter lengths (measurements made in metaphase cells), and these cells showed reduced expression of genes involved in regulating telomere length, e.g., *dyskerin*, *Terf1*, *Rtel1*, and *Tinf2*. These observations of Simeonova *et al.* (2013) seem to explain why somewhat paradoxically, increased transcriptional activity of p53 is associated with tumor suppression, but at the same time leads to features of accelerated ageing and premature death. However, the data do not "prove" a direct causal relationship between p53-mediated telomere attrition and premature ageing.

The prevailing view in telomere biology is that the two chromosome ends are symmetrical with both ends terminating in a single-strand 3′ G-rich extension (Nelson and Shippen, 2012). However, the processes that lead to 3′ overhangs at the two ends differ (Kazda *et al.*, 2012; Nelson and Shippen, 2012). The polarity of DNA-strand synthesis has led to the proposal that the formation of telomere caps at one end of the chromosome is a product of leading strand replication, whereas the telomere caps at the other end of the chromosome are formed by the lagging strand replication mechanism. Leading strand synthesis is continuous and creates a full-length daughter strand with a blunt-ended telomere. A 3′ G-rich overhang is created at this end by resection of the end of the 5′ C-rich strand by nucleolytic processing. At the other chromosomal end, lagging strand synthesis does not involve continuous DNA replication. RNA primers are laid down as part of the replication process and when the last RNA primer is removed from the extreme 5′ end of the daughter strand, an unreplicated gap is created resulting in a 3′ G-overhang. Hence, vertebrate telomeres at the two ends of the chromosome are structurally identical and presumably functionally identical even though the creation of a single-strand G-overhang at each end is the result of differing mechanisms. The presumption has been that the telomeric single-strand G-overhang is a critical structural feature to protect the chromosome terminus from illicit DNA repair activities.

Two main protein complexes, shelterin and CST, bind to the G-overhang. However, these complexes also have components that bind to the telomere duplex and it is thought that these components recruit nucleases which are responsible for resecting part of the 5′ C-rich tract (Nelson and Shippen, 2012). Hence, it is proposed that components of shelterin and CST regulate 5′ C-rich strand resection at the "leading-end" telomeres.

Chow *et al.* (2012) have examined the steps involved in the formation of telomere G-rich overhangs. Leading strand synthesis creates a 3′ G-rich telomere daughter strand at one chromosome end, while lagging strand replication forms a 5′ C-rich telomere daughter strand at the other chromosome end. These investigators used human foreskin fibroblasts and looked at telomere formation kinetics during the cell cycle. Adult human duplex telomeres contain about 5–15 kb of DNA nucleotide repeats; TTAGGG in the 3′ strand and corresponding AATCCC nucleotide repeats in the 5′ strand. The 3′ strand ends with a single-strand overhang (G-rich overhang) of variable length. Chow *et al.* (2012) looked at the mechanisms responsible for the size of the 3′ G-rich overhangs in the absence of telomerase activity. Duplex telomeres replicate during the S phase of the cell cycle, but there is a delay of about 2 h before significant numbers of telomeres appear. Overhangs approached mature sizes within about 2 h of DNA replication in these fibroblasts. On average, leading strand synthesis overhangs were about 30 nucleotides in length, while lagging strand overhangs were about 100 nucleotides long. Further analysis revealed that lagging strand daughter (telomere) overhangs reached mature size within 1 h of DNA replication, whereas leading strand daughter (telomere) overhangs required an additional hour to reach mature sizes. In the case of the lagging-end telomere, the 3′ G-rich overhang is created by removal of the RNA primer at the terminus of the 5′ daughter strand together with inability of the replication machinery to fill in the gap between the final RNA primer and the extreme end of the 5′ strand. The 3′ G-rich overhang of the leading end telomere is formed by resection of the terminal part of the 5′ C-rich strand. The data showed that the creation of mature leading-end 3′ G-rich overhangs took about 1 h longer than the creation of mature lagging-end overhangs. Chow *et al.* (2012) raised an interesting question based upon their observations. Given the long delay in processing leading and lagging strand telomere overhangs (leading more than lagging), these overhangs presumably remain in an unfolded state for at least 1–2 h following DNA replication. Once mature, a G-rich overhang can be inserted into a double-strand region to form a lasso-like telomeric T-loop secondary structure which hides the free end of the DNA. What mechanism prevents these single-strand unfolded overhangs from eliciting a DDR during this maturing time period?

Part of the answer to this question can be found in the observations of Kazda *et al.* (2012). These investigators looked at the structure of telomeres in the angiosperm plants. *Arabidopsis*, an angiosperm plant, has two distinct populations of telomeres terminating with either a long 3′ overhang or a blunt end. In this plant, lagging strand replication would form a telomere with a 3′ G-rich overhang at one end of the chromosome, while leading strand synthesis would result in a blunt-ended telomere at the other end. The proposal is that the DNA-binding protein Ku (Fisher and Zakian, 2005) loads onto and remains associated with the blunt-ended telomere, forming a physical cap that prevents nucleolytic resection of the 5′ C-rich strand. Hence, in *Arabidopsis*, chromosomes have asymmetrical ends — one end has a telomere with a 3′ overhang whereas the other end has a blunt-ended telomere. In addition, Kazda *et al.* (2012) point out that in human cells, mature leading strand 3′ G-rich overhangs take about 2 h to form after DNA replication (Chow *et al.* 2012). Perhaps the Ku DNA-binding protein is involved in temporarily "protecting" this chromosome end from eliciting a DDR while the blunt-ended telomere is being processed (by C-strand resection) to form a mature 3′ G-overhang. The data of Kazda *et al.* (2012) demonstrate that telomeres do not require a single-strand overhang to form a functional capping structure and underscore evolutionary flexibility in solving the chromosome end protection problem.

Plants encode the CST complex and homologs of some of the vertebrate shelterin components (Nelson and Shippen, 2012). CST binds to single-strand telomeric DNA and participates in the regulation of C-strand resection. In *Arabidopsis*, binding of CST and shelterin to the chromosome end with the telomeric overhang protects this end, while binding of the protein Ku to the blunt-ended telomere protects that end of the chromosome. Why the chromosome end problem was solved, at least partially by blunt-ended telomeres in angiosperm plants, but exclusively by telomeres with a single-strand DNA overhang in vertebrates is not clear. Nelson and Shippen (2012) speculated that the extended proliferative capacity of blunt-ended telomeres may not be an attractive choice for vertebrates such as mammals which have a potential for developing cancerous transformation of cells.

Telomeres also display a higher order of structural organization which has important functions with respect to the maintenance of telomere integrity and length (Gonzalez-Suarez and Gonzalo, 2008). As already noted, the 3′ G-rich overhang can fold back and interpose itself into the duplex region of the telomere, forming a structure known as a T-loop whose arrangement hides the free DNA end. This structure is stabilized by a number of DNA-binding proteins (components of the shelterin complex) and this arrangement helps to protect the chromosome end (Gonzalez-Suarez and Gonzalo, 2008). One of the major mechanisms maintaining telomere length is the telomerase complex which consists of TERT and an associated RNA molecule (TERC). TERT adds nucleotide repeats using TERC and the 3′ overhang as a primer. Most somatic cells lack sufficient telomerase activity to offset the telomere attrition that occurs with each cell division. When telomere length falls below a critical threshold value, the chromosome end becomes unprotected and elicits a cell-cycle arrest known as replicative senescence. An alternate mechanism of teleomere length (ALT) maintenance has also been described which relies upon recombination among telomeric sequences to maintain telomere length. Chromatin structure around the duplex telomere is also very important. DNA is wound around histone protein complexes to form nucleosomes and this arrangement also occurs with telomeric DNA, although the nucleosomal distribution differs between subtelomeric and telomeric regions. The amino terminal tails of the histones are subject to extensive post-translational modifications; acetylation/deacetylation, methylation/demethylation, phosphorylation/dephosphorylation. These modifications are classified as epigenetic marks and they, along with the binding of specific proteins (e.g., heterochromatin protein 1, HP1), can alter the packing state of the chromatin. For example, a heterochromatin state is one of dense chromatin packing. In mouse embryonic cells with a knockout of two specific methyltransferases, histone di- and tri-lysine methylation is reduced, less HP1 is bound, chromatin structure becomes less compact, and abnormally long telomeres (compared to wild-type cells) are observed (Garcia-Cao *et al.*, 2004). Hence, telomere length can be regulated by epigenetic-induced changes in chromatin structure. It is not clear whether the epigenetic-induced lengthening of telomeres is due to increased accessibility of the telomerase complex to the less compacted teleomeres or to the induction of the ALT mechanism (Gonzalez-Suarez and Gonzalo, 2008). However, an important point is that these studies have been performed in murine cells and the results may not be completely transferable to other mammalian cells such as those of humans.

In eukaryotic cells, nuclear architecture is maintained by attachments to a nuclear matrix, consisting of the nuclear lamina and an internal fibrogranular network known as the inner nuclear matrix. In mammals, the lamin A, B, and C proteins are involved in forming the nuclear lamina. Telomeres show a preferential location within the nucleus which is apparently the result of important interactions between telomeres and lamin proteins. This topic will be discussed in a subsequent section on **lamins, telomeres, and cellular senescence**.

In vertebrates, telomere elongation is regulated by the cell cycle; the telomerase complex adds nucleotide repeats to the 3′ G-rich single-strand during the S phase and into M phase (Armanios and Blackburn, 2012). In yeast, the rate of telomere shortening in the absence of telomerase is about 3–5 nucleotides per division and about 50–100 nucleotides per division in cultured telomerase-negative human cells (Chow *et al.*, 2012). The loss of telomere length with each cell division is the result of inability of the lagging strand replication mechanism to copy the end nucleotides on the 5′ strand in the case of telomeres generated by the lagging strand, and incomplete 5′ strand resection in the case of telomeres generated by the leading strand. In addition, most somatic cells lack sufficient telomerase to counteract telomere shortening (Box *et al.*, 2008). Human cancer cells have high levels of telomerase and can add about 60 nucleotides to every chromosome per cycle (Chow *et al.* 2012). However, the limitation of the telomerase synthetic capacity in most human cells favors telomere shortening as the default state and means that telomere shortening generally accompanies cell proliferation (Armanios and Blackburn, 2012). Telomerase preferentially elongates the shortest telomeres, and telomerase only elongates a few telomeres in any given cell cycle or generation (Armanios and Blackburn, 2012). In normal human somatic cells, telomere shortening with cell division sets a limit on the cells' proliferative capacity since replicative senescence occurs when a critical telomere length is reached; this process is considered to act as a tumor suppression mechanism. In the germline, telomere length is maintained by an active telomerase system. During development with tissue growth, telomerase activity compensates for telomere length erosion during consecutive rounds of cell division (Bekaert *et al.*, 2004). However, once tissue differentiation has occurred, telomerase activity is downregulated in most somatic cells.

Measuring telomere length is not without difficulties. For example, a human interphase cell would have 92 telomeres and each telomere tends to have its own unique length. Hence, a cellular telomere length profile would reflect both the mean length plus a measure of length distribution. In most studies, it is the mean telomere length that is measured. Studies in both laboratory animals and humans indicate that telomere length is a heritable characteristic (Armanios and Blackburn, 2012). Offspring of parents with mutations in telomerase genes have shorter telomeres than age-matched controls even though these offspring do not have a telomerase gene mutation; the heritability of telomere length is estimated to be between 36–84%.

The generation of the telomerase complex involves an incomplete functional role of the spliceosome. Telomerase is comprised of a catalytic subunit (TERT) and an RNA template (telomerase RNA, TERC) for telomere repeat synthesis. Box *et al.* (2008) looked at the generation of telomerase RNA in the yeast *Schizosaccharomyces pombe*. In *S. pombe*, the gene encoding telomerase RNA is known as *ter1*. Normally, when the spliceosome is removing an intron from an RNA molecule, two consecutive cleavages take place. First, the spliceosome cleaves the intron at the 5′ss allowing the free 5′ phosphate to bond with the branch point (forming a lariat-like structure) which is close to the 3′ss. The intron is cleaved at the 3′ss, and the exons previously joined by the intron are spliced together. In the case of telomerase RNA, the process is different. The first cleavage occurs at the 5′ss of the intron, joining exon 1 and exon 2. This first cleavage step generates the proper 3′ end of exon 1. However, the second cleavage step does not take place probably because the distance between the branch point and the 3′ss is unusually long (Bonnal and Valcarcel, 2008). Exon 1 corresponds to the mature form of telomerase RNA. If the second cleavage step had taken place, the intron joining exons 1 and 2 would have been removed and the spliceosome would have spliced exons 1 and 2 together, forming a non-functional form of telomerase RNA. As pointed out by Bonnal and Valcarcel (2008), this type of incomplete activity of the spliceosome could be a much more general mechanism and may be exploited to generate alternative RNA transcript ends.

Telomeres and ageing

In proliferating human somatic cells, telomere attrition of between 50 and 200 nucleotides occur with each division. Somatic cells can undergo 50–80 doublings before telomeres reach a critically short length. At this stage, the cell either undergoes apoptosis or enters a stage of permanent proliferative arrest called replicative senescence. The evidence indicates that the length of the shortest telomeres in a cell (not average telomere length) determines when replicative senescence or apoptosis occurs (Salomons *et al.*, 2009). As already noted in the section on **telomere biology**, measuring telomere length is complicated by the heterogeneity in telomere length within a given cell. Lansdorp *et al.* (1996) assessed telomere length in human metaphase cells; fetal liver, adult bone marrow, and chronic myeloid leukemia (CML) cells. Each metaphase chromosome has four telomeres located at the ends of sister chromatids. Telomeres situated on sister chromatids from the same side of the chromosome had closer lengths than chromatid telomere pairs from opposite chromosome sides. In a given metaphase cell, there was considerable variation in chromatid telomere length between individual chromosomes. In addition, there was considerable variation in chromosome total telomere length (sum of the four chromatid telomere lengths) between tissues (fetal liver, bone marrow, and CML) and between cells within the same tissue. Give this variability in telomere length across chromosomes, cells, and tissues, most studies report mean chromosome total telomere length as well as some measure of telomere length distribution.

Somatic cells downregulate telomerase expression prior to birth, whereas adult stem cells continue to constitutively express low levels of telomerase sufficient to substantially delay cellular senescence (Boccardi and Herbig, 2012). Stem cells divide asymmetrically where one daughter cell is another stem cell and the other daughter is a "progenitor" cell that differentiates into an adult cell of a fixed tissue type. Since telomere attrition occurs as a result of cell division, non-dividing cells should maintain a stable telomere length. Given that telomere shortening occurs primarily in proliferating somatic cells, e.g., skin, gastrointestinal tract, and bone marrow, what is the relationship between whole-animal ageing and lifespan and telomere length? This relationship has been examined in several natural experiments.

Raices *et al.* (2005) examined this relationship in the nematode, *Caenorhabditis elegans*. Fully developed adult *C. elegans* hermaphrodites consist of 959 non dividing somatic cells. The telomere repeat sequence in *C. elegans* (TTAGGC) is very similar to that in mammals, TTAGGG. Using a single strain, Raices *et al.* (2005) measured the variation in telomere length between clones of individual worms. Telomere lengths in most clones were in the range of 2.5–3 kb, but some clones had telomeres as long as 9 kb. When individual clones were followed for 10 generations, telomere length for each clonal population remained relatively constant, indicating that telomere length had a strong heritable component. Telomere length was not only variable between individuals in the same strain, but was also variable between different strains. Across seven different strains of *C. elegans*, mean telomere length varied from 2.6–15.9 kb, yet all seven strains had the same mean lifespan. The investigators used both RNA interference techniques and specific genetic mutations to create strains of worms with average lifespans varying from about 15 to 42 days. In these worms, there was no correlation between lifespan and average or maximum telomere length. In fact, mean telomere length was about the same for all these strains of *C. elegans*. Hence, this study demonstrated that in a post-mitotic organism such as *C. elegans*, telomere length and lifespan were independent of each other.

Salomons *et al.* (2009) looked at telomere length in the erythrocytes (nucleated in birds) of a population of free-living corvids. For each sample, a telomere length distribution was measured from which mean, minimum, and maximum length values were obtained. Cross-sectional data showed that the fastest rate of telomere shortening occurred early in life with the rate of shortening decreasing with advancing age, a pattern that has also been observed in humans. Longitudinal data demonstrated that within individual birds, there was a significant shortening of telomeres with age, with the longest telomeres shortening at twice the rate as the shortest telomeres. The fastest rate of telomere shortening was observed in the year preceding death (for those birds that died within the observational period). Hence, this study revealed that the rate of telomere shortening was a function of telomere length and that the rate of telomere shortening appeared to be a predictor of an individual's survival probability.

The observation that telomere length (or rate of shortening) correlated with survival prospects was also documented by Foote *et al.* (2011). These investigators measured mean telomere length in nucleated erythrocytes from the long-lived seabird, the Southern giant petrel. DNA sampling was performed on 47 adult birds in the years 1999 and 2000, and subsequently, the study subjects were checked for survival eight years later. The cross-sectional data indicated that in adult birds there was no significant change in mean telomere length with age, unlike the data presented by Salomons *et al.* (2009) in corvids. Of the original sample, 30 (64%) died over the eight-year period. Mean telomere length, measured at the beginning of the eight-year period, was significantly shorter in those that had died compared to those still alive.

There is also evidence in natural populations that telomere length can be influenced by environmental perturbations. Mizutani *et al.* (2013) measured telomere length in blood samples from long-lived black- tailed gulls over a five-year period. The sample size consisted of 25 individual birds: nine birds were measured twice, 12 were measured three times, two were measured four times, and two were measured five times. There was no relationship between telomere length and chronological age although adult birds had shorter telomeres than chicks. The telomere measurements were made over the years 2007 to 2011. Two significant environmental events occurred within this time period. From the summer of 2009 to the spring of 2010, El Nino conditions provided ideal weather and foraging conditions for these gulls, a situation that would have mitigated the physiological stress associated with the foraging for food. In March 2011, the great earthquake and tsunami in Japan took place. As a result of this severe environmental event, coastline foraging was unavailable and gulls were forced to forage at sea, which is associated with higher energetic demands owing to the need to travel farther in search of food. There were distinct changes in mean telomere length over these time periods. Between May 2009 and May 2010 (El Nino conditions), telomere length tended to remain unchanged or to increase. The mean change in telomere length over this one-year period was 0.38 kb. Between May 2010 and May 2011, the period encompassing the earthquake and tsunami, telomere length tended to decrease with the mean change in length being –0.65 kb. Although not conclusive, these observations suggested that significant environmental changes can alter telomere dynamics (relative rates of telomere loss and extension). Favorable environmental changes tended to promote telomere length maintenance or extension, whereas significant adverse environmental changes tended to promote telomere shortening. Hence, environmental perturbations appear to modulate telomere length dynamics depending on the nature of the environmental perturbation. The underlying mechanism by which environmental fluctuations alter telomere dynamics is unknown.

The results of these natural experiments indicate that telomere length can be altered by environmental perturbations and that telomere length appears to correlate with survival probabilities. Whether this correlation reflects a causal relationship is not known. These results beg the question as to what would be the biological effects of inducing an increase in telomere length in adult animals. To answer this question, de Jesus *et al.* (2012) measured the effects of telomerase gene therapy in altering telomere length in adult and old mice.

These investigators used an adeno-associated virus vector carrying the murine reverse transcriptase gene (*mTERT*). This vector was injected into the tail vein of two cohorts of mice, one and two years old. One-month post-injection heart, lung, liver, kidney, brain, and muscle tissues were assayed for mTERT mRNA and protein as well as telomerase activity. The tissues of both groups of mice showed large increases in mTERT mRNA and protein and in lung, liver, kidney, and heart tissues, the investigators demonstrated increased telomerase activity. Compared to controls, the *mTERT*-injected mice showed greater bone mineral density, better preservation of subcutaneous fat, lower insulin levels, higher levels of IGF-1, and better neuromuscular coordination and memory function. In addition, lifespan was extended by 24% in the one-year-old cohort and by 13% in the two-year-old group. Telomere length was measured in the cerebral cortex (brain), myocytes (heart), lung, kidney (medulla), hepatocytes (liver), and muscle fibers, and in mice injected with *mTERT*, these tissues showed longer telomeres as well as a decreased percentage of short telomeres compared to these same tissues in control mice. Recently, it has been observed that TERT acts as a modulator of Wnt signaling (Park *et al.*, 2009). *mTERT*-injected mice showed increased tissue expression of the Wnt target genes *B-catenin* and *cyclinD1*. Finally, as an additional control, the investigators injected mice with a catalytically inactive form of mTERT (mTERT-DN). The tissues of these mice did show increased levels of mTERT-DN mRNA and increased expression of cyclin D1, but there was no change in telomere length, bone mineral density, insulin levels, or animal lifespan compared to controls.

The experimental observations of de Jesus *et al.* (2012) are very interesting but not straightforward in their interpretation. The action of telomerase in elongating telomeres is regulated by the cell cycle. Telomere repeats are added during S phase and into M phase (Armanios and Blackburn, 2012). Telomerase preferentially elongates the shortest telomeres, and telomerase even at wild-type levels generally elongates only a few telomeres in any given cell cycle or generation (Armanios and Blackburn, 2012). In essence, telomere elongation occurs during the process of cell division and therefore would not necessarily occur in quiescent non-dividing cells. What is not clear in the study by de Jesus *et al.* (2012) is the mechanism underlying the telomere elongation they observed in post-mitotic (non-dividing) tissues such as of the brain, kidney, and heart. One possibility they considered was that these tissues have become repopulated with cells that had longer telomeres. It is also possible that the biology of murine telomeres is unusual and that these experimental results cannot be extrapolated to other mammals. This possibility is discussed in the following sections. In addition, telomerase appears to have other important functions besides telomere elongation. Telomerase is also known to be a transcriptional regulator of Wnt signaling target genes (reviewed in the next section). Also, there are complex interactions between telomeres, lamins, progerin, and the process of alternative splicing, which was detailed in the previous sections. All of these considerations render the interpretation of the study by de Jesus *et al.* (2012) difficult.

Telomerase and Wnt signaling

The Wnt family of secreted proteins is critical for determining cell fate during embryonic development and in adult stem cell niches (Berndt and Moon, 2013). The *Wnt* gene was identified in 1982 and early studies in fruit flies and *Xenopus* defined a highly conserved signaling pathway referred to as the (canonical) Wnt cascade. In fact, the name 'Wnt' alludes to a gene called 'wingless' because a mutation in this gene causes fruit flies to lack wings. The human genome contains 19 *Wnt* genes and Wnt proteins contain many conserved cysteins (Clevers and Nusse, 2012). Wnt proteins become modified by the attachment of lipids in the endoplasmic reticulum; this lipid modification is important for the secretion of the protein from the cell as well for its signaling functions. At target cells, Wnt proteins interact with a transmembrane receptor complex. A key protein component of the Wnt signaling cascade is B-catenin. Upon Wnt protein–receptor binding, the ubiquitination of B-catenin is inhibited and destruction of newly synthesized B-catenin does not occur. The B-catenin translocates to the target cell nucleus to activate genes (Clevers and Nusse, 2012). In the Wnt "off" state, target genes are bound by transcriptional repressors, whereas in the Wnt "on" state, B-catenin replaces transcriptional repressors and recruits activators, including histone modifiers such as BRG1 and CBP. Most *Wnt* target genes are tissular or developmental stage specific and include Axin 2 and cyclin D1. Wnt signaling is crucial for embryonic development, stem cell maintenance, and stem cell activation. The role of Wnt signaling in maintaining stem cells has garnered the most interest. The Wnt/B-catenin pathway maintains pluripotency and self-renewal of embryonic stem cells and recent observations suggest that stem cell division is influenced by the spatial orientation of the cells relative to the source of Wnt protein (Berndt and Moon, 2013). Aberrant Wnt signaling is associated with developmental abnormalities as well as cancer.

Park *et al.* (2009) described the influence of telomerase in modulating Wnt signaling. Telomerase is known to support stem cells by maintaining telomere lengths and thus sustaining proliferative capacity. However, other observations indicate that telomerase had a more direct effect on stem cells which was independent of its catalytic role of adding telomere repeats. In murine embryonic stem cells, Wnt 3A-induced Axin 2 expression was significantly reduced in *TERT*-deleted cells. *TERT* deletion also reduced basal expression of Axin 2 which was rescued by overexpression of *TERT*. Inhibiting *TERT* altered *Xenopus* embryonic development, but these phenotypes could be rescued by a catalytically inert *TERT* mutant. In the absence of a Wnt protein, the promoter regions of *Wnt* target genes are occupied by repressor complexes. In the context of a Wnt protein, B-catenin displaces these repressors and recruits proteins that confer transcriptional activation. These proteins include histone modifiers such as BRG1 and CBP/p300 as well as proteins that mediate the initiation and elongation of mRNA transcripts by RNAP II. Telomerase represents another co-factor recruited to the promoter region of *Wnt* target genes and also acts in conjunction with B-catenin and the histone modifier BRG1. Park *et al.* (2009) suggested that co-factors may act in a context-specific manner and that telomerase acts to regulate Wnt-determined progenitor cell fate, i.e., proliferation or differentiation. In essence, telomerase and Wnt signaling play interconnected and coordinated roles in the regulation of progenitor cell proliferation (symmetric or asymmetric) and differentiation.

Comparative mammalian telomerase biology

Three papers have reviewed the comparative biology of telomerase activity and telomere length in mammals, namely Gorbunova and Seluanov, 2009; Gorbunova *et al.*, 2008; Seluanov *et al.*, 2007. The mouse, the traditional mammal for biomedical research, differs dramatically from humans in terms of its telomere biology. The majority of normal human somatic cells (with the exception of stem cells) have no detectable telomerase activity and human fibroblasts demonstrate replicative senescence after about 60 population doublings. In contrast, the adult mouse expresses telomerase activity in most of its somatic tissues and murine fibroblasts in culture do not experience replicative senescence. About 90% of mice in captivity die of cancer.

Seluanov *et al.* (2007) measured telomerase activity in six tissues (heart, liver, spleen, kidney, skin, and lung) and telomere length was measured in the liver of 15 adult rodent species. These rodents have a maximum lifespan ranging from 4 to 28 years and body mass ranging from 20 to 55,000 g. The data were corrected for shared phylogenetic ancestry. The only two rodents that showed nearly complete somatic repression of telomerase activity, similar to humans, were the two largest: the beaver (20,250 g) and the capybara (55,000 g). Total telomerase activity, a composite measure from all tissues, showed a significant negative correlation with body mass but no correlation with maximum lifespan. Across these species, the larger the species size, the lower the telomerase activity. The six tissues were also analyzed separately across these species; liver, spleen, and kidney telomerase activities showed a negative correlation with body mass, whereas heart and lung telomerase activities were not correlated with body size. Mean telomere length in liver tissue across these 15 species did not correlate with either body mass or maximum lifespan. The lack of a relationship between telomere length and lifespan is consistent with the findings in *C. elegans* (Raices *et al.*, 2005) discussed in a previous section on **telomerase and ageing**. Hence in rodent species, reduced telomerase activity appears to have evolved in larger but not in longer-lived species. By contrast, telomere length has evolved independently of either body mass or lifespan.

Gorbunova and Seluanov (2009) reported data on telomerase activity and telomere length in other mammals. In non-human primates, telomerase activity was repressed in most somatic tissues similar to humans, and mean telomere lengths were somewhat longer than those of humans. In addition, proliferating cells in non-human primates also displayed replicative senescence. In ungulates such as the cow, sheep, and horse, telomere biology was similar to that of the human. Pigs appeared to be an exception. Somatic tissues expressed significant telomerase activity and cultured porcine fibroblasts did not show replicative senescence. In dogs and cats, somatic tissues showed little telomerase activity. Skin fibroblasts in four species of rabbit showed telomeres of heterogeneous length and had no detectable telomerase activity. Gorbunova and Seluanov (2009) summarized data on telomerase activity in somatic tissues from 32 species of mammals including rodents. They made the following general observations: (1) mammals with a body mass of less than 2 kg expressed telomerase activity in somatic tissues and proliferating somatic cells did not display replicative senescence; and (2) animals with a body mass of more than 5 kg had reduced telomerase activity in somatic cells and these cells underwent replicative senescence after a certain number of population doublings. Animals with a body mass between 2 and 5 kg represented an intermediate group with some showing telomerase activity in somatic tissues and others none.

The co-evolution of an inverse relationship between animal size and somatic cell telomerase activity is partly understandable in terms of cancer defense mechanisms. Evolutionary increases in body mass (and hence more cells) lead to an increased risk of cancer. To counteract that risk, large species have evolved repression of telomerase activity in somatic cells (Seluanov *et al.*, 2007). However, this model does not encompass a number of exceptions (Gorbunova and Seluanov, 2009). For example, the long-lived rodent, the naked mole rat, has a lifespan in excess of 20 years and yet has high levels of telomerase in its somatic tissues. The laboratory mouse and naked mole rat have comparable levels of telomerase activity yet the naked mole rat has a lifespan about seven times that of the mouse. At the other end of the size spectrum, marine mammals such as whales generally have a low incidence of cancer despite their very large mass (Singer, 2011). Large mammals would have reduced levels of telomerase, but is this adaptation sufficient to explain the low rates of cancer in these very large mammals? Most likely these mammals have evolved with additional tumor-suppressing mechanisms.

Regardless, interpreting and understanding the experimental actions of telomerase gene therapy as reported by de Jesus *et al.* (2012) is not straightforward when viewed within the context of comparative mammalian telomere and telomerase biology and the known non-telomere maintenance functions of telomerase.

Lamins, telomeres, and cellular senescence

Studies of mutant lamins such as the protein progerin have highlighted the role of lamins in the process of cellular senescence. The lamina is part of the nuclear envelope and consists of a meshwork of filament proteins known as lamins and lamin-associated proteins. Mammalian cells express two major B-type lamins, and the gene *LMNA* encodes for lamins A and C. B-type lamins remain farnesylated while lamin A, as already discussed, is cleaved by a zinc metalloprotease which removes both the farnesyl group and the 15 amino acids at the C-terminus. Lamins A and C are the main contributors to nuclear stiffness and also form a structural bridge between the lamina and cytoskeleton (Vlcek and Foisner, 2007). In addition, lamins interact with DNA, the histones found in chromatin, and histone-modifying enzymes. The latter interaction allows lamins to epigenetically modify gene expression. Lamin A binds to a variety of transcription regulators and thereby can regulate the expression of a number of target genes. Lamins are also involved in the DNA repair mechanism perhaps by acting as a scaffold for the assembly of proteins that orchestrate the repair process (Vlcek and Foisner, 2007). Hence, lamins appear to have widespread functions within the nucleus ranging from maintenance of nuclear structural properties to regulation of gene expression and repair of DNA damage.

The cell's nucleus has compartmentalized features. The nuclear periphery is involved in activities such as gene transcription and DNA repair (Gonzalez-Suarez *et al.*, 2009). Lamins as components of the nuclear lamina provide scaffolding for this compartmentalization of nuclear functions. Some of the novel roles of lamin A in nuclear functions were investigated by Gonzalez-Suarez *et al.* (2009) using two strains of MEFs; wild-type *LMNA+/+* and *LMNA−/−*, where *LMNA* is the gene encoding lamin A and lamin C. In *LMNA−/−* metaphase cells, telomeres were preferentially localized to the nuclear periphery compared to metaphase wild-type *LMNA+/+* cells. Telomere length (measured in metaphase cells) also differed between the two cell types. The telomeres in *LMNA+/+* cells had a longer mean length than the telomeres of *LMNA−/−* cells. This observation appears to be inconsistent with the data of Huang *et al.* (2008), who described shortening of telomere length in adult human fibroblasts overexpressing wild-type lamin A. Perhaps the discrepancy is due to the different cell types used, murine embryonic fibroblasts versus human adult fibroblasts, and to the different expression levels of lamin A in the two experimental setups. However, the two studies do indicate a role for lamin A in regulating telomere length. Gonzalez-Suarez *et al.* (2009) observed reduced chromatin histone epigenetic markers (lysine methylation) in *LMNA−/−* cells compared to *LMNA+/+* cells although the investigators did not think that these chromatin changes were responsible for the short telomeres seen in *LMNA−/−* cells. *LMNA−/−* cells also showed signs of genomic instability with an increase in number of telomere signal-free chromosome ends and an increase in chromosome breaks compared to *LMNA+/+* cells. The protein p53 is a crucial mediator of the cellular responses to DNA damage, including the senescence response (Campisi, 2005). The protein p53 is a tumor suppressor and functions as a transcription regulator. Dysfunctional telomeres which are sensed as DNA damage activate the p53 pathway. The transcription of p53-dependent genes induces a senescent-like growth arrest (Campisi, 2005). The protein 53BP1 binds to the tumor suppressor protein p53 and may control its phosphorylation (Wang *et al.*, 2002). In keeping with the observation that *LMNA−/−* cells showed evidence of DNA damage, Gonzalez-Suarez *et al.* (2009) demonstrated that *LMNA−/−* cells had lower levels of the protein 53BP1 than *LMNA+/+* cells. This difference in 53BP1 levels between the two cell types was not due to a difference in transcript levels but rather attributed to the increased degradation of 53BP1 in *LMNA−/−* cells. Hence, lamin A appears to be necessary for the stabilization of the protein 53BP1, and this component of the DDR pathway is disrupted in *LMNA-/-* cells. 53BP1 also functions to promote repair of DNA breaks via non-homologous end-joining (NHEJ) by blocking the action of enzymes that create single-strand DNA at the site of breaks (Lukas and Lukas, 2013).

Laminar biology

On the basis of recent experimental observations, it is becoming clear that lamins and the nuclear lamina subserve multiple functions with respect to the regulation of gene expression. The nuclear periphery is demarcated by two layers of nuclear membranes separated by a perinuclear space. The inner nuclear membrane is lined by the nuclear lamina, a meshwork of intermediate filamentous proteins. Lamins are the major components of the nuclear lamina and these proteins are classified into types A and B. The gene *LMNA* encodes for the lamins A and C which are formed by alternative splicing. The gene *LMNB1* encodes lamin B1, while the gene *LMNB2* encodes lamins 2 and 3 (Ho and Lammerding, 2012). A-type lamins are found in differentiated tissue and in some adult stem cells (mesenchyme and hair cells) but absent in early embryonic cells. B-type lamins are more ubiquitously expressed and at least one B-type lamin is present in all cell types during development (Meshorer and Gruenbaum, 2008).

Shevelyov and Numinsky (2012) reviewed the effects of lamins and the nuclear lamina on chromatin structure and gene expression. They noted that the radial positioning of chromosomes and chromosome loci in the interphase cell nucleus was programmed with gene-dense chromosomes localized close to the center of the nucleus and gene-poor chromosomes close to the periphery. Although the positioning of chromosomes varied between nuclei and probably over time in the same nucleus, gene-dense and gene-poor chromosome loci appeared to form distinct non-overlapping clusters. The positioning of chromosomes was determined by chromatin–lamina interactions. Sections of chromatin attached to the lamina have been denoted as lamina-associated domains (LADs). About half of the mammalian genome interacts with the lamina with these lamina–chromatin interactions being dynamic and not static. In addition, LADs turn out to be mostly gene-poor segments. Changes in chromatin–lamina interactions vary during the cell cycle (during mitosis, the lamina disassembles and then reassembles after cell division) and during development. In summary, chromosome loci located at the nuclear periphery are sites of LADs and sites of silent or poorly transcribed genes. The current model postulates that genes tethered to the lamina are exposed to multiple protein repressor molecules. These molecules include deacetylases and methyltransferases which establish histone epigenetic marks leading to gene repression. During cell differentiation, the expression of multigene regions can be regulated by translocation of these regions away from the periphery and their association with the lamina. Such a translocation would lead to transcriptional activation of these genes. The model suggests that chromatin domains "attached" to the lamina are deacetylated and compacted with silencing of the contained genes. Detachment of the chromatin domains and movement of the chromosome loci away from the periphery leads to histone acetylation and a more "open" gene active chromatin structure. Compactness of the chromatin determines accessibility of the genes to transcription regulators.

Lamins can regulate gene expression through additional mechanisms which involve direct interactions with various transcription factors (Ho and Lammerding, 2012). Most of these interactions involve A-type lamins. Lamins act as scaffolding for DNA replication factors and are also involved in DNA repair mechanisms, e.g., non-homologous end-joining (Redwood *et al.*, 2011).

Meshorer and Gruenbaum (2008) have reviewed evidence that lamins are involved in stem cell differentiation. Adult somatic stem cells are tissue specific and usually self-renewing. They are restricted in their differentiation potential and are normally maintained as a small reservoir to provide a source of tissue-specific replenishment particularly during damage repair. Mutations in lamin A can interfere with progenitor cell differentiation and this has been demonstrated in mesenchymal stem cells and stem cells located within the hair follicle. In both cases, mutant forms of lamin A induced either decreased stem cell proliferation or abnormal differentiation. The signaling pathways that seemed to be involved were the Notch signaling pathway in the mesenchymal stem cells and the Wnt signaling pathway in the hair follicle stem cells. The important observation is that these studies link lamin A biology with stem cell function, and abnormal stem cell function has been considered a contributor to the ageing process.

Van Bortle and Corces (2013) reviewed the current evidence that chromatin–lamin interactions have a key role in cell differentiation. During lineage-specific cell differentiation from embryonic stem cells through to terminally differentiated tissue cells, there occurs a gradual and cumulative restructuring of lamina–chromatin interactions. This reorganization of lamina–chromatin interactions results in the relocation of genes which leads to changes in their expression. Genes that are repositioned to the nuclear periphery become repressed or silenced, whereas genes translocated away from the periphery and into the nucleoplasm become transcriptionally active. The nature of the reorganization of lamina–chromatin interactions and the set of genes that are repositioned (toward or away from the nuclear periphery) are lineage-specific. However, there is good evidence that nuclear lamins play a major role in these lineage-specific cell fate decisions. These observations would provide a mechanism underlying the involvement of lamins in stem cell differentiation as reviewed by Meshorer and Grenbaum (2008). One example cited by Van Bortle and Corces (2013) underscores the role of lamins in the differentiation process. In *Drosophila*, the transcription factor Hb specifies early neuroblast differentiation. The gene encoding Hb is gradually repositioned to the nuclear lamina and transcriptionally silenced at a specific time in the neuroblast differentiation sequence. If lamin proteins are depleted, this differentiation stage would be extended in length because the relocation of the *Hb* gene to the periphery and its repression are impaired.

In cells from individuals with HGPS, the accumulation of the mutant lamin A protein progerin is associated with loss of the nuclear compartmentalization of active and inactive chromatin (McCord *et al.*, 2013). In their experiments, McCord and co-workers used three fibroblast cell lines: a cell line from a HGPS patient, a cell line from the father of the HGPS patient, and a cell line from an aged matched control subject. Across the genome of normal cells, they found large patches of the epigenetic marker H3K27me3 (trimethylation of histone H3 lysine 27) in gene-poor regions. Such patches were either decreased or lost in cells from the HGPS patient. They also found that mRNA levels of EZH2, a member of the polycomb repressive complex 2 which has methyltransferase enzymatic activity, were decreased in HGPS cells compared to normal cells. When they examined lamin–chromatin associations, they observed that in HGPS cells, these associations were reduced in the same gene-poor genomic regions that displayed decreased H3K27me3 levels. The spatial clustering in normal cells of gene-poor inactive chromatin (heterochromatin) at the nuclear periphery and gene-dense active chromatin in the nuclear interior was altered in HGPS cells. Electron micrographs showed a layer of heterochromatin attached to the nuclear lamina in cells from the father but a loss of this layer of heterochromatin in the cells from the HGPS patient. McCord and co-workers proposed a model in which progerin accumulation disrupted the normal nuclear lamina architecture. This disruption was associated with decreased chromatin–lamina associations and reduced levels of the epigenetic chromatin marker H3K27me3 (probably as a result of the reduced expression of EZH2). Progression of these changes resulted in a global loss of spatial chromatin compartmentalization in HGPS cells — changes that would probably be associated with transcriptional misregulation of multiple genes.

The experimental results of McCord *et al.* (2013) demonstrated a point of intersection between the premature ageing disorder of HGPS and the premature ageing (early onset dementia, immune dysfunction, premature menopause, and various endocrinopathies) observed in trisomy 21 (Down syndrome). Impairment of the self-renewal function of somatic stem cells is thought to underlie the premature ageing of Down syndrome (Souroullas and Sharpless, 2013). The point of intersection between these two premature ageing phenotypes involves the polycomb complex of proteins. There is experimental evidence that trisomy of the *Usp16* gene is responsible for reduced stem cell self-renewal in Down syndrome (Souroullas and Sharpless, 2013). The protein encoded by *Usp16* detaches a ubiquitin protein from histone H2A K (lysine) 119. This action counteracts the action of polycomb repressive complex 1 which silences genes by monoubiquinating histone H2A at lysine 119 (Wang *et al.*, 2004). Briefly, the proposed model for early ageing in Down's syndrome is as follows (Souroullas and Sharpless, 2013). Polycomb proteins by the addition of methyl groups or ubiquitin proteins to histones, repress the expression of genes such as *CDK2NA*, which encodes tumor suppressor proteins, as well as repress other genomic loci. Repression of these genes allows for the self-renewal of somatic stem cells. Trisomy of *Usp16* results in the removal of ubiquitin proteins from histones counteracting the repressive actions of the polycomb proteins. The re-activation of genes such as *CDK2NA* promotes premature senescence in somatic stem cells leading to the premature ageing of Down syndrome. The important point is that in both HGPS and Down syndrome there appears to be impaired polycomb function, with the downregulation of polycomb activity in HGPS leading to reduced histone H3 methylation, and the counteraction of the polycomb-mediated ubiquitination of histone H2A by trisomy *Usp16* in Down syndrome. Hence, variations in the function of polycomb complexes serve as a bridge between two premature ageing disorders which on first blush appear to be very dissimilar.

HGPS affects diverse body physiology and systems including growth, skeletal development, fat, skin, hair, and the cardiovascular system. However, individuals with HGPS show no impairments in cognitive functioning, and organs such as the liver, kidney, lung, brain, gastrointenstinal tract, and bone marrow appear to be unaffected (Baek *et al.*, 2013). The prevalence of cancer, cataracts, and cognitive impairment is not higher in those with HGPS. The lack of involvement of certain tissues in HGPS has been a puzzle (Baek *et al.*, 2013). However, recent data appear to supply at least some of the pieces to this puzzle. One of them is the documented link between mutations in A-type lamin and stem cell functioning. There is evidence that the target cell in HGPS is the mesenchymal stem cell (Meshorer and Gruenbaum, 2008; Csoka *et al.*, 2004), and this observation gives at least a partial explanation for the tissue specificity observed in the HGPS phenotype since most of the affected tissues in HGPS are of mesenchymal origin.

A second piece to the puzzle relates to the respective roles and tissue distributions of A-type and B-type lamins. The B-type lamins are expressed in nearly every cell type and during all stages of development and therefore have been considered crucial for embryogenesis. However, *LMNB1*-deficient mouse embryos survived through development to birth although they died shortly after birth. Many of the tissues appeared normal although the cranium was misshapen and brain development was abnormal (Young *et al.*, 2012). *LMNB2*-deficient mice displayed only abnormal brain development, specifically in the cerebral cortex and cerebellum and these mice also died shortly after birth. Additional studies have indicated that the abnormal brain development in the *LMNB2*-deficient mouse and the *LMNB1*-deficient mouse was due to defective glial-directed neuronal migration during embryogenesis (Young *et al.*, 2012). The extent of defective brain development is more severe in the *LMNB1*-deficient mouse than that observed in the *LMNB2*-deficient mouse. Although B-type lamins are crucial for brain development, they are not necessary for the development of skin, hair, nails, and the liver and the proliferation of keratinocytes. These experimental observations can be reconciled by the different tissue distributions of type A and type B lamins. Lamins A and C, in contrast to B-type lamins, are found at very low levels during early development but are expressed at high levels in most differentiated cells (Young *et al.*, 2012). Lamin C does not possess a Ca1a2X motif and therefore is never farnesylated. It is still not clear why mammals produce both lamin A and lamin C since studies in mice that were engineered to synthesize either lamin A or lamin C exclusively have not revealed unique functions for either of these two protein isoforms. Also, the physiological function of post-translational modifications of prelamin A is unclear. Mice engineered to synthesize mature lamin A directly, thereby bypassing the production of prelamin A and its processing, were healthy without obvious phenotypic alterations. Young *et al.* (2012) pointed out that the murine brain (neurons and glial cells) contained abundant amounts of lamin C, whereas lamin A is restricted to meningeal cells and vascular endothelial cells. In peripheral tissues such as those of the heart, kidney, and liver, lamins A and C are found in roughly equal amounts. The low level of lamin A in the brain was the result of downregulation by a brain specific microRNA (miR-9) which binds to the 3′-UTR (untranslated region) of prelamin A. MicroRNAs regulate gene expression post-transcriptionally by binding to mRNA and reducing mRNA translation (initiation and elongation) or inducing deadenylation and degradation of the target mRNA (Filipowicz *et al.*, 2008). In the case of *LMNA*, its expression was downregulated by miR-9 as reflected by reduced levels of prelamin A transcripts and the lamin A protein (Jung *et al.*, 2012). Downregulation of *LMNA* expression by miR-9 was specific since the transcript levels for other lamins, i.e., B-type lamins and lamin C, were not reduced (Jung *et al.*, 2012).

These experimental results raise some interesting questions. The observation in the mouse, that lamin C is highly expressed in the brain, implies it plays an essential role in this tissue. On the other hand, why the mouse has evolved a mechanism for limiting lamin A production in the brain is unclear.

Mechanical stresses applied to a tissue can cause deformation, and at the cellular level, these stresses can be transmitted through cell–cell and cell–extracellular matrix interactions and within the cell to the nucleus. Mechanical stresses have been implicated in the regulation of gene expression patterns as well as in determining the direction of stem cell differentiation (Bainer and Weaver, 2013). Tissues respond to these mechanical stresses by adjusting the stiffness or (micro)elasticity of the extracellular matrix and by altering the nuclear lamina such as stabilization of the integrity of the nuclear membrane. In essence, the stiffness and strength of a tissue should relate to the physical stresses acting on that tissue. Tissues such as those of the brain and fat, experience low stresses which probably explains why these tissues are soft. Adult bone is exposed to high stress which promotes its growth and stiffening, thus matching bone tissue properties with the high stress. These observations strongly imply the existence of a "mechanostat", a mechanism that senses mechanical stresses and initiates tissue and cellular adjustments to these stresses. Recently, Swift *et al.* (2013) suggested that lamin proteins play an important role within the "mechanostat" mechanism. Cells sense some aspect of the mechanical stress and the transfer of mechanical cues from the extracellular matrix to the cell nucleus induces changes in gene expression (Bainer and Weaver, 2013).

Swift *et al.* (2013), using a variety of human and mouse tissues, observed a positive and strong correlation between the ratio of lamin A to lamin B and tissue stiffness or microelasticity (*E*). Over a range of tissues with increasing values of *E*, from very low *E*(brain) to very high *E* (bone), the amount of lamin A increased by about 30-fold, whereas the amount of lamin B increased by about threefold. For very "soft" tissues such as those of the brain and bone marrow, the ratio of lamin A to B was slightly in favor of lamin B. However, as the microelasticity of the tissue increased, the importance of lamin A became dominant and the ratio of lamin A to B became progressively weighted toward lamin A, i.e., increased matrix stiffness enhanced the transcriptional activity of the gene *LMNA*. Hence, stiffer tissues such as those of the heart, muscle, cartilage, and bone would be associated with a high ratio of lamin A to B. This observation is consistent with the finding that *LMNA* gene mutations which cause premature ageing syndromes primarily affect stiffer tissues such as in the heart and large arteries, while sparing softer tissues such as in the brain and bone marrow. Swift *et al.* (2013) also found a strong positive correlation between the abundance of the extracellular matrix protein collagen type 1 and the value of *E*.

To further assess the effects of extracellular matrix stiffness, they cultured mesenchymal stem cells on soft and stiff gels. Cells grown on stiff gels, compared to soft gels, showed a different phenotype — a more spread-out shape with prominent stress fibres, higher levels of alpha smooth muscle actin, and nuclei that were smoothed out and flattened rather than wrinkled as observed on soft gels. Lamin A levels were higher in mesenchymal stem cells cultured on stiff gels and lamin A phosphorylation (which promotes lamin disassembly and turnover) was about 30% less when mesenchymal stem cells were grown on a stiff matrix. The degree of elasticity or stiffness of the extracellular matrix can direct stem cell differentiation probably by inducing distortions in nuclear architecture with resulting changes in gene expression patterns. The induction of tissue-specific stem cell differentiation by the degree of the extracellular matrix stiffness is modulated by the abundance of lamin A. For example, increased matrix stiffness and increased levels of lamin A maximized osteogenesis in mesenchymal stem cells grown in an osteogenic induction medium, whereas low matrix stiffness and reduced levels of lamin A maximized adipogenesis in mesenchymal stem cells cultured in an adipogenic induction medium. Conversely, a soft matrix suppressed osteogenesis even in the presence of an osteogenic induction medium, and abundant lamin A and a stiff matrix suppressed adipogenesis even in the presence of an adipogenic induction medium and low levels of lamin A. The degree of matrix stiffness (microelasticity) regulated the transcriptional activity of *LMNA* and also influenced the phosphorylation of lamin A. Cells grown on stiff gels compared to soft gels showed an increased expression of lamin A as well as a reduced phosphorylation of lamin A threonines and serines. Reduced phosphorylation would promote assembly of lamin A homodimers into higher-order filaments.

Lamin A regulates a number of genes, one of which is the transcription factor, serum response factor (SRF), and through SRF, lamin A targets genes such as alpha smooth muscle actin (*ACTA2*) and a number of cytoskeletal genes involved in stem cell differentiation. Swift *et al.* (2013) demonstrated that the transcription of the gene *LMNA* was regulated by all-*trans* retinoic acid (RA) and an antagonist of RA denoted AGN. RA is a factor involved in vertebrate morphogenesis, growth, cellular differentiation, and tissue homeostasis (Mark *et al.*, 2006). RA acts by binding to the RA nuclear receptor, RARG, with the complex then binding to sites in the promoter region of the *LMNA* gene. RARG can reside in the nucleus or cytoplasm dependent on conditions such as whether a ligand is bound and whether the receptor is heterodimerized with RXR receptors (Han *et al.*, 2009). In mesenchymal stem cells grown on soft gels, RARG was mostly nuclear in location (nuclear to cytoplasmic ratio 3:1), but increases in the stiffness of the matrix and increases in the amount of lamin A were associated with significant increases in the nuclear to cytoplasmic ratio of RARG by as much as fourfold. Hence, the nuclear localization of RARG was a function of both lamin A abundance and degree of matrix stiffness. Whether the only mechanism by which matrix stiffness regulates *LMNA* transcriptional activity is through changes in the nuclear/cytoplasmic distribution of RARG is not clear. In mesenchymal stem cells grown on soft gels, RA had essentially no effect on lamin A expression since *LMNA* transcriptional activity was low and more RARG was cytoplasmic in location. In mesenchymal stem cells cultured on stiff gels, *LMNA* transcriptional activity is high and almost all RARG was residing in the nucleus. Under these conditions, RA suppressed *LMNA* expression, while the antagonist AGN increased *LMNA* expression. In keeping with this observation, RA suppressed osteogenesis in mesenchymal stem cell grown on stiff gels by repressing lamin A expression and reducing lamin A levels, while an antagonist had the opposite effect, i.e., enhanced osteogenesis.

Furthermore, Swift *et al.* (2013) developed the following model: increases in mechanical stress are associated with increases in extracellular matrix stiffness. Changes in the degree of matrix stiffness are transmitted to the cell nucleus. The "transmission" of increased matrix stiffness to the nucleus promotes increased lamin A expression as well as suppression of lamin A phosphorylation. Thus, increased matrix stiffness leads to an increased abundance of lamin A and a greater assembly of lamin A dimers into filaments. Reduced matrix stiffness was associated with low levels of lamin A and greater phosphorylation leading to lamin A filament disassembly. In addition, increased matrix stiffness promotes increased expression of extracellular collagen type 1. The degree of matrix stiffness and the level of lamin A also regulate the transcription of the gene *LMNA* by determining the distribution of RARG between the cytoplasm and nucleus. High matrix stiffness and increased levels of lamin A promote the localization of RARG to the nucleus. RA binding to the receptor represses lamin A expression in the context of high matrix stiffness, while an antagonist AGN increases *LMNA* expression. Finally, lamin A regulates a number of genes involved in the differentiation of stem cells. This regulation can occur by several mechanisms: via the action of transcription factors such as SRF and by the interaction of chromatin with the nuclear lamina to form LADs which are generally regions of gene repression.

The observations reported by Swift *et al.* (2013) help to unify some of the diverse functions of lamins. Lamin A appears to function as an intermediary in the mechanisms linking (1) mechanical stresses applied to a tissue, (2) the cellular responses which modify matrix stiffness appropriate to those stresses, and (3) the mechanisms linking mechanical stresses to the induction of tissue-specific differentiation of stem cells.

Comparative lamin biology

The comparative properties of lamin proteins have been reviewed in several reports: Dittmer and Mistelli, 2011, Erber *et al.*, 1999, and Peter *et al.*, 1989. Lamins are found only in metazoans and have not been detected in single-cell eukaryotes such as yeast. The evidence is that a B-type lamin was the ancestral form, and hence, B-type lamins appeared before its A-type counterpart. The A-type lamins are restricted to amphibians, fish, birds, and mammals; chicken lamin A shares 91% amino acid sequence similarity with human lamin A and 87% amino acid sequence similarity with *Xenopus* lamin A. Humans, amphibians, and birds share the same lamin A Ca1a2X motif, which consists of cysteine–isoleucine–serine–methionine (CISM). Since the Ca1a2X motif directs farnesylation (Rusinol and Sinensky, 2006), the prelamin A in mammals, birds, amphibians, and probably fish most likely undergoes similar post-translational processing to mature lamin A in these different vertebrate groups (Peter and Stick, 2012).

Peter and Stick (2012) have reviewed existing data with respect to the evolution of the lamins. Lamins are founding members of the multigene family of intermediate filament (IF) proteins. All IF proteins share common structural features: a central rod domain capable of forming specific coil arrangements flanked by a short N-terminal and a longer C-terminal domain. Lamins have three additional characteristic features associated with their specific nuclear localization: a nuclear localization signal (NLS) segment within the C-terminus end, a Ca1a2X motif at the very end of the C-terminus, and a cyclin-dependent kinase 1 (CDK1) phosphorylation site on the N-terminus side of the central rod. Phosphorylation and dephosphorylation at the CDK1 site triggers disassembly and reassembly of lamin filaments, respectively, during cell division. The NLS segment directs the import of the lamin into the nucleus, while the Ca1a2X motif is involved with farnesylation and carboxymethylation which add hydrophobic groups to the lamin. It is believed that an archetypal lamin progenitor gave rise to cytoplasmic IFs by the "loss" of the NLS segment and the Ca1a2X motif. Lamins are found only in metazoans; invertebrates possess a single lamin gene encoding a B-type lamin, whereas vertebrates have multiple lamin genes. Lamin A is restricted to vertebrates. During vertebrate evolution, it is believed that the prototype lamin type B gene, which contained nine introns, underwent two rounds of gene duplications. These duplication events were associated with the gain of a vertebrate specific intron which was positioned between introns 1 and 2 of the prototype gene. Other structural changes included adding an extra intron to the linker 1 region of the rod domain, giving rise to the *LMNB2* gene and inserting a new exon into intron 9 to extend the tail domain giving rise to the *LMNA* gene. In essence, the gene *LMNA* arose by insertion of a new exon into the last intron of a B type progenitor gene. Mammals possess three lamin genes: *LMNB1*, *LMNB2*, and *LMNA*. *LMNA* produces two lamins, A and C, through alternative splicing. Lamin C lacks the Ca1a2X motif and hence does not undergo farnesylation.

The investigators noted that for mammalian species, lamin A sequences are very similar; 97–98% of residues are identical in the dog, pig, and human. One particular substitution, however, is associated with very different contextual results. In canine and porcine lamin A, there is a serine in position 602, while in human lamin A, glycine occupies position 602. A mutation in the human *LMNA* gene, resulting in a substitution of serine for glycine at position 602 of the lamin A protein, leads to insulin resistance (Young *et al.*, 2005). Given that lamin A is almost identical in the dog, pig, and human, it is difficult to understand how a serine at 602 could be the "wild type" in the dog and pig and yet be associated with a "disease" phenotype in the human. This observation is consistent with an epistatic effect (Breen *et al.*, 2012; Wagner, 2012). The substitution of a serine for glycine at position 602 in the mammalian *LMNA* gene requires a specific genomic background as found in the human but not in the dog or pig in order to result in a "diabetic"-like phenotype. This natural experiment indicated that whether the occurrence of a serine in position 602 of the lamin A protein was neutral or represented a deleterious mutation was dependent on the genomic context or background in which that substitution occurs — an epistatic effect. Perhaps epistatic factors are involved in the phenotypic expression of other mutations in the human *LMNA* gene which are associated with features such as premature ageing.

Cardiovascular disease

Accelerated vascular remodeling is a prominent feature of individuals with HGPS and generally is the cause of death (stroke, myocardial infarction) by age 13 (Gordon *et al.*, 2012; Olive *et al.*, 2010; Pollex and Hegele, 2004). Olive *et al.* (2010) performed a detailed morphological examination of the cardiovascular system in two deceased individuals with HPGS: a nine-year-old girl and a 14-year-old boy. Both had a C>T nucleotide substitution at position 1824 in exon 11 of the *LMNA* gene. The two patients with HGPS had evidence of advanced vascular morphological changes that had many features of atherosclerosis. The coronary lesions showed a dense fibrotic structure, but foci of calcification, foam cells, and necrotic areas were also present. The aorta in the subjects with HGPS displayed intimal and adventitial thickening with about a 50% loss of medial smooth muscle cells. The one structural feature that appeared to be different in the HGPS subjects was the extent of vascular adventitial thickening. In both the aorta and coronary vasculature, the degree of adventitial thickening was greater in those with HGPS compared to non-HGPS subjects of various ages. The investigators described the vascular changes in HPGS subjects with the predominance of adventitial fibrosis as analogous to a chronic injury. Most cells (medial smooth muscle cells, intimal plaque cells, adventitial fibroblasts) stained positive for progerin. For example, in HGPS coronary arteries, 68% of plaque cells, 91% of cells in the media, and 77% of cells in the adventitia contained progerin. In keeping with the observation that normal cells do utilize the cryptic 5′ss in the *LMNA* gene and produce low levels of progerin, coronary arteries of non-HGPS subjects also stained positive for progerin. The staining level was extremely low, amounting to less than 0.02% of cells. In non-HGPS individuals, the percentage of cells staining positive for progerin was much greater in the adventitial layer than for intimal plaques or the media. Also, for all three layers, there was a significant increase in the percentage of cells staining for progerin with the age of the donor. For example, in the adventia, the percentage of cells staining for progerin increased from 0.1% to 2% over the age range of one month to 97 years old for control subjects. Finally, the two subjects with HPGS had none of the classical risk factors for atherosclerosis. Both were normotensive throughout life and serum lipids for both were essentially within normal limits (Olive *et al.*, 2010; supplement material).

Summary and perspectives

The unusual ageing features of HGPS and the discovery that HGPS is the result of a rare point mutation in the *LMNA* gene (which encodes lamins A and C), have spurred investigations into the role of lamins A/C in the "normal" ageing process. The responsible HGPS point mutation results in the increased utilization of a sheltered 5′ss within exon 11 of the *LMNA* gene. Utilization of this splice site leads to the production of a truncated form of lamin A called progerin. However, it has been well established that normal cells also use this sheltered splice site in the *LMNA* gene. Hence, low levels of progerin are a normal cellular constituent. Comparative data suggest that epistatic factors are involved in the phenotypic expression of *LMNA* gene mutations although the nature of these epistatic mechanisms is unknown.

Lamins A/C are filamentous proteins that form part of the nuclear lamina. Lamins subserve important functions such as maintaining the structural integrity of the nucleus as well as regulating the compartmentalization of nuclear functions. Lamins interact with DNA, chromatin histones, and histone-modifying enzymes. Lamins bind to transcription regulators and can alter gene expression patterns not only through this binding function, but also epigenetically by modifying chromatin histones tethered to the nuclear lamina through the recruitment of histone-modifying enzymes (deacetylases and methyltransferases). Lamins are also involved in the DDR. Additional studies have linked lamins with telomere biology. Progerin can induce telomere damage and shortening as can overexpression of wild-type lamin A. Hence, lamins play a role in telomere biology, by regulating telomere integrity and length. Shortened telomeres appear to activate cellular use of the sheltered 5′ss, leading to the production of more progerin. Hence, there appears to be a feedback loop with short telomeres inducing the production of progerin and progerin inducing telomere loss. The wide-ranging set of nuclear functions that lamins A/C regulate will clearly not be "correctly" regulated by mutant forms of lamin A/C such as progerin.

If other aspects of telomere biology are included, the web of interactions involving lamins becomes even more extensive. As noted by Blackburn and

Epel (2012), telomere shortening (loss) has been associated with chronic stress. Cortisol, one of the stress hormones, can contribute to telomere shortening by reducing the activity of telomerase (Blackburn and Epel, 2012). There is also evidence that an active versus a sedentary lifestyle can slow down telomere attrition with ageing (Stromberg, 2013). Hence, chronic stress appears to induce telomere shortening and telomere shortening will in turn induce the cellular production of more progerin.

In summary, there is a very complex web of interactions between lamins, normally produced mutant forms of lamins (progerin), telomere integrity and shortening, chronic stress, DDR, and cellular ageing. In HGPS, the most extreme manifestation of the accelerated ageing phenotype is severe vascular remodeling which is usually the cause of early mortality in these individuals. These severe vascular structural changes occur in the absence of an elevated arterial pressure or elevated serum lipid levels. The details of the mechanistic links between telomere biology, lamin functions, progerin production, and extensive vascular and connective tissue structural remodeling remain to be worked out. However, this multifaceted web of interactions opens a new window through which one can view a different set of mechanisms underlying the ageing phenotype.

mTOR Pathway

In a recent review article, Johnson *et al.* (2013) summarized observations implicating the mTOR (mammalian target of rapamycin) pathway as a key determinant of the ageing phenotype. In the abstract of their review, the authors stated: "Many experts in the biology of ageing believe that pharmacological interventions to slow ageing are a matter of 'when' rather than 'if.' A leading target for such interventions is the nutrient response pathway defined by the mechanistic target of rapamycin (mTOR)." The mTOR network is a major determinant of ageing and lifespan, but this statement by the authors of this review is misleading. The ageing process is as complex as the living state itself. In fact, the two are clearly intertwined and inseparable. To imply that one selected pathway holds the key to ageing and that the ageing process can be significantly altered by manipulating discrete steps in this single pathway is not consistent with a wealth of biological and genomic data (Singer, 2011; 2013). Within the context of evolution, natural selection only "sees" the phenotypic expression of the underlying genetic variation. Traits are selected which allow the whole animal to adapt to ecological pressures. Evolution does not work to optimize single systems or individual biological processes. By analogy, ageing can be considered as the mirror image process within this same context. No single pathway or biological process is "responsible" for the ageing phenotype. The inseparable nature of ageing and the "living state" is illustrated by the following example. Telomeres perform a critical function in protecting chromosome ends from being treated as sites of DNA damage. This function allows correct DNA replication to occur at the time of cell division. Yet, by the very nature of the telomere structure, DNA damage is difficult to repair when it occurs in this DNA region (Singer, 2013). Also, the loss of telomere length with successive cell divisions sets a limit on the number of possible cell divisions that can occur and when a critical telomere length is reached a senescent state is triggered in the cell. This example illustrates that ageing and the living state cannot actually be treated as separable issues. In this example, ageing represents the trade-off between the structure of telomeres that has evolved to solve the chromosome end problem and functional consequences of that structure, namely, the attrition of telomere length with cell division and the inability to completely repair DNA breaks within the telomere region. Evolutionary solutions to ecological pressures by their very nature contain ageing features. Yet, the implication of the statement of Johnson *et al.* (2013) appears to be that ageing is a "disease" which can be pharmacologically managed.

The role of the mTOR pathway in the ageing process has been reviewed by Johnson *et al.* (2013) and Kaeberlein and Kennedy (2011). The antifungal agent rapamycin was first discovered in 1970 and details concerning its mechanism of action first became clarified in the 1990s. In 2006, rapamycin was observed to extend the lifespan of yeast and in 2009, this compound was shown to extend the lifespan of mice. mTOR is a serine /threonine protein kinase and a member of the phosphatidylinositol-3-OH kinase (PI3K)-related family. The protein mTOR is encoded by the gene *MTOR* and functions in two distinct complexes, mTORC1 (rapamycin-sensitive) and mTORC2 (rapamycin-insensitive). Much more is known about the upstream regulation and downstream targets of mTORC1 than those of mTORC2. mTORC1 is a multi-protein complex containing mTOR and five or six other proteins. Rapamycin inhibits the activity of mTORC1 by binding to a cytosolic protein (FKBP12). The rapamycin–FKBP12 unit interacts physically with mTORC1, disrupting the interaction between mTOR and another mTORC1 protein constituent called Raptor. This disruption leads to the inhibition of the activity of mTORC1 (Johnson *et al.*, 2013). mTORC1 has a number of upstream regulators. Insulin and other growth factors activate mTORC1 through PI3K and AKT (protein kinase B) signaling. The mechanism responsible for this activation involves two other proteins known as tuberous sclerosis complex 1 and tuberous sclerosis complex 2 (Hay and Sonenberg, 2004). Amino acids work via a different pathway to induce a conformational change in the mTORC1 complex that activates the complex and promotes phosphorylation of downstream targets. mTORC1 is inhibited by AMP-activated protein kinase (AMPK) which is responsive to low ATP levels. AMPK is activated by cellular stressors that reduce ATP generation such as nutrient starvation (glucose deprivation), hypoxia, and metabolic poisons (Inoki *et al.*, 2012). In addition, reduced p70S6 kinase activity can activate AMPK (Kaeberlein and Kennedy, 2010). Two important downstream outputs of mTORC1 are regulation of protein synthesis and regulation of autophagy.

mTORC1 promotes global mRNA translation and protein synthesis by phosphorylating p70S6 kinase and eukaryotic translation initiation factor 4E binding protein 1 (4E-BP1). Activation of p70S6 kinase results in the phosphorylation of ribosomal protein S6 (rpS6) and other components of the translational machinery. Phosphorylation of 4E-BP1 decreases its binding to the eukaryotic translation initiation factor which then allows the recruitment of ribosomal components and initiation of cap-dependent translation. In essence, phosphorylation of p70S6 kinase directly promotes mRNA translation, whereas phosphorylation of 4E-BP1 releases an inhibition on mRNA translation.

Autophagy is the process whereby cellular constituents are degraded by a membrane (autophagosome) lysosomal system and the degradation products recycled (Marino *et al.*, 2008). mTORC1 negatively regulates autophagy through inhibitory phosphorylation of ULK1 which prevents the formation of a complex required for initiation of autophagy.

The observation that rapamycin, which inhibits the activity of mTORC1, can extend lifespan in yeast, fruit flies, and mice (Johnson *et al.*, 2013) indicates that mTOR is an important modulator of the ageing process. Inhibition of mTORC1 activity would reduce its downstream outputs (i.e., decrease protein synthesis and promote autophagy) and this reduction in downstream outputs would appear to be the mechanistic link by which mTORC1 modulates the ageing process.

As noted by Johnson *et al.* (2013), there is considerable crosstalk between the mTOR pathway and the insulin/insulin-like growth factor 1 (IGF-1) network. The insulin/IGF-1 signaling network has also been implicated in modulating the ageing process. In addition, several recent papers have reported experimental observations linking the mTOR pathway and nuclear lamins, thereby uncovering an even more extensive web of biological processes underlying the ageing phenotype.

Ramos *et al.* (2012) examined the activity of the mTOR pathway in mice in which exons 8–11 in the *LMNA* gene were deleted (*LMNA*−/− mice). These mice developed dilated cardiomyopathy and skeletal muscle dystrophy. At four weeks of age, these mice displayed increased phosphorylation of mTOR, rpS6, p70S6 kinase, and 4E-BP1 in heart tissue and increased phophorylation of rpS6 and 4E-BP1 in skeletal muscle, compared to control *LMNA*+/+ mice. Non-diseased tissue such as that of the liver did not show these changes. Echocardiographic studies highlighted left ventricular dilatation and impaired cardiac function in *LMNA*−/− mice. Moreover, *LMNA*−/− mice showed evidence of impaired skeletal muscle function. When these mice were fed rapamycin, they displayed improved cardiac and skeletal muscle function as well as reduced cardiac size. *LMNA*−/− mice fed rapamycin had a 23% increase in mean lifespan. The improved cardiac function with rapamycin treatment was associated with a significant reduction in the extent of phosphorylation of mTOR, p70S6 kinase, and rpS6 compared to untreated *LMNA*−/− mice. However, skeletal muscle in *LMNA*−/− mice, in contrast to cardiac muscle, did not show decreased phophorylation of mTOR signaling components with rapamycin treatment. Cardiac muscle but not skeletal muscle in *LMNA*−/− mice showed evidence of decreased autophagy-mediated degradation compared to control *LMNA*+/+ mice. Treatment of *LMNA*−/− mice with rapamycin resulted in an increase in autophagy-mediated degradation in the heart. These observations indicate that in *LMNA*−/− mice, the activity of the mTORC1 signaling pathway is increased and that this hyperactivity contributes to the cardiac and skeletal muscle dysfunction in these mice. The mechanistic link(s) between impaired production of lamin A/C due to a deletion within the *LMNA* gene and increased activity of mTORC1 signaling was not addressed by these investigators.

Marino *et al.* (2008) also looked at the crosstalk between lamins and the mTOR pathway, but used a different rodent model for their experiments. These investigators studied a mouse model with a mutation in the gene encoding the metalloprotease Zmpste24. This enzyme cleaves the 15 amino acids at the C-terminus including the attached farnesyl group from lamin A. Mice lacking the *Zmpste24* gene present with an early ageing phenotype consisting of alopecia, growth retardation, muscle weakness, bone fractures, and early death (Ibrahim *et al.*, 2013; Johnson, 2013). These mice showed an increased level of autophagy in skeletal muscle, the liver, and the heart compared to wild-type littermates. This increased autophagy was associated with evidence of decreased activity of mTOR signaling as manifested by reduced phosphorylation levels of downstream mTOR targets, 4E-BP1 and p70S6 kinase. The mTOR pathway is regulated by growth factors via PI3K/AKT kinase signaling and by AMPK in response to nutrient signals and ATP levels. The depression of the mTOR pathway in these mice was due to the upregulation of AMPK signaling. Blood glucose concentrations and blood levels of leptin and insulin were significantly reduced in these mutant mice, whereas adiponectin blood levels were significantly increased. Decreased blood glucose levels and increased blood adiponectin levels can induce increased AMPK activity. Hence, mice with a mutation in the gene encoding the protease Zmpste24 displayed premature ageing, had nuclear lamina structural abnormalities due to abnormal post-translational processing of lamin A, and also showed increased autophagic proteolysis. The increase in autophagy was due to the depression of mTOR signaling activity brought about by increased AMPK activity. Marino *et al.* (2008) pointed out that depressed activity of the mTOR pathway, with an increase in autophagy, and reduced insulin levels were features usually associated with lifespan extension. These features in the Zmpste24-null mouse which showed a premature ageing phenotype appear to be paradoxical. In this mouse model, with abnormal lamina A post-translational processing, it is quite possible that a chronic upregulation of autophagy and chronic hypoinsulinemia may actually contribute to the ageing process. These experimental observations underscore that trade-offs contribute to the ageing phenotype. Depression of mTOR signaling activity and hypoinsulinemia which have been considered "anti-ageing" processes appear to contribute to the premature ageing in this mutant mouse model.

Recently, Ibrahim *et al.* (2013) used the Zmpste24-null (Zmpste24−/−) mouse model to explore additional interrelationships between mTOR signaling and lamin biology. These investigators bred mice with a double mutation: homozygous null for Zmpste24 and homozygous for a hypomorphic (reduced gene function) allele for the gene *ICMT* (Icmt/hm). ICMT (isoprenylcysteine carboxyl methyltransferase) carries out the carboxymethylation of the terminal cysteine of the farnesylated lamin A. ICMT expression and activity levels were 70–90% lower in mice homozygous for the hypomorphic allele compared to wild-type mice. As already noted, Zmpste24−/−, ICMT+/+ mice showed a premature ageing syndrome, whereas Zmpste24−/−, Icmt hm/hm mice showed much fewer ageing features. Zmpste24−/− mice deficient in ICMT had a stronger grip, longer survival, and greater normal body weight than Zmpste24−/−, ICMT+/+ mice. In hepatocytes, prelamin A was localized to the nuclear rim in Zmpste24−/−, LCMT+/+ mice, but was abundant in the nucleoplasm in Zmpste−/−, Icmt hm/hm mice. Zmpste24−/−, LCMT+/+ fibroblasts proliferated slowly and senesced prematurely compared to Zmpste24−/−, Icmt hm/hm cells. Cell lysates from Zmpste24−/−, Icmt hm/hm mice showed activation of mTOR signaling (increased phosphorylation of downstream targets) compared to Zmpste24−/− , LCMT+/+ mice. Enhanced mTOR signaling was due to activation of the AKT kinase, and the investigators observed that prelamin A was required for the increased AKT/mTOR signaling accompanying ICMT deficiency. Inhibitors of AKT but not mTOR (rapamycin) blocked the proliferation of Zmpste24−/−, Icmt hm/hm fibroblasts. Finally, the investigators demonstrated that reducing ICMT expression (induced by RNA interference technology) in HGPS fibroblasts increased the proliferation rate and delayed senescence. In normal fibroblasts, inhibiting the expression of ICMT had no effect on cell proliferation. In summary, Ibrahim and co-workers showed that mice lacking the *Zmpste24* gene developed a premature ageing phenotype that could be significantly attenuated by simultaneously inhibiting the enzyme ICMT. Reduced carboxymethylation of the farnesylated lamin A activated AKT-mTOR signaling, which appeared to be the pathway responsible for reversing the manifestations of early senescence.

The studies by Ramos *et al.* (2012), Marino *et al.* (2008), and Ibrahim *et al.* (2013) demonstrated that there is considerable crosstalk between alterations in lamin A biology and the mTOR pathway. In the study of Ramos *et al.* (2012), mice with a deletion in the *LMNA* gene and hence impaired production of lamin A showed increased mTOR signaling activity, whereas in the study of Marino *et al.* (2008) and Ibrahim *et al.* (2013), mice with impaired post-translational processing of lamin A (absent Zmpste24 activity) showed decreased mTOR signaling activity. Marino *et al.* and Ibrahim *et al.* implicated at least two possible causes for decreased mTOR signaling in Zmpste24-null mice: reduced AKT kinase activation by the carboxymethyl moiety of the farnesylated lamin A and activation of AMPK by an undefined mechanism. There are some inconsistencies between these studies. Ramos *et al.* observed that LMNA-null mice had increased mTOR signaling, whereas Ibrahim *et al.* found that lamin A was required for the increased AKT/mTOR signaling associated with ICMT deficiency. However, since lamins interact with DNA, transcription regulators, and histone-modifying enzymes, there are a number of mechanisms by which lamins could alter mTOR signaling activity. The important observation is that the various biological mechanisms that have been associated in the ageing process cannot be considered in isolation. Based on this short review, it is evident that nuclear lamins, telomere biology, alternative splicing, and the mTOR pathway are all interconnected via multiple functional interactions. The very biological functions that sustain the living state are inherently "imperfect" and as a consequence, induce ageing.

Molecular Clocks

The importance of circadian molecular clocks in animal physiology has been well documented. Disruptions in the functioning of these clocks have been associated with impaired cognitive functioning and neuropsychiatric behaviors (Singer, 2011; Singer, 2013). In addition, there is evidence of ageing of molecular clock functions (Singer, 2013). Although the basic machinery of molecular clocks has been well described, the finer details concerning how clock genes are regulated are becoming clarified (Masri and Sassone-Corsi, 2013). It is currently believed that about 10% of the genome in a given cell is subject to circadian oscillation.

A heterodimeric transcription factor formed by the proteins CLOCK and BMAL1 binds to a regulatory sequence (known as an E-box motif) in the promoter region of target genes. These genes include those that encode for the proteins PER1, PER2, PER3, CRY1, and CRY2. The period (PER) and cryptochrome (CRY) proteins translocate to the nucleus and displace CLOCK and BMAL1 from their binding sites, thus repressing their own transcription. PER and CRY proteins are degraded which derepresses CLOCK- and BMAL1-mediated transcription of its target genes (including those encoding PER and CRY proteins), hence beginning a new cycle.

Koike *et al.* (2012) examined the temporal details of this cyclic mechanism. First, a transcriptionally poised state exists in which both activators (BMAL1 and CLOCK) and repressors (PER and CRY) occupy regulatory sequences on the target genes. Subsequently, several steps occur. The binding of repressor proteins decreases, a co-activator protein (p300) binds to the regulatory region, and histone epigenetic marks are established. These steps lead to the activation of transcription of the target genes by the bound BMAL1–CLOCK protein complex. Target genes include those that encode for the proteins PER and CRY proteins. In the final repressive state, BMAL1 and CLOCK occupancy decreases and these proteins are replaced by the repressor proteins (PER and CRY) and transcription decreases. Koike *et al.* (2012) observed some interesting features of this cyclic clock mechanism when they examined the mRNA transcriptional activity of 3408 clock-controlled genes. Among these genes, the cycles were driven primarily by transcriptional activity in 458 of them; i.e., the cyclic mRNA levels of these genes reflected cyclic transcriptional regulation. In 1,579 genes, the cyclic levels of mRNA were the result of post-transcriptional regulation and not due to cyclic transcriptional activity, e.g., cyclic changes in mRNA splicing. In 913 genes, pre-mRNA levels showed cyclic transcriptional activity, but mature mRNA levels showed much dampened oscillations. Hence, transcriptional regulation of cycling mRNA transcripts occurs in less than 20% of clock-controlled genes. Post-transcriptional processes contribute significantly to the generation of steady-state cycling mRNA levels. On a genome-wide basis, binding of RNAP II to DNA was cyclic as was the establishment of epigenetic histone marks (methylation/acetylation) at gene promoters. These cyclic changes were detected for thousands of expressed genes whether or not mRNA cycling was detectable. The biological significance of these cyclic rhythms in RNAP II binding and regulation of histone epigenetic marks was unclear given that these genome-wide cycles did not appear to correlate with cyclic levels of mRNA.

More details concerning the post-transcriptional regulation of the clock mechanism are slowly being unravelled (Doherty and Kay, 2012). Lim and Allada (2013) studied the clock system of *Drosophila*. The gene *TYF* ('twenty-four') encodes a protein which is required for the efficient translation of the core clock protein PER mRNA. Lim and Allada (2013) found two proteins that specifically associated with the 290-amino acid domain of TYF at the C-terminus: ATX2 and PAPB (polyadenylate-binding protein). ATX2 is the fly homolog of the mammalian *ATAXIN-2* gene, ATAXIN-2 being an RNA-binding protein. Polyglutamine expansion of human ATAXIN-2 is associated with neurodegenerative diseases such as amyotrophic lateral sclerosis and spinocerebellar ataxia type 2. ATX2 has a polyadenylate-binding-protein-interacting domain by which ATX2 and PAPB associate. TYF and PAPB bind to separate regions of ATX2. TYF then associates with clock gene mRNAs particularly PERIOD and TIMELESS. ATX2 also binds to PERIOD mRNA and this binding activity depends on its interaction with PABP. In the study, ATX2 was required for TYF activation as well as recruitment of PAPB to the TYF containing protein complex. In summary,

ATX2 formed a complex with TYF and recruited PAPB to this complex. This indicates that ATX2 was crucial to the function of TYF by coordinating the formation of an active complex (ATX2–TYF–PABP–PERIOD mRNA) for the translation of PERIOD mRNA.

Duong *et al.* (2011) used murine liver and lung tissue to examine the mechanisms by which PER proteins inhibit transcription. These investigators isolated PER complexes which contained a number of proteins such as PER, CRY, and RNA-binding proteins. One particular RNA-binding protein was PSF (PTB-associated splicing factor), which functions as a transcriptional co-repressor by recruiting histone deacetylase complexes. PER complexes interact with CLOCK–BMAL1 dimers bound to the promoter region of clock-controlled genes (including the genes encoding for PER and CRY proteins themselves). By virtue of its constituent PSF, the PER complex recruits a deacetylase complex to the promoter locus which deacetylates histones 3 and 4, and through this chromatin modification represses CLOCK–BMAL1 transcriptional activity. Many of the proteins associated with molecular clocks function as epigenetic chromatin modifiers (Masri and Sassone-Corsi, 2013). The CLOCK protein acts as a transcription factor and as a histone acetyltransferase (HAT). The co-activator protein p300 also functions as a histone acetyltransferase. According to Masri and Sassone-Corsi, CLOCK and BMAL1 promote transcription by modification of chromatin histones at the promoter loci of clock-controlled genes. One can envisage a cycle in which CLOCK–BMAL1 dimers promote transcription by acetylating specific histone residues, while PER complexes containing PSF repress transcription by deacetylating specific histone residues (Duong *et al.*, 2011).

In summary, the molecular processes underlying the cyclic expression of clock-controlled genes include transcription-factor regulation of RNAP II recruitment, chromatin histone epigenetic modifications (e.g., acetylation and deacetylation), and changes in post-transcriptional alternative RNA splicing. The data of Koike *et al.* (2012) indicate that cyclic changes in RNAP II recruitment/initiation and cyclic changes in the establishment of histone epigenetic marks are very pervasive in thousands of expressed genes across the genome, although many of these genes do not display cyclic mRNA expression. This observation reveals that additional, though as yet undefined, regulatory mechanisms control whether these underlying widespread cyclic changes are converted into actual cyclic gene expression.

The observation that post-transcriptional regulation of cyclic mRNA expression (probably through changes in alternative splicing) is a crucial mechanism underlying circadian clock function provides another example of intersecting biological processes. As discussed in a previous section, lamin A and telomere integrity and length can regulate alternative splicing patterns. Lamin A and/or telomere-induced alterations in alternative splicing could result in changes in the cyclic expression of those clock-controlled genes regulated by post-transcriptional processes.

The functional organization of the central and peripheral circadian clocks has recently been reviewed by Masri and Sassone-Corsi (2013). The central clock in the suprachiasmatic nucleus (SCN) of the hypothalamus is entrained by light via the retinohypothalamic tract and directs circadian rhythms in peripheral tissues using humoral cues and synaptic processes. In this manner, the central clock communicates internal time to all organ systems in the body. Conversely, peripheral tissues release various hormones and metabolites in a circadian pattern and these oscillating factors (e.g., glucose, insulin, leptin, glucocorticoids, and mineralocorticoids) feed back to the SCN. How these peripheral signals are sensed by oscillators in the SCN is unknown, but the important observation is that the central clock and peripheral clocks maintain a two-way system of communication.

A key feature of the SCN clock is its entrainment by light. Light signaling received by the retina is transduced to the SCN and drives resetting of the clock by a mechanism thought to involve the rapid transcription and translation of the repressor proteins PER1 and PER2 (Cao and Obrietan, 2010). One key pathway underlying light entrainment is the mitogen-activated protein kinase (MAPK) pathway, and one of the routes by which this pathway regulates light entrainment is through mTOR signaling (Cao and Obrietan, 2010). Examination of the brain tissue comprising the SCN has revealed high levels of expression of the mTORC1 effectors which regulate mRNA translation, namely p70S6 kinase and 4E-BP1. The phosphorylation of these mTORC1 pathway effectors was triggered in mice by light exposure during the early or late part of night. This light-induced phosphorylation of p70S6 kinase and phosphorylation of 4E-BP1 could be suppressed, at least temporarily, by the intracerebroventricular infusion of rapamycin. The upstream regulator of the mTOR pathway was MAPK signaling since this pathway was activated by light. Hence, the overall mechanism appeared to be the light activation of the MAPK pathway which then activates mTORC1. Additional evidence in support of the involvement of mTOR signaling in SCN light entrainment was the observation that light-induced PER1 and PER2 protein expression could be suppressed by the administration of rapamycin. These observations indicate that mTOR signaling regulates central clock physiology and also provide a link between molecular clock functioning and the biology of lamins.

One final note: the complexity of molecular clocks and in fact their ultimate function remain areas of intense research (Asher and Schibler, 2011). Between 2–15% of all expressed genes appear to manifest circadian cycles. A large fraction of these genes encode enzymes and regulators of carbohydrate, cholesterol, and lipid metabolism. Crosstalk between the central clock in the SCN and peripheral clocks is quite complex as previously noted. In the liver, for example, the fraction of cyclic genes controlled by systemic signals and local oscillators is 14% and 86%, respectively (Asher and Schibler, 2011). The phases of many peripheral clocks can also be entrained by feeding-fasting cycles. Although the importance of molecular clocks is well established, the observation that "cyclic" genes in the liver of mice with a double Cry1/Cry2 knockout (KO; and hence "clock-less") can still be phase-entrained by feeding rhythms is difficult to understand. In fact, clock-less laboratory rodents appear perfectly viable from a metabolic perspective (Asher and Schilber, 2011). These observations in rodents with genetic disruption of core clock genes point to redundancies in the circadian/molecular clock system that have not yet been elucidated. In addition, the clock system can also be plastic and display modified molecular mechanisms in species that engage in around-the-clock activities (Bloch *et al.*, 2013). Furthermore, recent experimental studies of the circadian clock neuronal network in *Drosophila* have underscored a complexity in this network not previously appreciated (Yao and Shafer, 2014). The network was found to consist of multiple independent oscillators and behavioral rhythms emerged from the interactions of these independent oscillators rather than from a single group of master pacemakers.

Summary and Perspectives

The initial observation that the premature ageing phenotype of HGPS was associated with mutations in the gene encoding lamin A/C underscored the link between lamin biology and the ageing process. This link was strengthened by the documentation that normal cells produce, although in low amounts, the mutant lamin A protein progerin that accumulates in HGPS. Using this link as a starting point, one can trace a vast web of interconnections between lamin biology and a number of other biological processes that have been implicated in ageing. These processes include: regulation of telomere length and integrity, mTOR signaling pathway, alternative splicing patterns, DNA replication and repair, and the function of molecular clocks. The interconnections between these multiple processes can result in situations in which a mechanism that was pro-survival under one set of circumstances can become an "ageing" process under a different set of circumstances. For example, the observations of Marino *et al.* (2008) (previously discussed) indicate that in Zmpste24-null mice, the mTOR signaling pathway is downregulated with the result that autophagic activity is increased. As pointed out by these investigators, the autophagic pathway which is pro-survival in the short term can be detrimental when chronically activated as occurs in these mice with impaired lamin A/C maturation due to absence of the metalloprotease Zmpste24. Regardless, this web of interconnections between various biological processes accentuates that the living state and the ageing process are tightly intertwined; there are no specific "longevity" genes or "ageing" genes. The biological processes that maintain the living state are the very same processes that underlie the ageing process. However, this discussion has not included the profound influence of environmental cues on the course of ageing.

CHAPTER 2

Ageing and Environmental Cues (Experiences)

In this discussion, environmental cues are considered to be any experience that an individual animal undergoes. These experiences would include the type and quantity of food intake, level of physical activity, and the nature of social interactions with other animals to name a few. The terms 'environmental cues' and 'experiences' are treated as interchangeable. When considering environmental cues, it is important to remember that in vertebrates such as mammals, exposure to environmental begins *in utero* during the period of embryogenesis and fetal maturation. Each newborn human has a unique genome and epigenome (Singer, 2013). For example, McVean and the 1000 Genomes Project Consortium (2012) examined over 1,000 human genomes and found that individuals possessed more than 2,500 non-synonymous SNPs and more than 150 loss of function variants at evolutionarily conserved sites. Many of these variants were found in untranslated regions, non-coding RNA sequences, and transcription factor-binding motifs — sites involved in regulatory functions. The unique genome and epigenome of an individual will shape how this particular individual responds to specific experiences.

In his book *The Blue Zones*, Buettner (2008) described geographical areas in which there were a large number of long-lived humans, many of them centenarians. These areas have been referred to as 'blue zones.' The four areas discussed in the book included: the Barbagia region of Sardinia, Okinawa, Loma Linda California, and the Nicoya Peninsula region of Costa Rica. Buettner and colleagues interviewed many inhabitants of these blue zones and tried to uncover common factors responsible for their longevity. Given the diversity of these geographical areas, it is unlikely that the longevity of the resident populations has a common genomic basis. These people do not share a collection of "longevity" genes. The commonalities among these groups were all environmental cues as previously defined. In the last section of the book, Buettner distilled out some common characteristics based on the data collected in these blue zones. Recommendations built upon these common features included: regular low-intensity physical activity as an integral part of one's daily schedule, not overeating, having a plant-based diet with limited ingestion of meat, daily but modest intake of alcohol, having a sense of purpose, reducing stress through breaks in the work schedule and through social activities, engaging in spiritual or religious activities, making family a priority, and associating with individuals with a similar value system, i.e., like-minded individuals. As one reads about the individuals living in these blue zones, it is apparent that they have a relatively simple, active life driven by a sense of purpose and a devotion to family. These recommendations all focus on environmental cues which broadly comprise the nature and quantity of food intake, regular physical activity, and developing a social support network which helps to alleviate stress and create a "purpose" in life.

The importance of environmental cues in shaping lifespan as typified by the longevity of blue zone residents described by Buettner is supported by other studies.

Eaton *et al.* (1988) collected data on the incidence of "chronic degenerative diseases" in early non-industrialized societies and compared this data with disease incidence in current affluent industrialized societies. The profound influence of environmental cues (nutrition, physical activity, exposure to toxins such as alcohol and tobacco) on disease incidence is striking. In particular, the very low incidence of "chronic degenerative diseases" (obesity, diabetes mellitus, atherosclerosis, hypertension, and cancer) in "primitive" non-industrialized societies increased dramatically when these societies adopted a Western lifestyle whether by migration or acculturation. Although these "primitive" societies have a low life expectancy, age per se is not the primary determinant underlying the differences in disease incidences. Modern humans live longer but age quite differently than humans of pre-industrialized times.

A more recent genome-wide association study (GWAS) also underscored the importance of environmental cues in shaping the ageing process. Beekman *et al.* (2010) compared the occurrence of "disease"-associated SNPs in older individuals compared to younger controls. One population consisted of 723 individuals (males and females) of mean age 94 years (range 89–104 years) and 721 younger controls of mean age 52 years. A second population of elderly individuals consisted of 979 individuals of age 85 years and older together with 1,167 younger controls of mean age 41 years. The number of SNPs statistically associated with coronary artery disease, heart failure, cancer, and type 2 diabetes was compared between the older and younger populations. There was no difference in the total number of disease-associated SNPs between the long-lived individuals and the younger controls. There was also no difference between the elderly cohort and younger controls when number of disease-associated SNPs for specific diseases was compared. Although not conclusive, the results of this study are consistent with the construct that long-lived individuals do not appear to possess a genomic background characterized by a lesser burden of SNPs statistically associated to cardiovascular disease, cancer, and type 2 diabetes mellitus.

Meyer *et al.* (2012) sequenced the genome of a *Denisovan hominin* who lived 40,000 years ago, using DNA extracted from a finger bone. Data related to the ageing and lifespan pattern characteristic of Paleolithic era humans have been extensively collected and analyzed by Gurven and Kaplan (2007). As expected, infant and child mortality were extremely high among our ancestors, but those surviving to 45 years of age had an expected lifespan of an additional 15–20 years. The major causes of mortality were illness (respiratory, gastrointestinal, and other infections) and accidents and violence. These two categories accounted for over 90% of deaths. So-called degenerative causes accounted for only 9% of deaths. However, this category was very poorly documented and often death due to "old age" was attributed to a degenerative cause. Hence, the causes of mortality in our Paleolithic ancestors were very different from the causes of mortality in our modern society. Yet, the individual *Denisovan* genome sequenced by Meyer *et al.* (2012) contained 28 high-frequency SNPs that have been identified by GWAS to be associated with specific "disease" phenotypes in the modern human. In 11 of these SNPs, the *Denisovan* allele was the risk allele for the associated disease phenotype. These 11 "disease" phenotypes included: bipolar disorder and schizophrenia, altered bone mineral density, coronary disease, migraine, myocardial infarction (early onset), ovarian cancer, QT interval "disorder," and certain hematological and biochemical traits. The important observation is that although the genome of this *Denisovan* individual contained DNA sequence variants distinct from the modern man, the genome also contained SNPs considered as risk factors for several "modern diseases," diseases that appear to have been rare during Paleolithic times.

These case studies of longevity of blue zone inhabitants, comparable burden of "disease"-associated SNPs in very elderly and young individuals, and the extremely low incidence of "chronic degenerative diseases" in early non-industrialized societies, although "disease"-associated SNPs were present in the *Denisovan* genome, underscore the poor correlation between genes and human longevity and ageing trajectories.

For environmental cues (experiences) to alter biological processes underlying ageing, they must ultimately modulate the expression of genes. Regulation of gene expression has become an extensive area of research and in the next sections, selected aspects of gene regulatory mechanisms will be reviewed.

Regulation of Gene Expression: Epigenetics

In the section on **laminar biology**, the role of chromatin–lamina interactions in the regulation of gene expression was discussed. Chromatin domains bound to the nuclear lamina are subject to the establishment of epigenetic repressive marks resulting in repression of the transcriptional activity of the associated tethered DNA sequences. Hence, the formation and release of chromatin–lamina binding and the spatial compartmentalization of active and inactive chromatin represent one level of gene expression regulation. As discussed in a previous section on **alternative splicing**, this alternative splicing process represents another mechanism by which gene expression can be regulated; a single gene can encode for multiple proteins. Finally, recent observations have confirmed what had been previously suspected — protein translation and mRNA transcription can be regulated independently.

Khan *et al*. (2013) quantified protein levels and mRNA transcript levels in lymphoblastoid cell lines from individual humans, chimpanzees, and rhesus macaque monkeys. The expression levels of proteins and corresponding mRNAs from over 3,000 orthologous protein-coding genes were measured in the three species. Khan *et al.* identified 1,151 protein-coding genes that were differentially expressed between the chimpanzee and human at the mRNA and/or protein expression level: 580 genes showed divergent mRNA transcript but not divergent protein expression levels between the chimpanzee and human, 336 genes showed divergent expression levels at the protein level only, and 235 genes showed divergent expression levels at both the mRNA transcript and protein levels. Among genes which showed divergent expression (between the chimpanzee and human) at either the mRNA transcript or protein expression level, the inter-species variation was in general much greater for genes divergent at the mRNA transcript level than for genes divergent at the protein level. These observations suggest that protein translation was under greater evolutionary constraint than mRNA transcription. The different evolutionary constraints with respect to translation and transcription were illustrated by specific gene examples across the three species: human, chimpanzee, and rhesus monkey. The gene *GRB2* showed a pattern of human lineage-specific increase in protein expression levels but no interspecies divergence for mRNA transcript levels; within-species variation was low for both protein levels and transcript levels. This pattern was consistent with human-lineage directional selection for protein translation regulation. The gene *PSMD1* showed a pattern of human lineage-specific decrease in mRNA transcript levels but no interspecies divergence in protein expression levels. This pattern was consistent with a human lineage-specific directional selection for mRNA transcription regulation.

The observations of Khan *et al.* (2013) indicated that across primate species, orthologous coding- gene protein translation and corresponding mRNA transcription can be separately regulated and that protein translation is the more tightly regulated process. Some of the factors involved in the regulation of protein translation include microRNAs and RNA-binding proteins (Filipowicz *et al.*, 2008; Vogel, 2013). The human genome encodes about 800 microRNA genes and possibly 1,000 RNA-binding proteins. These factors, plus others, could be responsible for regulating protein translation such that across primate species, protein translation from orthologous coding genes is "buffered" from large variations in the transcription of corresponding mRNAs from these genes. As noted by Khan *et al.* (2013), the protein products encoded by genes are probably much more relevant to the link between phenotypes and genotypes than the corresponding mRNA transcript levels of these genes. A single mRNA transcript can be translated many thousands of times. Hence, mechanisms must be in place to separately regulate transcription and translation such that large interspecies variations in orthologous gene transcript levels do not result in comparable variations in translated protein levels. These experimental data confirm the presence of separate but coordinated regulatory processes for gene mRNA transcription and corresponding protein translation.

In this section, the focus will be on epigenetic mechanisms and their role in regulating gene expression patterns. The basic level of gene expression regulation starts at the nucleosome itself, a complex consisting of a segment of DNA wrapped around an octamer of histone proteins. Recent experimental studies have shed light on the stacking modes of nucleosomes and the role of linker histones and linker DNA (Travers, 2014). These observations are important pieces in understanding the packing of chromatin.

Bintu *et al.* (2012) studied the nucleosomal elements controlling gene transcription by RNAP II. The three elements examined were the histone tails, specific histone–DNA contacts, and the underlying DNA sequence; the three nucleosomal "domains" considered were entry, central, and exit. RNAP II cannot mechanically detach DNA from the histones. In front of the nucleosomal barrier, the enzyme pauses, often backtracks, and advances only when the DNA spontaneously unwraps from the surface of the core histone octamer. The dynamics of RNAP II, i.e., pause density and duration in each domain, were uniquely controlled to different extents by the histone tails, histone–DNA contacts, and the DNA sequence. The histone tails primarily gate access of RNAP II into the entry domain of the nucleosome by regulating the wrapping and unwrapping rates of DNA around the histone proteins. Post-translational modifications of histone tails (e.g., acetylation of lysine residues) can alter the "compactness" of the histone core and thus the initiation of transcription. Specific DNA–histone interactions appear to regulate the rate of RNAP II transcription in the central domain of the nucleosome, while the nature of the underlying DNA sequence regulates RNAP II transcription at the exit domain. In the central domain, the rate of transcription is a function of the number and duration of polymerase pauses plus any backtracking of RNAP II. The number and duration of pauses is regulated by the kinetics of DNA–histone wrapping and unwrapping. By modeling the transcription process in terms of three specific spatial domains (entry, central, and exit), the investigators were able to assess the relative effects of alterations in histone tails, DNA–histone interactions, or the underlying DNA sequence on the rate of

RNAP II-mediated RNA transcription as determined by the number and duration of polymerase pauses.

The modification of histone tails by acetylation or deacetylation of component amino acid residues is one of the epigenetic "marks" associated with changes in gene expression. The other epigenetic mark is methylation of DNA, which occurs predominantly at cytosines that are adjacent to guanines (CpG: a C and G next to each other joined by a phosphodiester bond). Epigenetics is defined as changes in the expression of genes in the absence of underlying changes in DNA sequences, and epigenetic marks are the chemical flags that mark the modifications on the histone proteins or DNA. In order to produce a diverse set of cell types under the condition that each cell has the same genetic material, cells must be capable of expressing or repressing a given set of genes to generate a neuron, a hepatocyte, or a hematocyte. These complex gene transcription programs initiated during cellular differentiation and division are epigenetically regulated and ultimately ensure that a given cell lineage can be maintained through multiple rounds of cell division or prolonged life in the case of non-dividing cells (Day and Sweatt, 2011).

Cytosine methylation is carried out by methyltransferase enzymes. Most CpGs (about 70–80%), except for CpG islands, are methylated in mammals. CpG dinucleotides do not have a uniform distribution across the genome and DNA areas rich in CpG dinucleotides are referred to as CpG islands (Schubeler, 2012). CpG islands are found close to gene promoter regions and are mostly unmethylated. In general, CpG methylation is associated with gene repression, for example, the methylation of promoter regions of imprinted genes. As pointed out by Schubeler (2012), methylation of CpG promoter islands can block binding of a transcription factor. Epigenetic marks (both DNA and histone protein methylation) can be transmitted across somatic cell generations and hence are somatically heritable (Hathaway *et al.*, 2012).

A description of the DNA methylation of autosomal CpGs has been reported by Ziller *et al.* (2013). These investigators studied 42 whole genome bisulfite sequencing data sets comprising a range of human cell and tissue types. DNA methylation is essential for normal cell development. About 70–80% of all CpGs are methylated, but for most sites, the methylation pattern is stable. However, about 22% of autosomal CpGs show a dynamic variable methylation pattern and most of these CpGs are downstream from transcription start sites. These dynamic variably methylated CpGs co-localize with gene regulatory elements, particularly enhancers and transcription factor-binding sites. It would appear that this fraction of CpGs which displays dynamic changes in methylation state do so as part of coordinated regulatory programs.

Demethylation occurs in the germline (primordial germ cells) and in the early embryo following fertilization and prior to implantation (Faulk and Dolinoy, 2011; Hackett and Surani, 2013a). In primordial germ cells, global DNA demethylation is essential for removal of parent-of-origin-dependent genomic imprints and following this epigenetic demethylation, global *de novo* remethylation takes place (Hackett and Surani, 2013b). In the embryo, *de novo* remethylation begins at the time of implantation. Epigenetic regulation is critically important during development since it forms the underpinnings of cell identity and differentiation (Champagne, 2010; Hackett and Surani, 2013a). Immediately after fertilization, there is widespread demethylation of DNA in the male pronucleus. Zygotic genes are silenced through chromatin-mediated transcription suppression, and imprinting of both maternal and paternal genes also occurs. Hence, epigenetic mechanisms are critical determinants of the transition from the zygote to embryo to fetus. In the adult mammal with ageing, somatic cells undergo gradual DNA demethylation coupled with hypermethylation at normally unmethylated CpG islands (Faulk and Dolinoy, 2011). Recently, a family of enzymes (ten-eleven translocation, Tet) have been described which oxidize methylated cytosine nucleotides and appear to be involved in the process of DNA demethylation (Guibert and Weber, 2012; Schubeler, 2012).

Recently, the epigenetic reprogramming of parental genomes in the zebrafish at the time of fertilization and early zygote development was described and the process differs significantly from that occurring in the mammal (Hackett and Surani, 2013b). In the rodent, parental genomes undergo either passive DNA demethylation (maternal) or active conversion to 5-hydroxymethyl cytosine (paternal), which results in a hypomethylated epigenome in the blastocyst distinct from both maternal and paternal original gametic methylomes at the time of fertilization. Subsequently, the genome undergoes *de novo* methylation at the time of implantation to reset the epigenome for organismal development. In the zebrafish, the paternally inherited methylome is stably inherited and does not undergo any changes at least through early development, whereas the maternally inherited methylome undergoes reprogramming with demethylation of hypermethylated regions and *de novo* methylation of hypomethylated areas. In this loss and gain of methylation patterns, the maternal methylome is reset to that of the male and the overall methylome is now competent for development of the zygote. These contrasting processes raise a number of possibilities. First, the observation that the zebrafish paternal methylome is stably maintained through fertilization and at least early development implies that the paternal germline could be a potential route for transgenerational epigenetic inheritance. Secondly, what roles do the initial parental methylomes, prior to their reprogramming, play in the process of development?

In addition to DNA methylation, histone modifications represent a second class of epigenetic marks. Histone modifications such as trimethylation of lysine 27 and lysine 9 on histone H3 are associated with target gene repression, whereas trimethylation of lysine 4 and lysine 36 and acetylation of lysine 27 on histone H3 are associated with increased gene activity (Carone and Rando, 2012). Hathaway *et al.* (2012), using murine embryonic stem cells and embryonic fibroblasts, looked at the kinetics and heritability of histone epigenetic marks. These investigators measured the expression of the gene *Oct4* by replacing the first exon with an enhanced green fluorescent protein (GFP) reporter. *Oct4* encodes a transcription factor that is critical for self-renewal and pluripotency in stem cells. Upon cellular differentiation (e.g., in murine embryonic fibroblasts), *Oct4* expression is silenced by a number of mechanisms which include trimethylation of lysine 9 on histone H3, HP1 (heterochromatin protein 1) binding, and DNA methylation. Hathaway *et al.* (2012) engineered embryonic stem cells and embryonic fibroblasts containing the *Oct4–GFP* fusion gene with a modified promoter region. The modified promoter region had two new transcription factor binding sites in an arrangement that allowed the investigators to control the recruitment of either the repressor protein HP1 or the transcription activator VP16. Furthermore, the investigators could wash out (remove) these proteins from the promoter region, thus controlling the length of the exposure time. HP1 specifically binds to methylated lysine 9 on histone H3 and can form oligomers which bridge adjacent nucleosomes. HP1 also recruits lysine 9 histone H3 specific methyltransferases to its binding site which results in linear spreading of methylation to adjacent nucleosomes.

In embryonic stem cells, targeted by the repressor protein HP1, gene expression was completely repressed within five days. This repression was associated with a decline in lysine 27 acetylation and lysine 4 trimethylation on histone H3 and an increase in lysine 9 trimethylation. The lysine 9 methylation spread over the five days forming a domain of 10 kbp, 5 kbp on either side of its initial binding locus. To test the stability of the epigenetic marks, binding of HP1 to the promoter region for either seven days or 4.5 weeks (followed by washout of HP1) was compared. Levels of lysine 9 trimethylation were similar after seven days and 4.5 weeks, but the level of DNA methylation (CpG sites) at the promoter region was much higher after 4.5 weeks of HP1 recruitment compared to seven days. Four days after washout of HP1 from cells targeted by this protein for 4.5 weeks, the level of the repressive mark methylated lysine 9 was significantly reduced and 20% of the cells displayed gene reactivation. Six days after washout of HP1, about 30% of cells showed gene reactivation. Gene repression was lost if the cells were treated with a DNA methyltransferase inhibitor, indicating that this repressive state was controlled by DNA methylation. In murine embryonic fibroblasts (differentiated cells), the gene *Oct 4* is normally transcriptionally silent and chromatin in the area of the gene has high levels of the repressive marks trimethylated lysine 9 and lysine 27 on histone H3. The transcription activator VP16 was recruited to the promoter region of these embryonic fibroblasts. Within five days of recruitment, about 10% of cells showed *Oct4* expression and this activation of the *Oct4* gene was associated with changes in histone epigenetic marks: increased levels of methylated lysine 4 and acetylated lysine 27 on histone H3 and low levels of methylated lysine 9. Subsequently, HP1 was recruited to VP16-activated murine embryonic fibroblasts. The result was repression of *Oct4* expression within five days and reversal of the histone epigenetic marks: reduction in methylated lysine 4 and acetylated lysine 27 and an increase in methylated lysine 9. This observation suggests that active transcription is necessary for the formation of repressive marks. In a confirmatory experiment, murine embryonic fibroblasts were activated by the recruitment of VP16 which resulted in the removal of repressive epigenetic marks and expression of *Oct4*. After washout of VP16, *Oct4* expression was essentially eliminated but re-establishment of a repressive mark (trimethylation of lysine 9 at histone H3) did not occur.

In differentiated cells, the epigenetic mark, methylated lysine 9, can be transmitted through numerous cell divisions in the absence of the initial stimulus. Murine embryonic fibroblasts were activated by the recruitment of VP16, and subsequently, HP1 was also recruited to these same cells. After seven days, both VP16 and HP1 were washed out. The level of the repressive mark methylated lysine 9 was stably maintained in these cells for at least eight days after washout across multiple cell divisions and this level remained unchanged even when DNA methylation was inhibited. Hence, murine embryonic stem cells and differentiated embryonic fibroblasts differ in at least two properties: the epigenetic mark methylated lysine 9 is stably transmitted across cell divisions in embryonic fibroblasts but not embryonic stem cells, and the level of the repressive mark was controlled by DNA methylation in embryonic stem cells but not embryonic fibroblasts.

These elegant experiments of Hathaway *et al.* (2012) underscored the balance and reversibility of histone epigenetic marks associated with gene repression or activation. These experiments also demonstrated epigenetic differences between cell types including the interplay between DNA methylation and histone marks. In differentiated cells, the establishment of repressive marks appeared to require active transcription and these marks can be stably transmitted across cell divisions.

The mechanisms responsible for transmission of chromatin marks across cell divisions are becoming better understood. Jacob *et al.* (2014) examined the re-establishment of one such mark, H3K27me1 (monomethylated lysine 27) during the S phase of the cell cycle in the plant, *Arabidopsis thiala*. This epigenetic mark functions to prevent over-replication of the heterochromatin regions of the genome as well as to prevent the transcription of certain repetitive sequences located within these regions. Eukaryotic chromosomes contain two distinct chromatin domains: euchromatin as the predominant domain and heterochromatin as the smaller domain (Leach *et al.*, 2000). These regions are usually defined cytologically by the intensity of staining and how condensed the chromatin appears. Heterochromatin regions are generally areas of suppressed gene expression and are found around the centromere and at telomeres. Heterochromatin is enriched in highly repetitive, noncoding satellite sequences and is usually the last portion of the genome to be replicated during the S phase.

In *A. thialana*, methyltransferases (ATXR5 and ATXR6) maintain the chromatin mark H3K27me1 during DNA replication. Histone H3 has two variants, H3.1 and H3.3, and ATXR5/6 have a much greater activity for nucleosomes containing the H3.1 variant compared to nucleosomes containing the H3.3 variant. Jacob *et al.* (2014) proposed the following model based upon their experimental observations. The protein CAF-1 acts as a histone chaperone and incorporates the H3.1 variant into nucleosomes during S phase of the cell cycle. ATXR5/6 methyltransferases are recruited to the replication fork during S phase and specifically monomethylate lysine 27 (K27) at newly incorporated H3.1 histones thus re-establishing this particular chromatin mark. Since ATXR5/6 only methylates K27 on H3.1 and not H3.3, transcriptionally active H3.3-enriched regions are protected against K27 monomethylation and gene activity repression during DNA replication.

The regulatory repertoire of the genome includes promoters and enhancers (discussed briefly in the **Introduction**) and these distinct genomic regulatory regions can be identified by different epigenetic markings on the nucleosome histones in which they reside (Schubeler, 2007). In essence, these epigenetic markings convey information as to the function of these regulatory DNA sequences. Promoters are located at the 5′ ends of genes and include the transcription start site and a short adjacent upstream region; promoters function as a point of assembly of the transcriptional machinery and the initiation of transcription (Heintzman *et al.*, 2007). The promoter is responsible for the basal level of transcription. Enhancers function to increase the rate of transcription of a target gene and can be located close to the target gene or some distance away, either upstream or downstream. The mammalian genome contains around 20,000 protein-coding genes and more than 1,000,000 enhancers with, on average, about four enhancers contacting an active gene per cell type (de Laat and Duboule, 2013). Active promoter and enhancer regions are both highly accessible to digestion with nucleases (DNase I hypersensitive) but they differ in their associated histone markings. Active enhancers show H3K4 monomethylation whereas active promoters show H3K4 trimethylation (Heintzman *et al.*, 2007; Schubeler, 2007). Both sites show H3/H4 acetylation (e.g., H3K27ac).

Much has been learned about enhancers (de Laat and Duboule, 2013) and in the **Introduction**, the evolution of enhancers involved in the lineage-specific embryogenesis of the human limb was discussed. Enhancers are DNA regulatory elements with sequences for binding of specific protein factors that are central to developmental transcription regulation and ensure the correct spatial-temporal expression of genes (Eccleston *et al.*, 2013). A number of techniques have been used to screen for functionally relevant enhancers: binding site of p300, epigenetic marks such as histone methylation (H3K4me1), and acetylation (H3K27ac) and DNase I hypersensitivity. Enhancers that display only the mark H3K4me1 are said to be in a "poised" state, whereas enhancers bearing the marks H3K4me1 and H3K27ac are considered to be active. Enhancers appear to act through the formation of a DNA loop which brings the enhancer into physical proximity with the promoter of the target gene (de Laat and Duboule, 2013). Two different types of enhancer–promoter loops have been described. In one type, tissue-specific enhancer–promoter contacts are formed *de novo* during differentiation depending on the set of transcription factors acting at that locus. In the second type, a preformed enhancer–promoter structure already exists and on differentiation, transcriptional activation just requires the recruitment of tissue-specific transcription factors. In only about 7% of cases do enhancers target the nearest gene; most enhancer–promoter contacts are long range. For example, an enhancer of the gene *IRX3* promoter is located about 500 kb away in the first intron of the *FTO* gene (Gorkin and Ren, 2014). At least 10% of distant enhancers contact multiple genes. Likewise, many genes contact multiple enhancers. Mammalian chromosomes are structurally subdivided into topological domains of about 1 Mb with flanking boundaries that hamper DNA contacts across these domains. Long-range enhancer–promoter contacts appear to be organized within these topological domains. Even this brief review highlights the important role that enhancers play within a very complex and dynamic regulatory landscape.

The formation of DNA loops is involved in other aspects of the transcription process. Genes have both promoter sequences which are involved in transcription initiation and terminator sequences which are involved in stopping transcription. It has been documented that DNA loops can form to physically connect a gene promoter with its terminator (Hampsey, 2012). Promoters are inherently bidirectional (Wei *et al.*, 2011) and the loop arrangement by bringing promoter and terminator regions into close proximity facilitates transcription in the sense direction, i.e., formation of mRNA, and represses antisense transcription, i.e., formation of a non-coding RNA (Hampsey, 2012). Hence, another functional role of DNA loops is to facilitate the repression of non-coding RNA transcription from bidirectional promoters (Hampsey, 2012). The critical role of DNA loops in gene transcription is not surprising given the observation that the basic structure of the chromosome consists of a linear array of chromatin loops which have been longitudinally compressed (Kleckner *et al.*, 2013).

Similar to the study of Cotney *et al.* (2013), Attanasio *et al.* (2013) examined the role of enhancers in the development of craniofacial morphology in the mouse. Murine embryonic craniofacial tissue was screened for enhancers by identifying genomic loci enriched for p300 binding. Enhancers were defined as enriched loci at least 2.5 kb from the nearest transcription start site. Candidate enhancers were a median distance of 44 kb from the nearest transcription start site with 38.4% in introns and 54.7% in intergenic regions. Candidate craniofacial enhancers were tested for morphogenic activity through reporter gene transgenesis in murine embryos. In addition, specific enhancers of genes known to be involved in craniofacial morphogenesis were studied. For example, the murine gene *Msx1* was found to be regulated by five distinct enhancers. These enhancers acted in a complex modular fashion to regulate the spatial expression of this gene. To further explore the role of enhancers, the investigators developed mice with selective deletions of three specific enhancers. The three enhancers chosen had close association with three genes documented to be involved in craniofacial embryogenesis. The three enhancers were located 190, 235, and 350 kb from their respective genes. Each specific enhancer deletion resulted in decreased expression of its respective target gene in those anatomic regions in which that enhancer was normally active. To study the craniofacial morphologic phenotypes resulting from these enhancer deletions, the investigators established three cohorts, each consisting of about 30 mice homozygous for one of the deletions. These cohorts were compared to a cohort of wild-type mice. Each specific enhancer deletion produced morphologic changes that were not confined to a single anatomic feature but involved multiple regions of the skull. This observation indicates the highly integrated nature of craniofacial development. This study illustrates that enhancers, by virtue of their regulation of the spatial expression of "developmental" genes, contributed significantly to the normal variation of craniofacial development.

Much of the mammalian genome serves to regulate and ensure the correct spatial–temporal expression of genes. This genomic regulatory function is critical for processes as diverse as cell differentiation and the encoding of environmental cues or experiences into memories. Epigenetic mechanisms, such as histone methylation/acetylation and DNA methylation, are an important part of the regulatory network, and the establishment of these epigenetic marks modulates the activity of genomic promoters and enhancers. Epigenetic histone marks can alter the packaging of chromatin and through this mechanism control the access and binding of various transcription factors to promoter and enhancer regions. Methylation of DNA cytosines is in general associated with repression of gene expression. The interactions between epigenetic marks, chromatin packing, promoter and enhancer activity, and gene expression are just one feature of the regulatory landscape.

Several classes of RNA molecules have been identified to be important in the regulation of gene expression. One class is called long noncoding RNA (lncRNA), which is defined as an RNA strand greater than 100 nucleotides in length (Lee, 2012). Some lncRNAs are involved in epigenetic regulation and the best-studied example is that of the lncRNAs which are involved in inactivation of one of the X chromosomes in female mammals (Lee, 2012). These lncRNAs direct epigenetic complexes such as the polycomb repressive complex 2 to specific genomic locations, resulting in the silencing of one of the X chromosomes through the establishment of repressive epigenetic marks such as trimethylated lysine 27 at histone H3 and DNA methylation. Several properties of lncRNAs make them suitable mediators of epigenetic regulation of gene expression. First, a lncRNA can target an epigenetic complex (such as the polycomb repressive complex 2) which does not itself possess a sequence-specific DNA-binding motif, to a specific genomic location since the lncRNA is tethered to the genome by a RNAP II complex at its site of transcription. Secondly, lncRNAs have a fast turnover (Lee, 2012). Recently, circular or loop RNAs have been described (Ledford, 2013). Circular RNAs possess binding sites for microRNAs which are short RNA segments that block gene expression by binding to and preventing the translation of messenger RNAs. The interplay between microRNAs and circular RNAs in the regulation of mRNA translation is quite nuanced and becoming better defined (Kosik, 2013).

This brief survey of selected aspects of gene expression regulation is meant to emphasize both the complexity of this biological process as well as the multiple sites at which regulation can occur. For example, histone epigenetic marks can modulate the kinetics of RNAP II activity within the nucleosome as well as the accessibility of various repressive and activation factors to the DNA. Histone epigenetic marks are functionally intertwined with the methylation state of DNA. Histone modifications and the level of DNA methylation are potentially reversible processes that can be stably transmitted across somatic cell divisions. Being potentially reversible, these mechanisms represent ideal candidates to mediate the link between environmental cues and changes in gene expression patterns.

Epigenetics and ageing

In a very interesting report, Fraga *et al.* (2005) measured in a cross-sectional study the epigenetic differences accruing during the lifetime of monozygotic (MZ) twins. MZ twins constitute a natural experiment for studying the influence of environmental cues on the epigenetic contribution to phenotypic differences between genetically identical individuals. The subjects of the study were 40 twin pairs, 16 male and 24 female with an age range of 3–74 years. Peripheral lymphocytes were used as the genomic source and global epigenetic marks were measured. These marks included global histone H4 and H3 acetylation and DNA 5-methyl cytosine content. A well-defined epigenetic mechanism is that of X chromosome inactivation in females occurring during development (Haque *et al.*, 2009). About 14% of female MZ twins had a different X chromosome methylation pattern between siblings, suggesting that epigenetic differences between MZ siblings could arise early on during development. Thirty-five percent of MZ twin pairs showed different global levels of epigenetic marks (H3 and H4 acetylation and DNA methylation) between siblings which varied according to the age of the twins. The youngest pairs were epigenetically similar, whereas the oldest pairs were epigenetically quite distinct. In addition, twin pairs who spent less of their lifetime together and/or had a different natural health or history of medical events also showed the greatest differences in levels of epigenetic marks.

The differences in global epigenetic patterns displayed by twin pairs were not spread uniformly across the genome but had a regional distribution. For example, regional differences in DNA methylation were located in all telomeres and a few selected gene-rich sites. Differences in regional DNA methylation were minimal in twin pairs that were younger, had similar lifestyles, and had spent more of their lifetimes together, whereas twin pairs who were older, had different lifestyles, and had spent less of their lives together showed distinct differences in DNA methylation levels in these same genomic regions. The methylation status of about 1,800 unselected CpG islands, which are predominantly found in gene promoter regions, were also compared in the younger versus older twin pairs. Older twin pairs showed 2.5 times as many DNA methylation differences in the surveyed CpG islands as the younger twin pairs. Differences in levels of global and regional epigenetic marks between twin siblings corresponded to differences in gene expression. Very young twin pairs who showed little differences in global levels of epigenetic marks showed very similar expression profiles, whereas older twin pairs who showed distinct differences in global epigenetic patterns also showed extremely different expression profiles in terms of both gene expression increase and decrease. There were four times as many differentially expressed genes in older twin pairs compared to younger twin pairs. The genes linked to the surveyed CpG islands which showed more methylation differences in older twin pairs also showed greater expression differences in these older twin pairs. The investigators confirmed their observations made in peripheral lymphocytes by documenting similar differences in epithelial buccal cells and intraabdominal fat and skeletal muscle biopsies in these twin pairs. In these three tissues, marked epigenetic differences were found in older MZ twins who had different lifestyles and who had spent less of their life together.

These observations indicated that the pattern of epigenetic marks in MZ twin pairs diverged as these twins aged and experienced different lifestyles and "medical" events. Although not conclusive, these observations are consistent with the construct that these epigenetic differences in genetically identical individuals were the result of the cumulative effects of environmental cues (influences) that these individuals experience as they age. The accumulation of such effects with ageing could explain the epigenetic differences between older MZ twin pairs. An alternative explanation is that these epigenetic differences were due to small "mistakes" in transmitting epigenetic marks across cell divisions or in maintaining stable epigenetic marks in differentiated cells. However, this explanation does not account for the correlation between lifestyle, "medical" events, and the time twins spend apart and the extent of epigenetic differences. The epigenetic differences between MZ twin siblings were associated with differences in gene expression patterns. Since MZ twins are genetically identical, these differences in epigenetic marks furnish a mechanistic link between environmental cues (experiences) and phenotypic differences.

The results of a recent study reported by Ouellet-Morin *et al.* (2013) were consistent with the observations of Fraga *et al.* (2005). Ouellet-Morin and co-workers studied epigenetic changes associated with bullying victimization in 28 pairs of monozygotic twins aged 12 years. Each twin pair was discordant for bullying, i.e., only one member of each twin pair had been subject to bullying. The occurrence and intensity of bullying were assessed when the cohort of twins was seven, 10, and 12 years of age. Buccal cell samples served as the DNA source and the methylation status of the serotonin transporter gene (*SERT*) was measured at ages five and 10. At age 12, the twin pairs were given a standardized psychosocial stress test and salivary cortisol level was measured after this test. At age 10, the mean extent of methylation of *SERT* DNA averaged across the 12 CpG sites was higher in the bullied compared to non-bullied co-twins. This difference was primarily due to a difference in the extent of methylation at one specific CpG site — CpG 8. At the age of five years, prior to bullying victimization, both groups had the same level of DNA methylation. Between ages five and 10, the non-bullied group of co-twins did not show any increase in *SERT* DNA methylation, while the bullied group of co-twins showed an increase in the extent of DNA methylation primarily due to methylation changes at CpG 8. These investigators also observed a negative correlation between post-stress test cortisol levels at age 12 and *SERT* DNA methylation at CpG 8 at age 10; i.e., twins with higher methylation levels at CpG 8 at age 10 years exhibited lower cortisol responses to the psychosocial stress test.

The data of Ouellet-Morin *et al.* (2013) showed that monozygotic twins exposed to different life experiences, such as one twin being subject to bullying victimization, manifested a divergence in *SERT* gene DNA methylation measured between ages five and 10. The extent of *SERT* gene CpG 8 methylation was also associated with an alteration in functioning of the hypothalamic–pituitary–adrenal (HPA) axis. Co-twins that had experienced bullying displayed a blunted cortisol response to stress compared to non-bullied co-twins. The importance of these observations of Ouellet-Morin and co-workers is supported by the data of Takizawa *et al.* (2014) that individuals subjected to bullying during childhood exhibited a greater incidence of poor general health, mental health, and cognitive functioning later in adulthood and middle age.

Kaminsky *et al.* (2009) also examined DNA methylation patterns in MZ and DZ (dizygotic) twins. The tissues used in this study included white blood cells, buccal epithelial cells, and that of the rectum, and in each tissue, the methylation status of about 6,000 unique DNA loci were measured. In MZ twins, the genome-wide DNA methylation pattern showed considerable variation between co-twins in all three tissues. The investigators also compared in MZ co-twins the variation in methylation of specific genomic elements: CpG islands, non-CpG sites, and promoter regions. In white blood cells, CpG islands and promoter regions showed less variation in methylation between MZ co-twins than non CpG loci, while in buccal epithelial cells, there were no statistically significant differences in methylation variation of these three genomic elements between co-twins.

The investigators compared genome-wide methylation patterns between MZ and DZ twins. A classical technique for estimating heritability of a "trait" is to compare the within-group variation of that trait between MZ twins and DZ twins (Bell and Spector, 2011). Heritability of DNA (cytosine) methylation can be estimated from the data of Kaminsky *et al.* (2009). For white blood cells, heritability was very low, whereas for buccal epithelial cells, heritability was in the range of 0.3. However, in buccal epithelial cells, the difference in co-twin variation between MZ and DZ twins was attributable to the 10 dichorionic MZ twins; i.e., dichorionic MZ twins showed much less co-twin variation than DZ twins. On the other hand, the co-twin variation was similar in monochorionic MZ twins and DZ twins. The observation that MZ dichorionic twins display much more co-twin epigenetic similarity than monochorionic MZ co-twins indicated that intra-uterine environmental cues can significantly modulate the establishment of epigenetic marks.

Boks *et al.* (2009) examined DNA methylation patterns in twins and healthy controls. They measured the methylation status of 1,505 CpG sites from more than 800 genes. Twin samples were used to obtain estimates for heritability. Approximately, 23% of the CpG sites displayed significant heritability. In Table 2 of their paper, they list the heritability of 25 CpG sites, where the values range from 0.57 to 0.94. Boks *et al.* (2009) described significant correlations between the levels of methylation at specific CpG sites and nearby SNPs, indicating that genomic sequence variation can control the level of local DNA CpG methylation. Boks *et al.* (2009) also reported effects of age and gender on level of CpG methylation.

Wong *et al.* (2010) performed one of the few longitudinal studies on epigenetic variation in twins. The subjects in this study were 46 MZ and 45 DZ twin pairs. Three candidate genes were selected for CpG methylation analysis: dopamine receptor 4 gene (*DRD4*), *SERT*, and monoamine oxidase A gene (*MAOA*). Buccal cells were used as the tissue source for DNA and samples were taken when the twins were five and 10 years old. The analysis measured the extent of methylation across a number of CpG sites within the 5′ promoter region of each gene. For all three genes, there was a wide range of changes in average DNA methylation between ages five and 10. In some children, the extent of methylation increased while in other cases the extent of methylation decreased. For CpG methylation of the genes *DRD4* and *SERT*, the MZ and DZ within-pair correlations were statistically the same, indicating little evidence of heritability. *MAOA* is located on the X chromosome, and for male twins, there was a difference in MZ–DZ within-pair correlations suggesting a heritable component. Further analysis suggested that for *DRD4*, the strong and similar within-pair correlations in the change in DNA methylation (between ages five and 10) among MZ and DZ twins indicated a significant environmental influence of factors shared by children growing up in the same family. On the other hand, for *SERT*, the weak but similar within-pair correlations in the change in DNA methylation in MZ and DZ twins indicated a significant influence of environmental factors unique to each child in the family. For *MAOA*, the CpG methylation data was more complex since this gene resides on the X chromosome which introduces significant sex differences. The data for the methylation pattern of this gene suggested very little heritability and no common environmental influences.

Talens *et al.* (2012) in a recent study collected both cross-sectional and longitudinal data on epigenetic variation in MZ twins. Genomic DNA was extracted from whole blood. The study subjects included 230 MZ pairs 18 to 89 years of age. The investigators measured global genomic DNA methylation as well as locus-specific methylation (promoter regions) close to seven genes: three non-imprinted and four imprinted. The range of absolute within-pair differences in percent DNA methylation was significantly higher in old as compared to young MZ pairs for global methylation as well as for six of the seven specific loci. Twin pairs were divided into four age groups: 18–25 years old, 26–50 years old, 51–75 years old, and 75–88 years old, and the range of absolute within-pair differences in global methylation and that of five loci were measured for the four groups. The range of absolute within-pair differences in global methylation showed no significant age related change, whereas three of the five loci showed increasing within-pair discordance with increasing age of the twin pairs. In a longitudinal study, 19 elderly twin pairs (73–82 years old) were followed for 10 years; global methylation and methylation of five loci were measured at the start and after 10 years. The range of absolute within-pair differences for global methylation and methylation at three of the loci showed an increase over the 10-year period. In the age-stratified MZ pairs, the contributions of familial (genetic and common environment) versus unique (individual) environmental effects on age-related DNA methylation differences were assessed. The age-related increase in methylation variation was generally attributable to unique environmental factors except for one locus for which familial factors appeared to play a more important role.

Hannum *et al.* (2013) measured genome-wide methylation patterns in relation to ageing rates using a cross-sectional study design. A preliminary survey of over 400,000 CpG sites indicated that the methylation status of about 70,000 (15%) of these sites showed a significant correlation with chronological age. The primary study was done using whole blood as the tissue but similar results were obtained using breast, kidney, lung, and skin samples. Methylation status was recorded as a fraction between 0 and 1, which represented the frequency of a given CpG site being methylated across the population of blood cells taken from a single individual. It is interesting to look at the methylation status for some of the specific CpG sites surveyed. For the site cg16867657, young individuals had a mean methylation fraction of about 0.55, while old individuals had a mean methylation fraction of 0.75. For the site cg24724428, mean methylation fraction increased from 0.1 at 30 years of age to 0.3 at 70 years of age. On the other hand, CpG sites cg27367526 and cg2230534 showed a decrease in mean methylation fraction with age, and in addition, for both of these CpG sites, there was a correlation between the numerical value of the methylation fraction (extent of methylation) and SNPs which mapped to the same genomic region. This association between the extent of methylation at a given CpG site and nearby SNPs is similar to the observations reported by Boks *et al*. (2009).

Based on the correlation between changes in the extent of DNA methylation and chronological age, Horvath (2013) developed an algorithm, using the methylation status of CpG sites, for computing an individual's chronological age (Gibbs, 2014). Horvath assembled a DNA methylation data set from 7,844 non-cancer samples which included methylation levels in 51 different tissues and cell types. He then screened more than 21,000 CpG sites for their correlation with chronological age and ended up with a selection of 353 CpGs in his "age estimator." Genes that co-located with these 353 CpG sites were primarily involved with processes related to cell death and survival, cellular growth and proliferation, tissue development, and cancer. Of these 353 CpG sites, 193 showed increased methylation with age, while 160 showed reduced methylation with age. Further analysis indicated that the 193 CpG sites which became hypermethylated with age were generally associated with "poised" promoters, whereas the 160 CpG sites that became hypomethylated with age were predominantly associated with "weak" promoters or strong enhancers. This set of 353 CpGs predicted chronological age with an accuracy of about a 96% (Gibbs, 2014; Horvath, 2013).

The association of changes in CpG methylation status with age appears to be well established, but the basis of this association has not been defined. A stochastic process such as epigenetic drift with advancing age could be responsible, but I believe that there is strong evidence that these age-related changes in DNA methylation reflect the effects of environmental cues (experiences). Another question that has not been answered is the impact of these methylation changes on gene expression patterns. Horvath looked at this question for the set of 353 CpG sites by matching each site with its corresponding gene and then plotting mRNA data against CpG methylation data. He did not find any significant relationship between age effects on mRNA levels (gene expression) and age effects on DNA methylation levels for his set of 353 CpG sites.

A cautionary note has to be considered when interpreting these epigenome-wide studies (Callaway, 2014). Many of these studies used whole blood as the DNA source. However, whole blood contains many cell types and each cell type has its own unique DNA methylation pattern (Jaffe and Irizarry, 2014). Jaffe and Irizarry demonstrated that age-related changes in the extent of DNA methylation correlated strongly with changes in the proportions of certain cell types with age; i.e., changes in whole-blood DNA methylation status with ageing actually reflected to a large extent changes in the proportions of certain blood cell types with ageing. These investigators pointed out that this confounding problem was not restricted to whole blood but would apply to any DNA tissue source that contained a mixture of cell types each with its own characteristic DNA methylation signature.

The data on CpG methylation in twins and singletons indicate that the contribution of heritability and of environmental cues on the methylation status of CpG sites varies across these sites. This variation appears to be tissue specific. In addition, the methylation status of many of these sites changes with age; some sites show an increase in the extent of methylation and others show a decrease. Very young MZ twins display a similar co-twin epigenetic pattern which tends to diverge as the co-twins age. This divergence is greater between co-twins who have spent less time together and who differ in their health and history of "medical" events. The observations of Boks *et al.* (2009) and Hannum *et al.* (2013) suggest that specific genomic polymorphisms (SNPs) can control the level of methylation at nearby CpG sites. Further, it hints at a possible genetic mechanism that could account for the heritability component of the methylation status of specific CpG sites. Other factors that can influence DNA methylation patterns include associated epigenetic histone marks and stochastic processes (Bell and Spector, 2011).

The interaction between environmental cues and epigenetic heritability is a complex one (Guerrero-Bosagna and Skinner, 2012). Epigenetic methylation marks (DNA and histone) can be faithfully transmitted through mitotic cell divisions (Bell and Spector, 2011; Guerrero-Bosagna and Skinner, 2012; Hathaway *et al.*, 2012), but the evidence for transgenerational transmission of epigenetic marks through germline replication (meiosis) is not as strong. Before looking at the evidence for transgenerational transmission of epigenetic marks, it is important to put this question in context. The epigenome functions to regulate the correct spatial–temporal expression of genes and hence is crucial for organismal development and cell differentiation. These functional aspects of the epigenome are encoded in the genome and must be transmissible across generations. There are now experimental observations pointing to a mechanism by which genomic variants in non-coding regulatory regions can encode variations in histone epigenetic marks.

A nucleosome consists of about 146 nucleotides of DNA wrapped around eight histones. Histone amino acid tails project from the nucleosome and are subject to chemical modifications, i.e., formation of epigenetic marks such as methylation, acetylation, and phosphorylation. Furey and Sethupathy (2013) pointed out that histone tail modifications are highly variable in the human population and that they are heritable across generations. It appears that DNA structural variants are associated with both histone tail modifications and gene expression variation.

McVicker *et al.* (2013) looked at these associations using lymphoblastoid cell lines (LCLs) as the test system. Four histone H3 modifications were measured: H3K4me1 (monomethylation of lysine 4), H3K4me3 (trimethylation of lysine 4), H3K27ac (acetylation of lysine 27), and H3K27me3 (trimethylation of lysine 27). These marks are generally associated with different genomic states: H3K4me3 is associated with active promoters, H3K4me1 is associated with active enhancers, H3K27ac is associated with active promoters and enhancers, and H3K27me3 is associated with gene silencing by the polycomb repressive complex 2. Allelic imbalance, which is a difference in the expression between two SNP alleles, can be used for identifying *cis*-acting regulatory SNPs (Milani *et al.*, 2007) and McVicker *et al.* (2013) developed a "combined haplotype test" that incorporated allelic imbalance. They applied this test to thousands of SNPs located in non-coding regulatory genomic regions. With this method, they identified high-confidence SNPs that were associated with quantitative trait loci (QTL) for specific histone marks and for RNAP II occupancy at transcription start sites. The final list included 27 distinct QTLs for H3K4me1, 469 for H3K4me3, 730 for H3K27ac, 118 for RNAP II occupancy, and two for H3K27me3. Many of these QTLs overlapped QTLs associated with deoxyribonuclease (DNase I)-hypersensitive sites. These sites are a marker of open chromatin packing states with active regulatory regions such as promoters or enhancers. They also measured expression QTLs (eQTL) and found that many of the histone mark QTLs and RNAP II QTLs were close to eQTL transcription start sites. McVicker *et al.* also identified a number of high-confidence transcription factor-binding sites that contained SNPs. These SNPs modulated the extent of transcription factor occupancy at these sites, and a positive correlation between increased transcription factor occupancy and increased levels of nearby activating histone marks, and between RNAP II occupancy at transcription start sites and decreased levels of the repressive mark H3K27me3, was found.

In summary, these investigators were able to associate genomic regulatory structural variations (SNPs) with QTLs for different histone modifications and RNAP II occupancy at transcription start sites. Many of these QTLs overlapped DNase I hypersensitivity sites (enhancers and promoters) and were close to eQTL transcription start sites. In addition, structural variants (SNPs) in transcription factor-binding sites were causally related to differences in transcription factor binding which resulted in changes in the levels of chromatin marks in nearby histones. These observations of McVicker *et al.* (2013) and others are consistent with the following model (Furey and Sethupathy, 2013). Specific structural variants (SNPs) that are located in noncoding regulatory regions can increase the recruitment and binding of transcription factors to DNA promoters and enhancers. Transcription factor binding leads to histone tail modifications that create a chromatin environment which facilitates the expression of nearby gene(s). Conversely, other SNPs at the same genomic loci do not promote transcription factor binding and thus create a chromatin environment less conducive to gene expression. This model provides a mechanism by which regulatory DNA structural variants influence variation in histone epigenetic marks and thereby influence gene expression patterns. This model would also explain the genetic encoding of epigenetic mechanisms and account for the transgenerational inheritance of epigenetic mechanisms that are critical in organismal development and cell differentiation.

Since, the induction of epigenetic marks can also occur in the developing or mature organism through the effects of environmental cues (experiences); the more problematic question is whether these environmentally induced epigenetic marks can be transmitted across generations.

Examples of environmental cues inducing epigenetic changes which are subsequently transmitted across generations are given by Skinner (2011) and Guerrero-Bosagna and Skinner (2012). One of the best-studied examples is that of the effects of the endocrine disruptor vinclozolin (a fungicide commonly used in agriculture). When pregnant rats (F0) were given vinclozolin during a critical time window, epigenetic changes could be measured in the sperm of third-generation male rats (F3) (Guerrero-Bosagna *et al.*, 2010). This study assessed germline promoter region DNA methylation, and 48 different promoters were found to have a differential DNA methylation pattern in treated F3 generation sperm compared to control F3 generation sperm. This data demonstrated that an environmental cue during gonadal development and sex determination can induce gene promoter DNA methylation changes that can be detected in the F3 generation. F3 generation males displayed "disease-type" phenotypes and the implication is that the germline epigenetic changes transmitted from the F0 generation mother are involved in the expression of these phenotypes. This transgenerational alteration in the F3 generation sperm epigenome promoted transgenerational changes in the gene expression pattern of adult somatic cells and hence changes in adult animal phenotypes.

In addition, there are experimental observations confirming the inheritance of epigenetic marks across generations in plants. Silveira *et al.* (2013) studied the gene Qua-Quine Starch (*QQS*) in *Arabidopsis thaliana*. *QQS* encodes a short 59-amino acid protein which is involved in starch metabolism in leaves. This gene contains several tandem repeats in its promoter and 5′-UTR and these tandem repeats produce a number of short interference RNAs. The investigators found that the extent of DNA cytosine methylation of these repeat elements in the promoter and 5′-UTR of *QQS* was negatively correlated with gene expression and that these DNA methylation epigenetic marks were transferrable across multiple generations. Other experiments using *Arabidopsis thaliana* as the model organism have demonstrated that differentially methylated DNA regions and not specific DNA sequences were responsible for variations in flowering time and root length, and that these DNA methylation patterns were stably transmitted across eight generations (Pennisi, 2013a).

Other examples of transgenerational transmission of epigenetic marks have been reported by Kaiser (2014) and Dias and Ressler (2014). The Dias and Ressler study is discussed in **Chapter 5**.

However, the transmission of epigenetic marks across generations still remains a controversial issue. Can an environmental stress alter an organism's epigenetic marks and lead to a permanent trait change that is acted upon by natural selection? If indeed epigenetic marks can be stably inherited across generations, then the essential ingredients of Lamarkian evolution are in place — a very controversial notion (Bird, 2013). The role of epigenetic inheritance in modern evolutionary theory was recently discussed in two articles that appeared in the *Proceedings of the Royal Society B* (Dickins and Rahman, 2013; Suter *et al.*, 2013). A number of questions were raised in these two articles. Are all epigenetic states in effect genetically encoded? Can epigenetic variants ("epimutations") be transmitted across generations? Are the phenotypes associated with epigenetic variants subject to natural selection? There is currently no consensus with regard to the answers to these questions but a recent paper by Cortijo *et al.* (2014) addressed some of them experimentally.

Cortijo *et al.* (2014) used the model plant *Arabidopsis thaliana* and established a population of epigenetic recombinant inbred lines. These lines had almost identical DNA sequences but differed in DNA methylation patterns. Two complex traits were examined: flowering time (FT) and primary root length (RL). Linkage analysis detected differentially methylated DNA regions (DMRs) that acted as eQTLs for these two traits. eQTLs on chromosomes 1, 4, and 5 were associated with FT and eQTLs on chromosomes 1, 2 and 4 were associated with RL. The degree of DNA methylation in these eQTLs was assoctiated with variations in these two complex traits: a higher methylation state in the FT eQTLs was associated with a later flowering time and a higher methylation state in the RL eQTLs was associated with longer primary roots. The eQTLs on chromosomes 1, 4, and 5 accounted for 90% of the heritability of FT and the eQTLs on chromosomes 1, 2, and 4 accounted for 60% of the heritability of RL. These eQTLs had been inherited from the original founder parent line and had not been introduced during later generations of inbreeding.

The investigators searched for putative causal DMRs, within these eQTLs. Within the FT eQTLs, 325 candidate DMRs were identified that mapped to specific genomic sequences, including 44 genes. Within the RL eQTLs, 506 candidate DMRs were detected that mapped to 71 unique genes plus other specific genomic sequences. However, none of the mapped genes were obvious FT or RL genes, so these candidate DMRs were not the actual causal DMRs. The investigators noted however that their methodology could not detect if these candidate DMRs were in linkage with the causal DMRs. Cortijo and co-workers concluded that in *Arabidopsis*, DMRs could be stably inherited across generations independently of DNA sequence changes and that these inherited DMRs functioned as eQTLs. They speculated that these eQTLs had all the necessary features to be potential targets of natural selection. In essence, Lamarkian evolution appeared to occur in plants and the heritability of epigenetic marks might explain part of the "missing heritability" noted for complex traits. If a similar epigenetic basis for Darwinian evolution, independent of DNA sequence, occurred in vertebrates then the epigenetic effects of environmental cues on an individual of one generation could be passed to members of descendant generations. With respect to this mode of inheritance, Dias and Ressler

(2014) have described the passage of an olfactory sensory experience across generations in a mouse model and this observation is discussed in more detail in **Chapter 5**.

The experimental observations summarized in the previous section support the construct that epigenetic marks in both somatic cells and germline cells are susceptible to environmentally induced changes. Alterations in epigenetic marks can be transmitted both across somatic cell divisions and across generations. These altered epigenetic marks are associated with changes in the transcriptomes of adult somatic cells and by inference associated with changes in the expressed phenotypes.

Summary

The establishment and removal of repressive or activating epigenetic marks reflect a dynamic and balanced set of processes. The most extensively studied epigenetic mark is that of CpG methylation. Observations in twins and singletons indicate that a number of factors contribute to the methylation status of a given CpG site. These factors include heritability, environmental cues, and stochastic processes. Heritability could be determined by several genetic mechanisms. Specific genomic polymorphisms can regulate the methylation status of nearby CpG sites and some epigenetic marks appear to be transmitted through the germline across generations. For many CpG sites, the methylation status is not stable over the individual's lifespan, although the change with age can vary from a decrease to an increase in the extent of methylation. MZ twin studies suggest that these age-related changes are strongly influenced by environmental cues and are not simply the result of stochastic events. I believe the data are consistent with the construct that age-dependent changes in epigenetic marks reflect the cumulative influence of environmental cues (experiences) and that given the uniqueness of individual genomes, age-dependent changes in epigenetic marks will also vary significantly across individuals. The establishment and/or removal of epigenetic marks appear to be important mechanisms linking environmental cues to changes in gene expression patterns. In the next section, I will explore the interrelationships between specific environmental cues and changes in epigenetic marks although a qualifier needs to be added. There is still some reluctance within the scientific community to accept that epigenetic mechanisms underlie the apparent environmental responsiveness of the genome. Buchen (2010) has outlined some of the concerns that form the basis of this reluctance.

Environmental Cues: Specific Examples

In the preceding sections, the terms 'environmental cues' and 'experiences' have been used interchangeably. In many ways, 'experiences' is preferable since this term better conveys the complex, rich, and multi-nuanced features that constitute environmental influences.

As noted in the **Introduction**, ageing can be conceptualized at both the species and individual level (Singer, 2013) and experiences are an important modulator of ageing at the individual level. Two examples support this construct. MZ twins have identical genomes, yet co-twins show divergent phenotypes as they age (Haque *et al.*, 2009). As discussed in the previous sections, epigenetic mechanisms appear to underlie these age-dependent phenotypic changes. A second example involves lifespan measurements and ageing patterns in inbred strains of laboratory rodents (Hollander *et al.*, 1984). Although inbred strains of laboratory mice and rats are not genetically identical, the extensive inbreeding process results in reduced genomic diversity. As reviewed by Hollander *et al.* (1984), multiple "lesions" are observed in older inbred mice and rats and there is considerable individual variability in these age-associated lesions. Lifespan studies of cohorts of inbred rodents also show a marked individual variability. For example, for the inbred rat strain WAG/Rij, different female cohorts showed a variation in mean maximum lifespan varying from 37.2 to 44.5 months, a variance of seven months or about 20%. This marked variation in lifespan and ageing across individual animals must be environmentally mediated.

These two examples indicate that animals with either identical or very similar genomes show significant between-individual variation in the ageing process, which is the result of differing individual experiences during development (intrauterine) and throughout the postnatal period. In the next sections, specific experiences and their phenotypic results will be reviewed. Only a few experiences will be highlighted, since the objective is to emphasize the complexity of the interplay between experiences and the ageing process.

Adult neurogenesis: Voluntary exercise and enriched environment

Adult neurogenesis has been well documented across many species of invertebrates and vertebrates and appears to be a general phenomenon occurring in brain areas that are important for learning and memory processes (Barnea and Pravosudov, 2011). Adult neurogenesis is a multistep process consisting of proliferation of progenitor cells followed by differentiation and migration of these new neurons to their final destination. At this terminal location, these young neurons are integrated and connected into existing circuits. The mechanisms underlying adult neurogenesis have been most extensively studied in mammals and birds. In mammals, new neurons are added to specific brain regions: the dentate gyrus of the hippocampus and the olfactory bulb, whereas in birds new neurons are added to a wider selection of brain regions.

Recently, adult neurogenesis was confirmed in humans using a carbon dating technique (Spalding *et al.*, 2013). Progenitor cells were found in the subventricular zone (SVZ) of the lateral ventricles and migrated to the hippocampal dentate gyrus but, unlike the rodent, no new neurons were detected in the olfactory bulb of the human. Hence, in the human, neurogenesis is restricted to the dentate gyrus. There are distinct differences between adult neurogenesis in the human and in the rodent. In mice, about 10% of neurons in the dentate gyrus comprise a renewing fraction of cells that are added during postnatal life, whereas in the human, almost all of the dentate gyrus neurons are renewed after birth. Further, in the human, the dentate gyrus contains a mixture of neurons of different ages. Neurons formed during development are non-renewing and have a long lifespan. This declining population of neurons generated during development is gradually replaced by postnatally generated cells. Neurons formed postnatally have a half-life of about 7.1 years. In the rodent, adult-generated neurons are additive and the result is a net increase in the number of dentate gyrus neurons with age. In the human, there is a net loss of dentate gyrus neurons with age since the generation of new neurons does not keep up with neuronal loss. The rate of hippocampal neurogenesis declines with age, but this decline is less in humans than in the mouse. Spalding *et al.* (2013) made some comparisons between species with respect to the rate of neurogenesis. In the human, 0.004% of dentate gyrus neurons are exchanged daily. In a two-month-old mouse, the value is 0.03–0.06% per day and in a nine-month-old mouse, 0.003–0.006% per day. The value in the adult macaque monkey is 0.004–0.02% daily. Hence, humans, non-human primates, and the middle-aged mouse appear to have comparable rates of neurogenesis. New neurons have enhanced plasticity for a period of time after differentiation and under normal conditions, synaptic strength plasticity (measured as changes in long-term potentiation) in the dentate gyrus can be attributed to newborn cells (Kempermann, 2013). These two observations plus the fact that the dentate gyrus acts as a "bottleneck" in the circuitry underpinning learning and memory acquisition enable a small number of postnatally generated neurons to have a substantial influence on hippocampal function (Spalding *et al.*, 2013).

In mammals, new neurons are "born" in the SVZ of the lateral ventricles and in the subgranular zone (SGZ) of the hippocampal dentate gyrus (Zhao *et al.*, 2008). Neurons originating in the SVZ migrate a great distance to the olfactory bulb (olfactory bulb neurogenesis does not occur in humans, see previous section), whereas new neurons born in the SGZ migrate a much shorter distance into the granule cell layer of the dentate gyrus. These two brain regions, the olfactory bulb and dentate gyrus of the hippocampus, are the only confirmed sites of adult neurogenesis in mammals (Barnea and Pravosudov, 2011). In the SGZ, two types of adult neural stem cells (progenitor cells) have been identified: type 1 radial glia-like (RGL) type cells and type 2 progenitor cells which do not express glial protein markers. It is thought that type 2 cells arise from type 1 cells. A new neuron is formed when a RGL cell divides and one of the daughter cells differentiates into a young migrating neuron (Barnea and Pravosudov, 2011; Bonaguidi *et al.*, 2011). Although progenitor cells can be found in many brain areas, only those in the SVZ and SGZ give rise to new neurons. Hence, there must be special features associated with the microenvironment (neurogenic niche) of these two locations that promote neurogenesis. The anatomical and functional factors responsible for the unique features of these neurogenic niches are discussed by Zhao *et al.* (2008). In both the SVZ and SGZ, neurogenesis declines with age, although as discussed in a subsequent section, voluntary exercise can ameliorate the age-related decline in neurogenesis. Mammalian neurogenesis is highly regulated by neurotransmitters and neural peptides such as acetylcholine, cholecystokinin, dopamine, gamma-aminobutyric acid (GABA), glutamate, and serotonin to name a few, as well as a number of growth factors including the epidermal growth factor, fibroblast growth factor, and brain-derived neurotrophic factor (BDNF) (Zhao *et al.*, 2008).

In both mammals and birds, hormones are important regulators of neurogenesis. In birds, testosterone increases neuronal recruitment and survival and estrogens also have positive effects on neuronal migration and survival (Barnea and Pravosudov, 2011). Another hormone regulating neurogenesis is corticosterone (Cameron and Gould, 1994). Rats treated with injections of corticosterone showed a reduced number of new neurons in the SGZ (decreased proliferation of progenitor cells) compared to control rats. Adrenalectomy resulted in more new neurons in the dentate gyrus compared to sham-operated control animals.

In birds, the ventricular zone (VZ), a region lining the lateral ventricles, is the neurogenic niche (Barnea and Pravosudov, 2011). In the canary brain, migration of young neurons begins about three days after birth and is generally completed within about 20 days. The majority of new neurons do not survive and by day 40, only one-third of the initial cohort remains alive. In the bird, new neurons are recruited to brain areas involved in song control and vocal learning as well as areas involved in processing spatial learning. In general, recruitment of new neurons to these areas declines with age. During the migration process, new neurons undergo differentiation and maturation in terms of morphological and physiological characteristics. At the end of their migration, these young neurons become functionally integrated into the existing circuitry although as already noted most of these new neurons do not survive.

As pointed out by Barnea and Pravosudov (2011), the functional role of adult hippocampal neurogenesis has not been completely established although the prevalent view is that neurogenesis is causally linked to learning and memory. In fact, many aspects of hippocampal function are poorly understood. For example, the hippocampus is involved in spatial navigation which involves measures of both distance and time parameters (Buzsaki, 2013). Which hippocampal neuronal circuits encode distance and time parameters for navigational events (Buzsaki, 2013)? Do these same circuits subserve episodic recall of these navigational events which allows for cognitive functions such as prediction, planning and action (Buzsaki, 2013; Pastalkova *et al.*, 2008)?

The linkage between adult hippocampal neurogenesis and individual variation in behavior was explored by Freund *et al.* (2013) using a natural-type of experimental design. Forty highly inbred female mice at age four weeks were put into a large enriched enclosure for three months. The assumption was made that these highly inbred mice were genetically identical at baseline. The enclosure had an elaborate tracking system to continuously monitor the exploratory movements of individual mice. Although there was some individual variation in exploratory behavior at the start of the experiment, by the three-month time point, individual differences in exploratory behavior were considerable. The label bromodeoxyuridine (BrdU) was used to measure hippocampal neurogenesis at the start and termination of the experiment. This label is incorporated into newly dividing cells and hence is used as a marker of neuronal "birth." When all of the individual data points were plotted, there was a significant positive correlation between the number of new neurons in an individual mouse and the size of the area regularly explored by that animal. Animals that regularly explored large areas were consequently exposed to more cognitive challenges and had the most new neurons. This study connected structural and functional brain plasticity in a more natural environment (Bergmann and Frisen, 2013). Newborn neurons have special electrophysiological properties for about a month post-birth, and for those that survive beyond a month, their features become indistinguishable from older neurons (Bergmann and Frisen, 2013). The data of Freund *et al.* (2013) indicate that this pool of new neurons in a given individual animal is involved in the learning and memory acquisition of individual-specific behaviors.

Zhao *et al.* (2008) and Barnea and Pravosudov (2011) reviewed the evidence linking neurogenesis with learning and memory formation in mammals and birds. Most of the studies demonstrate correlations between neurogenesis and various cognitive functions, but few, if any, have documented a direct causal link between the two. In their review, Zhao *et al.* (2008) summarized a number of studies in mammals which have explored the relationship between SGZ neurogenesis (hippocampal dentate gyrus) and various cognitive functions; in general, increased neurogenesis is associated with improved learning and memory, whereas decreased neurogenesis (decreased proliferation and/or decreased neuronal differentiation and survival) is associated with reduced or impaired memory and learning.

In birds, neurogenesis has been associated with specific behaviors that depend on learning and memory functions (Barnea and Pravosudov, 2011). The song system of songbirds has served as a model for the process of vocal learning. Brain nuclei associated with this system are one of the primary destinations for new neurons migrating from the VZ. These areas include the high vocal center (HVC) which is involved in the control of vocal behaviors, the robust nucleus of the arcopallium (RA) which innervates the vocal organ, area X which is critical for the acquisition of song in juveniles, and the caudal neostriatum (NC), an area which stores specific auditory information. Food-caching birds need to develop spatial memories regarding the location of stored caches of food. Limited data show that food-caching birds have greater hippocampal neurogenesis than non-food-caching birds. Migratory adult birds have more intense hippocampal neurogenesis than non-migratory adults, consistent with their use of large-scale cognitive navigational maps. Finally, birds housed in large groupings with other birds showed more new neurons in specific brain regions than birds housed singly or in pairs.

These studies in mammals and birds strongly support the construct that adult neurogenesis is associated with the processes of learning and memory acquisition. The real unknown is the underlying mechanisms. How do new neurons facilitate learning and memory functions? Are new neurons involved and necessary for the acquisition of new memories? When new neurons are integrated into existing circuits, do they function in the same way as the old neurons?

Two important experiences, environmental enrichment and voluntary exercise have been shown to have significant effects on neurogenesis. Since most studies involving these two experiences have used laboratory rodents as the test subjects, it is important to define these experiences. Environmental enrichment includes a combination of inanimate and social stimulation. The stimuli are constituted by bigger housing cages with a running wheel and a few toys that are periodically changed to foster animal curiosity and exploration (Laviola *et al.*, 2008). Voluntary exercise is managed by giving the laboratory rodents free access to a running wheel. Another experience that has been noted to influence neurogenesis is that of social isolation which is manipulated by controlling the number of animals housed together.

Both environmental enrichment and voluntary exercise increased hippocampal neurogenesis in laboratory mice (Olson *et al.*, 2006). The influence of these two experiences appeared to be different. Environmental enrichment promoted the survival of newborn neurons, whereas voluntary exercise primarily increased progenitor cell proliferation and creation of new neurons. As previously noted, a DNA synthesis marker BrdU is used to define the birth date of a cohort of new neurons. Voluntary exercise leads to an increase in BrdU positive cells within 24 h following the last BrdU injection, whereas environmental enrichment leads to no change at this time point. By 28 days post-BrdU injection, both voluntary exercise and environmental enrichment induced a comparable increase in new neurons. Both of these experiences, voluntary exercise and environmental enrichment, enhanced learning and memory acquisition (Olson *et al.*, 2006).

Recently, an indirect measure of exercise-induced neurogenesis has been described in humans (Pereira *et al.*, 2007). The technique used was magnetic resonance imaging (MRI) and the justification for the use of this technique was the known coupling between neurogenesis and angiogenesis and between angiogenesis and the regional cerebral blood volume (CBV). Using mice as a test subject, the investigators found that in comparing exercising to non-exercising mice, there was a positive correlation between changes in dentate gyrus CBV (measured by MRI) and the number of BrdU-labeled neurons in the dentate gyrus. These changes in CBV and BrdU labeling were specific to the dentate gyrus. The investigators then applied this model to exercising humans. Eleven healthy humans completed a three-month aerobic program and MRI-produced CBV maps of the hippocampus were generated for these subjects before and after the exercise program. The dentate gyrus was the only subregion of the hippocampus that showed an increase in CBV following the exercise program. Declarative memory testing was performed in these human subjects and the subjects performed better in this cognitive function test after the exercise program compared to their pre-exercise performance. These observations in humans are consistent with the animal studies that showed an exercise-induced increase in neurogenesis in the dentate gyrus associated with an exercise induced enhancement in memory acquisition.

Van Praag *et al.* (1999) examined the effects of voluntary exercise on learning, synaptic plasticity, and neurogenesis in mice. Mice were divided into two groups: controls and runners (runners had free access to a running wheel). Spatial learning was tested using a Morris water maze (MWM). Runners showed enhanced learning and memory with respect to the water maze task and hippocampal slices from runners displayed greater synaptic plasticity (measured as the strength of dentate gyrus long-term potentiation) than the control group. The hippocampal dentate gyrus of runners contained more than double the number of BrdU-positive cells two to four months following BrdU injection than control mice. This study highlighted the connections between hippocampal dentate gyrus neurogenesis, synaptic plasticity, and learning and memory, but did not define the mechanistic links underlying these connections. In a comparable study, Farmer *et al.* (2004), using rats as the test subjects, observed that runners showed almost twice as many BrdU-positive cells in the dentate gyrus as control rats and also displayed greater long-term potentiation (following titanic stimulation) than non-runners. Farmer *et al.* (2004) also explored possible mechanisms underlying these observations. They found that mRNA levels for brain-derived neurotrophic factor, NMDA receptor NR2B subunit, and glutamate receptor 5 were increased in the dentate gyrus of runners compared to non-runners.

The effects of voluntary exercise on neurogenesis are quite nuanced given that at least over the short term, these effects are modulated by the social context in which the exercise occurs. Stranahan *et al.* (2006) exposed rats to running or no running (i.e., with or without access to a running wheel) while housed in groups of three or alone. Runners housed in groups and exposed to 12 days of running had more BrdU-positive cells in the dentate gyrus than non-runners housed in groups. In contrast, runners housed singly showed a decrease in the number of BrdU-labeled cells in the dentate gyrus compared to non-runners housed individually. These changes in the number of BrdU-positive cells in the dentate gyrus reflected changes in the proliferation of SGZ progenitor cells. Blood corticosterone levels were measured in these rats and runners were found to have greater levels at the beginning of the active phase (7 pm) than non-runners irrespective of the housing arrangement. At a later time point (11 pm), for both runners and non-runners, rats in group housing had lower corticosterone levels than rats housed singly. When rats were adrenalectomized with baseline corticosterone replacement, runners housed either individually or in a group showed increased neurogenesis compared to non-runners in either housing arrangement. Since glucocorticoids suppress neurogenesis, an unidentified factor must be present in runners housed in a group context which "overcomes" the suppressive effects of corticosterone on neurogenesis. The investigators speculated that this unidentified factor could be serotonin. Finally, Stranahan *et al.* (2006) observed that if rats housed singly exercised for up to 48 days, they did experience an increase in dentate gyrus neurogenesis compared to non-runners. This changeover from running having a negative effect to a positive effect on neurogenesis in rats housed individually occurred sometime between days 24 and 48 of running. These observations underscored the complex interactions that existed between exercise, social context, and adult neurogenesis.

The experimental data indicate that in the adult mammal, ongoing new neuron recruitment to the hippocampal dentate gyrus is linked to the process of task learning and memory acquisition. Experiences such as voluntary exercise and environmental enrichment which increase adult hippocampal neurogenesis also improve learning and memory functions. In the case of exercise, an important variable is the social context in which the exercise occurs. The changes in mRNA levels described by Farmer *et al.* (2004) implicate various trophic factors and neurotransmitters as mediators of the link between dentate gyrus neurogenesis and learning and memory and the proposed roles of these and other factors have been reviewed by van Praag (2008), Zhao *et al.* (2008), and Barnea and Pravosudov (2011). However, none of these studies addressed the following questions. What are the underlying genomic mechanisms that link the experience itself, e.g., environmental enrichment or voluntary exercise, to an increase in the creation of new neurons and their recruitment to specific brain areas? Once these new neurons have been integrated into existing circuits, how do they facilitate and enhance the learning of tasks and memory formation? There are experimental observations that touch upon the first question concerning the mechanistic links between the experience and the process of neurogenesis, but there is very little information as to what, if any, role these new neurons play in task-learning and memory acquisition. Recent observations, for example, have shed more light on hippocampal memory functions such as spatial memory, a function subserved by hippocampal place cells (Poucet and Sargolini, 2013). However, these studies have also raised a variety of new questions concerning spatial memory, since this memory function appears to be based upon a set of exquisitely complex interactions between grid cells in the medial entorhinal cortex and hippocampal place cells.

Social interactions, in addition to environmental enrichment, can influence adult neurogenesis as illustrated by the effects of social isolation in rodents (housed singly) on exercise-enhanced neurogenesis. Electric fish engage in long-term social interactions which are mediated by electro-communication behaviors (Dunlap *et al.*, 2008). These fish emit a weak electric signal from specialized neural tissue in the tail termed the electric organ. The signals from this organ are regulated by a hindbrain pacemaker nucleus. During social interactions, fish modulate the frequency and amplitude of the signals from the electric organ to form chirps. Chirps are generated by a bilateral diencephalic nucleus, the pre-pacemaker nucleus-chirping (PPn-C) that increases the firing frequency of the pacemaker nucleus. The PPn-C receives electro-sensory and visuo-sensory inputs and makes monosynaptic output to the pacemaker nucleus. Social interactions between fish enhance chirping behavior, increase neurogenesis in the ventricular germinal zone (VGZ) adjacent to the PPn-C, and increase the density of radial fibers that serve as a migratory pathway between the VGZ and the PPn-C. New neurons born in the VGZ are guided to the PPn-C by this fiber network.

Dunlap *et al.* (2008) examined what features of the social interaction between fish resulted in increased cell birth in the VGZ and increased glial fiber density. Four arrangements were created: isolated fish that received no electric or non-electric social stimuli; paired fish that could interact through electric, chemical, or visual stimuli but had no direct physical contact (these fish can be aggressive and injure each other); stimulus fish that were housed singly but in tanks equipped with wire that could transmit electric organ signals; and receiving fish that were housed singly in tanks that received the electric signals from the stimulus fish. Paired fish and fish that received the electric signal only (receiving fish) had higher number of BrdU-positive cells and a higher density of glial fibers in the VGZ adjacent to the PPn-C than isolated fish. These experimental observations indicate that electrical signals alone were sufficient to induce the increase in neurogenesis and radial glia fiber density induced by the social interactions associated with fish pairing. The investigators also created an electric stimulus to mimic the electric organ discharge, but this signal was ineffective in promoting increased neurogenesis and glial fiber formation.

These experimental observations define electrocommunication signals between fish as the minimal sufficient stimulus to promote increased neurogenesis and glial fiber formation in the VGZ. The VGZ and its connections to the PPn-C are regions of the brain that participate in socially induced changes in electrocommunication behaviors. In addition, sine-wave electrical stimuli that mimicked the electrocommunication signals had no effect on neurogenesis and glial fiber formation indicating that a natural communication signal including its dynamic modulations was necessary. In electric fish, the social interactions promoted by fish pairing can be reduced to a sensory electrocommunication signal. This signal would be inputted into the receiving fish through direct sensory pathways and result in increased activity of target neurons.

There is good evidence that the activation of neural stem cells can be influenced by the activity of surrounding mature neurons (Hsieh and Schneider, 2013). Song *et al.* (2012) examined the effects of neural activity on progenitor cell activation using a mouse model. Type 1 RGL (radial glia-like) progenitor cells were identified in the mouse SGZ. RGL progenitor cells can undergo symmetric self-renewal or asymmetric self-renewal with generation of astrocytes or neurons (Bonaguidi *et al.*, 2011). The activity state of these cells was found to be regulated by GABA. Song and co-workers engineered mice such that the GABA receptor in these RGL cells could be conditionally deleted. Such RGL cells went from a quiescent to an activated state within two days of induction of this deletion, activation being assessed by the number of RGL cells labeled with the DNA synthesis marker EdU. Clonal analysis revealed that these activated cells had multiple fates: symmetrical self-renewal, astrocytes, or neurons. Hence, GABA had a direct role in maintaining RGL progenitor cells in a quiescent state. These RGL cells were in close proximity to the terminals of PV+ (expressing the calcium-binding protein parvalbumin) interneurons which release GABA. PV+ interneurons are abundant in the hippocampus and receive excitatory inputs from the dentate granule cells and to a smaller extent from entorhinal cortical neurons. Activity of these interneurons is a niche component controlling the behavioral state of the RGL progenitor cells. Activation of PV+ interneurons reduced the number of EdU-positive RGL cells, whereas suppression of PV+ interneurons significantly increased the number of EdU-positive RGL progenitor cells. Song *et al.* (2012) also examined the effect of social isolation on this interneuron–RGL circuitry. In contrast to group housing, social isolation resulted in an activation of RGL cells with both symmetric self-renewal as well as creation of astrocytes. Activation of PV+ interneurons abolished the increase in RGL activity induced by social isolation. In summary, PV+ interneurons within the hippocampus regulate the functional state of RGL progenitor cells through the effects of GABA. These interneurons can be considered a niche component for RGL cells; activity of these interneurons with release of GABA maintains the RGL cells in a quiescent state, whereas reduced activity of these interneurons promotes proliferation of these RGL cells.

The experimental observations of Dunlap *et al.* (2008) and Song *et al.* (2012) suggest the framework of a conceptual model to account for the effects of experiences such as environmental enrichment, group social arrangements, and voluntary exercise on neurogenesis. Experiences can be broken down into sensory components which then "enter" the animal through neural pathways originating with visual, olfactory, auditory, pain, temperature, kinesthetic, etc. receptors. The type of sensory system activated is important and in invertebrates, the role of specific sensory pathways in determining organismal phenotypes has been well described (Linford *et al.*, 2011). The model animals used for these studies were *Caenorhabditis elegans* and *Drosophila melanogaster* and the phenotypes assessed were ageing pattern and lifespan. The sensory systems tested included olfactory, gustatory, and mechanosensory. As reviewed by Linford and co-workers, sensory signals in these model animals can significantly change lifespan and ageing patterns and part of the mechanism involved activation of certain pathways such as the insulin/insulin-like peptide signaling cascade. These types of observations linking sensory system activation with alterations in ageing and longevity have not been documented in vertebrates, but perhaps this is because the necessary experiments have yet to be performed. However, sensory pathways are the only portal of entry by which environmental cues (experiences) can enter an animal; hence, these invertebrate studies are relevant to our discussion since they underscore that the nature of the sensory system itself is important in determining the effects of environmental cues. The specific role of sensory systems in mediating the effects of environmental cues has generally not been appreciated.

Within the proposed framework, sensory pathway signaling leads to the activation of downstream neurons such as for example the PV+ interneurons in the hippocampus. These neurons are anatomically closely associated with neural progenitor cells and can regulate the functional behavior of these cells (Hsieh and Schneider, 2013). In this construct, the electrical activity of mature neurons can determine whether nearby progenitor cells are in a quiescent state or in an activated proliferative mode. The development of this model to include epigenetic mechanisms is discussed in the next sections.

Ma *et al.* (2009) explored the role of epigenetic mechanisms in the process of adult neurogenesis in the murine dentate gyrus. Electroconvulsive therapy (ECT) in mice significantly increased the proliferation of neural progenitor cells as measured by the greater number of BrdU-labeled cells in the dentate gyrus of ECT-treated mice compared to control mice. In addition, a single ECT treatment was sufficient to induce a large increase in the expression of the gene *Gadd45b* (growth arrest DNA damage gene) in mature dentate granule cells within an hour following the ECT session. Two other experiences associated with enhanced neurogenesis, spatial exploration of a novel environment and a voluntary seven-day exercise program also upregulated *Gadd45b* expression. The association between experience-mediated increased neurogenesis and upregulation of *Gadd45b* was direct since ECT or voluntary exercise induced much less neurogenesis in *Gadd45b* KO (knockout) mice compared to wild-type mice. In addition, neurons born following ECT induction showed attenuated morphological maturation in *Gadd45b* KO mice compared to wild-type. The Gadd45 protein family mediates DNA demethylation during cell differentiation and during the stress response (Niehrs and Schafer, 2012). *Gadd45* genes are also members of the immediate early group (IEGs) of genes (Liebermann and Hoffman, 2008). Following ECT, mature dentate neurons showed demethylation of CpG sites within specific regulatory regions: regulatory region IX of brain-derived neurotrophic factor (BDNF IX) and promoter B of fibroblast growth factor 1 (FGF1B). This ECT-induced DNA demethylation did not occur in *Gadd45b* KO mice. Demethylation of these regulatory regions was associated with increased expression of these genes. Hence, experiences such as ECT, spatial exploration of a novel environment, or voluntary exercise were associated with increased expression of *Gadd45b*, an IEG, in mature neurons and subsequently *Gadd45b*-mediated DNA demethylation of specific gene regulatory regions. This demethylation of specific regulatory regions resulted in increased expression of those genes, e.g., BDNF and FGF1. BDNF promotes dendritic growth and FGF1 possesses mitogenic activity. The increased gene expression led to enhanced neural progenitor cell proliferation with the birth of a greater number of neurons.

Zhao *et al.* (2003) also examined the role of epigenetic mechanisms in the process of adult neurogenesis and associated hippocampal function. One of the means by which DNA methylation regulates gene expression is through the binding of methyl CpG-binding proteins (MBDs) to methyl CpG sites. The binding of a MBD to methyl CpG sites can block transcription factor binding and can also recruit histone deacetylase complexes to the binding sites, leading to histone deacetylation and formation of a tighter chromatin structure that is repressive for DNA transcription. One of the family members of MBDs is MBD1, which binds to methylated gene promoters. MBD1 has been shown to directly regulate the methylation status of the *FGF2* gene promoter region in neural progenitor cells (Hsieh and Eisch, 2010). (FGF2 is a mitogenic factor for neural progenitor cells.) MBD1 has a wide spread tissue distribution, but in the brain it is found in the highest concentration in the hippocampus: the CA1 zone, dentate gyrus, and SGZ. MBD1 is found in neurons but not astrocytes. Humans and mice each have a single *MBD1* gene. Zhao and co-investigators created a MBD1-deficient mouse (*MBD1−/−*) and observed that these animals developed normally and appeared healthy as adults. BrdU was injected into adult mice to assess neurogenesis. One day post-injection, both *MBD1−/−* and wild-type mice had similar numbers of labeled cells in the dentate gyrus, indicating comparable progenitor cell proliferation rates. However, at four weeks post-injection, the dentate gyrus of *MBD1−/−* mice had about 55% fewer BrdU-positive cells than the dentate gyrus of the wild-type mice, indicating a lower survival rate for newborn neurons in *MBD1−/−* mice. This decrease in neurogenesis was associated with reduced synaptic plasticity (severely attenuated long-term potentiation) in the dentate gyrus of the *MBD1−/−* mice as well as impaired spatial learning compared to the wild-type mice. Hence, adult mice lacking a functional *MBD1* gene displayed decreased hippocampal neurogenesis and impaired learning and synaptic plasticity. These experimental observations demonstrate that normal hippocampal neurogenesis and hippocampal mediated cognitive functions require the action of the *MBD1* gene in regulating the DNA methylation status of gene promoter regions.

The observations of Dunlap *et al.* (2008), Ma *et al.* (2009), Song *et al.* (2012), and Zhao *et al.* (2003) can be combined to further develop the framework of this model. The specific objective experience is inputted through sensory pathways with the activation of downstream central neurons. Within these neurons, IEGs are upregulated and partly through the effects of IEGs, changes in epigenetic marks (DNA methylation/demethylation; histone protein acetylation/deacetylation/methylation/demethylation) are established. The IEG, zif268 (alternate names ERG-1, NGFI-A) is known to be upregulated during different forms of associative learning and in the induction of long lasting long-term potentiation (Davis *et al.*, 2003). These changes in epigenetic marks regulate the expression of specific genes within these neurons, which in turn determine levels of various growth factors and neurotransmitters. Through the intermediary of these factors, mature neurons can regulate the proliferative activity of neural progenitor cells. If the experience leads to an increase in progenitor cell proliferation with the birth of more neurons, then as a result of this increased number of young neurons being integrated into existing hippocampal circuits, the animal demonstrates an enhancement in synaptic plasticity and learning and memory functions. It is obvious that this framework begs many questions, not the least of which is the mechanism(s) by which new neurons actually enhance cognitive functions.

Social experiences during the life cycle

The important role of early childhood experiences in promoting normal cognitive development has been well documented (Nelson III *et al.*, 2007; 2013). The Bucharest Early Intervention Project compared three groups of children all less than 31 months of age exposed to different early childhood experiences. One hundred and thirty- six orphaned children were recruited from institutions within the Bucharest region. These children lived in highly structured, bleak environments with typically one adult to oversee the care of 12 to 15 children. Half of these children were randomized to remain in institutional care, while the other half were randomly assigned to foster care. A third "control" group consisted of 80 children who were living with their biological parents and had never experienced institutional care. The children were assessed using a set of cognitive function tests at baseline, 30 months, 42 months, and 54 months. The authors noted several findings. Children reared in institutions showed greatly diminished cognitive functioning, whereas children randomly assigned to foster care showed significant gains in cognitive performance compared to institutionalized children. There appeared to be a time-sensitive window within the first two years of life within which foster care exerted a maximal effect on cognitive development. Even though children raised in foster care had better cognitive functioning than institutionalized children, they did not perform as well as children living with their biological parents.

The results of studies such as that of the Bucharest early Intervention project indicate the complexities characterizing the interactions between experiences and animal biology. Some of these complexities have been discussed by Rutter (2012). What are the components that constitute experiences such as abuse, social neglect, and social disadvantage? How does the objective experience relate to the experience as perceived by the individual, the subjective experience? Given the uniqueness of each person, two individuals may be exposed to the same objective experience but perceive quite different subjective experiences. For example, comparable experiences may have a sensitizing effect in some individuals, a steeling effect in others, and no discernable effect in yet other individuals (Rutter, 2012). The highly nuanced nature of experiences is underscored by the observation that the same experience may have different effects depending upon the context in which that experience occurs (Lemos *et al.*, 2012).

Lemos *et al.* (2012), using a mouse model, examined changes in behavior induced by severe stress. The nucleus accumbens plays an important role in reward-type behaviors. Corticotrophin-releasing factor (CRF) is one of the regulators of behaviors that are mediated by dopamine within the nucleus accumbens. Two CRF receptors, CRFR1 and CRFR2 are widely distributed in the brain including the nucleus accumbens. Dopamine release can be evoked from neurons of the nucleus accumbens by either an electric impulse or by CRF. CRF injected into the nucleus accumbens increases dopamine release in a concentration-dependent manner. This CRF-induced dopamine release requires the presence of both receptor types since mice with a deletion of either receptor gene failed to release dopamine in response to CRF. Dopamine release within the nucleus accumbens induces place preference behavior in mice. When mice receive CRF directly into the nucleus accumbens (NA), they can be conditioned to prefer the test chamber that they were in at the time of the injection — a positive affective state. When a selective neurotoxin was used to prevent CRF-mediated dopamine release, place preference conditioning could not be induced. In a second part of these experiments, mice were subjected to severe repeated swim stress for two days. In these "stressed" mice, CRF was unable to induce dopamine release from the nucleus accumbens. This inability of CRF to induce dopamine release in stressed mice could be prevented by prior administration of a glucocorticoid receptor (GR) antagonist. Interestingly, this stress-induced loss of CRF-mediated dopamine release persisted for 90 days after stress exposure. When mice that had undergone the repeated swim stress were subsequently tested for conditioned place preference mediated by intra-NA CRF injections, they actually showed an aversion rather than a preference for the conditioned test chamber. This aversive state persisted for up to 90 days. Hence, a positive affective state or behavior becomes an aversive behavior if the animal is pre-conditioned with a severe repeated stressful experience. Mechanistically, the severe stress prevents CRF-mediated dopamine release from NA neurons, and converts a positive preference type behavior to an aversive negative behavior.

One of the inputs into the nucleus accumbens comes from neurons in the ventral tegmental area (VTA). In the mouse brain, the VTA is one of a group of subregions of the brain involved in reward and avoidance behaviors as well as regulating anxiety and fear (Johansen, 2013). These subregions contain intermixed cell populations that can produce or ameliorate anxiety. Although the observations of Lemos *et al.* (2012) document that severe chronic stress can convert a positive behavior to an aversive one, the underlying neural circuitry involved in these and related behaviors are only beginning to be defined (Johansen, 2013).

The mechanisms underlying the multifaceted interactions between an experience and the biology of an individual have generally been broadly defined under the rubric of gene–environment interactions. The primary class of gene–environment interactions thought to mediate the effects of social experiences is that of changes in DNA methylation and histone modifications (Champagne, 2010). Changes in epigenetic marks act as a transduction mechanism to convert the exposure to objective experiences into altered gene expression patterns and this mechanism is operative throughout all stages of the life cycle, both prenatal and postnatal. Several examples illustrate this point (Buchen, 2010; Champagne, 2010).

In rodents, postnatal maternal licking and grooming (LG) behavior can modulate the strength of the stress response in offspring through epigenetic modification of hippocampal GR expression. This is an example of developmental plasticity; the nature of maternal care during a critical developmental time window can modify offspring behaviors in later adult life. The adult offspring of mothers that provided increased levels of pup LG (high LG mothers) over the first week of life, manifested increased expression of hippocampal GRs, enhanced glucocorticoid feedback inhibition, decreased hypothalamic CRF expression, and more modest HPA (hypothalamic-pituitary-adrenal axis) stress responses compared to adult offspring reared by low LG mothers (Weaver *et al.*, 2007). These effects of maternal care (LG) involved a thyroid hormone-dependent increase in serotonin (5-HT) activity at 5-HT_7 receptors and the subsequent activation of cAMP and cAMP-dependent protein kinase A. This sequence of events led to increased expression of the gene encoding the transcription factor nerve growth factor- inducible protein A (NGFI-A). NGFI-A is a member of a family of transcription factors and is encoded by the immediate early gene, *ERG-1* (Swirnoff and Milbrandt, 1995). The *GR* gene contains nine exons: exons 2–9 code for the receptor protein while exon 1 is composed of multiple regulatory regions (11 such regions), each of which is capable of activating gene transcription. The activity of the various exon 1 promoters is thought to be tissue specific with evidence that exon 1_7 is more active in the brain (Zhang and Meaney, 2010). The exon 1_7 *GR* promoter region contains a NGFI-A binding sequence and differences in the methylation level of the 5′ CpG locus within this sequence regulate binding of NGFI-A to this sequence and subsequent transcriptional activity of the *GR* gene. At birth, the 5′ CpG locus is hypermethylated in the offspring of both high and low LG mothers. Over the first week of postnatal life, in offspring of high LG mothers, there is demethylation of the 5′ CpG locus in the exon 1_7 *GR* promoter region with increased binding of NGFI-A and increased *GR* gene expression. The experimental findings indicate that NGFI-A mediated the epigenetic reprogramming of the exon 1_7 promoter. This reprogramming included demethylation of the 5′ CpG locus as well as acetylation of histone H3, lysine 9 by NGFI-A-recruited CBP (CREB-binding protein), a histone acetyltransferase. In summary, the tactile sensation of high LG maternal behavior led to a series of events involving thyroid hormone and 5-HT which increased the expression of the *NGFI-A* gene. NGFI-A-mediated epigenetic modification of the exon 1_7*GR* promoter (CpG demethylation and histone acetylation) enhanced NGFI-A binding to the promoter. Enhanced binding of NGFI-A increased transcriptional activity of the *GR* gene, with increased expression of *GR* in the hippocampus and modulation of the

HPA stress response as noted at the beginning of this section. In sum, a social experience, enhanced LG maternal care during the first postnatal week, induces a stable persistent epigenetic reprogramming in the offspring which is associated with a phenotypic change in these offspring — modulation of the HPA stress response (Weaver *et al.*, 2007).

The extent of maternal LG can also epigenetically modify the expression of hypothalamic estrogen receptors in female offspring and by so doing, modulate the maternal care provided by these female offspring to their offspring (Champagne, 2010).

In adult rodents, exposure of animals to social defeat results in altered behaviors: reduced locomotion, decreased social behavior, increase in the stress response, and an overall depressive state (Champagne, 2010). One of the underlying changes measured in these animals was a reduction in the expression of *BDNF* in the hippocampus through the establishment of histone marks (increased histone H3 lysine 27 dimethylation) in the III and IV promoter regions of this gene in socially defeated males. These results were consistent with the construct that social experiences alter gene expression by means of epigenetic reprogramming.

As a general approach to the study of epigenetic mediated gene–environment interactions, Lam *et al.* (2012) measured DNA promoter methylation patterns in a community cohort of 92 individuals. The tissue sample used in this study was peripheral blood mononuclear cells. In somatic cells, as previously noted, DNA methylation occurs almost exclusively at CpG sites and these sites are non-randomly distributed across the genome. The investigators measured the methylation status of CpG sites in gene promoters. They classified genomic regions according to the density of CpG sites: high density (HD), intermediated density (ID), and low density (LD). HD regions were primarily hypomethylated whereas LD regions were primarily hypermethylated, although there was considerable variability in level of CpG methylation within these regions. For the whole cohort, specific CpG loci showed extensive variation in the level of methylation between individuals. A small number of CpG sites showed differences in methylation levels that correlated with gender and a smaller number of sites had different methylation levels that correlated with age. (However, the cohort had a very limited age range, 24–45 years.) Age- and gender-related changes in methylation levels were generally in the order of 5%. When socioeconomic status (SES) of the cohort was examined, there were a few sites that showed a difference in level of methylation associated with early-life SES but not with current SES. In addition, a subject's perceived stress and measurements of salivary cortisol correlated with variations in methylation levels at a small number of specific CpG sites. In 55 members of the cohort, the relationship between gene expression and DNA promoter methylation was assessed. Within an individual, there was a strong negative correlation between DNA promoter methylation and gene expression; highly methylated genes showed a low expression and vice versa. However, there were genes that did not fit this pattern and some genes with high expression had highly methylated promoters. Across individuals, about 90% of CpG sites showed the expected negative correlation between methylation level and gene expression, but about 10% of CpG sites had a positive correlation. The observations reported in this study were consistent with the model framework discussed in the section on **adult neurogenesis**. CpG loci methylation is quite variable across individuals, and there appears to be a general correlation between variations in the level of DNA methylation at specific CpG loci and age, gender, early-life SES, and level of stress (self-reported and cortisol production). These correlations are consistent with a mechanism whereby alterations of epigenetic marks mediate social experience-induced phenotypic changes.

In summary, the elements of the gene–environment interactions underlying the effects of social experiences on phenotypic expression are similar to those discussed in the section on **adult neurogenesis**. These elements include: input of the objective specific experience through various sensory pathways and activation of downstream neurons with activation of various IEGs. These IEGs mediate changes in epigenetic marks with modulation of gene expression patterns. In this model, IEGs (for example *Gadd45b* and *NGFI-A*) act to transduce the effects of experience-mediated activation of neurons into epigenetic reprogramming resulting in modulation of gene expression patterns. As a result of this sequence of events, an objective social experience can induce a stable phenotypic change.

The gene–environment interaction involves not only specific environmental cues (experiences) but may also require a specific allele of the gene under question. The effect of gene polymorphism in gene–environment interactions was examined by Caspi *et al.* (2003). The test subjects comprised a birth cohort (in New Zealand) of 1,037 children (52% male) that had been assessed at multiple ages and was still intact at the age of 26. Eight hundred and forty-seven members of the cohort were genotyped for the serotonin transporter gene (*SLC6A4*), the protein product of which regulates the reuptake of serotonin at brain synapses. This system is the target of serotonin reuptake inhibitor drugs (SSRIs). The promoter region of this gene contains a polymorphism which modulates transcriptional activity; the short (s) allele is associated with reduced transcriptional activity compared to the long (l) allele. The 847 members of the cohort were made up of 147 s/s homozygotes, 435 s/l heterozygotes ($n = 435$), and 265 l/l homozygotes. Stressful life events that occurred between ages 21 and 26 were assessed and correlated with past-year depression at the age of 26. There was a significant positive association between the number of stressful life events between ages 21 and 26 and depression outcomes at age 26 as a function of the promoter polymorphism of gene *SLC6A4*. The strength of the association was in the order s/s > s/l > l/l. The investigators also examined the relationship between childhood maltreatment between ages three and 11 and adult depression (between 18 and 26 years of age) as a function of promoter polymorphism. There was a positive correlation between 'no,' 'probable,' and 'severe' childhood maltreatment and adult depression, and the strength of the correlation was in the same order as noted above, s/s > s/l > l/l. These experimental observations confirm that social experience–gene interactions can result in specific phenotypic expressions (in this case depression), but add the wrinkle that the interaction is modulated by a particular allele of the gene. Here, the effect of the environmental cue (stressful life experience) in inducing a specific phenotype required a specific allele of the target gene. Individuals homozygous for the short allele of the serotonin transporter gene were much more likely to develop depression in response to adverse stressful life experiences than individuals homozygous for the long allele.

The effects of early childhood adverse experiences are not restricted to cognitive impairments as discussed with respect to the Bucharest Early Intervention Project, but also include an increased risk for development of "chronic degenerative diseases" in later (adult) life. Danese *et al.* (2009) looked at statistical associations between adverse psychosocial experiences in childhood and "diseases" such as depression, levels of inflammatory biomarkers, and clustering of metabolic risk markers in a birth cohort of 1,037 individuals. The adverse childhood experiences (assessed over the first decade of life) included: socioeconomic level (a measure of socioeconomic disadvantage), childhood maltreatment (maternal rejection, physical abuse, sexual abuse, harsh discipline, and two or more changes in a child's primary caregiver), and childhood social isolation. The main outcomes, measured at age 32, were adult major depression, level of C-reactive protein as a biomarker of an underlying inflammatory state (e.g., cardiovascular disease), and clustering of metabolic risk markers, e.g., obesity, high blood pressure, high levels of total and low density cholesterol, and high glycated hemoglobin. The investigators performed a multivariate analysis and included in this analysis established developmental risk factors such as family history of cardiovascular disease and depression, birth weight and childhood body mass index as well as current risk factors at age 32 years; socioeconomic status at age 32 years, current smoking activity, current physical activity, current diet, and current medications. The analysis of the data showed that at 32 years of age, 31.6% of the cohort cases with depression, 13.0% of cohort cases with an elevated inflammatory biomarker, and 32.2% of cohort cases with clustering of metabolic risk biomarkers could be attributed to adverse childhood experiences (socioeconomic disadvantage, maltreatment, or social isolation). In addition, the prevalence of each of these outcomes (depression, inflammation, and metabolic risk factors) in adulthood increased as a function of the number of adverse childhood experiences. The implication of the data is that these adverse childhood experiences can induce enduring consequences that affect nervous, immune, and endocrine/metabolic functioning in adulthood; this is another example of developmental plasticity.

Although the investigators did not propose a mechanism, the linkage between early childhood adverse experiences and adult onset impaired biological functions most likely would involve epigenetic reprogramming. One possibility is that these adverse childhood experiences epigenetically induced a long-lasting upregulation of the stress response with an increased production of glucocorticoids. Stress hormones can mediate adult onset biological changes such as insulin resistance, arterial vascular remodeling, and a resetting of arterial blood pressure levels as discussed by Singer (2013). The role of glucocorticoids and corticotropin-releasing hormone (CRH) in learning and memory functions is discussed in the next sections.

Immediate early genes (IEGs)

The term 'immediate early gene' originated from the field of virology. When a virus infects a host, viral genes are rapidly transcribed since the viral genome uses the pre-existing transcription machinery of the host cell, and this transcription occurs in the absence of *de novo* protein synthesis (Okuno, 2011). By analogy, cellular genes have been described which can be rapidly activated by extracellular signals without the need for *de novo* protein synthesis; i.e., the rapid upregulation of these genes utilizes existing regulators and is not dependent on the synthesis of new regulators. These genes have been classified as IEGs and these genes encode various products, including transcription factors, signaling molecules, and structural proteins. IEGs are rapidly upregulated following activation by cell signaling pathways and mRNA transcripts can appear within minutes.

As discussed in the previous sections, IEGs are involved in social experience-mediated modulation of hippocampal neurogenesis and hippocampal GR expression (with modulation of the stress response). These effects of IEGs are the result of epigenetic reprogramming of target genes. IEGs are also involved in a number of cognitive functions such as learning and memory as well as being involved in strength and structural synaptic plasticity (Korb and Finkbeiner, 2011). KO mice for the IEG *Arc* (coding for the activity-regulated cytoskeleton-associated protein) display normal short-term memory but are deficient in the formation of lasting memories (Korb and Finkbeiner, 2011). KO mice for the gene *NGFI-A* showed deficits in long-term memory formation and impairment of recognition memory, whereas mice with a deletion (localized to the brain) of the gene *FOS* had impairments in hippocampal-dependent spatial and fear memory (Okuno, 2011). During the formation of declarative long-term memory in macaque monkeys, the IEGs *BDNF* and *NGFI-A* are selectively induced in patch-like patterns in a specific region of the inferior temporal cortex (Okuno, 2011). Selective induction of the IEG *BDNF* (in the parietal association cortex) is observed in association with tool-use learning in the monkey (Okuno, 2011). The experimental observations summarized by Guzowski (2002) and Okuno (2011) underscored the multiple roles that IEGs play in high-level cognitive functions.

The role of IEGs in learning and memory is a highly conserved role of IEGs given that these functions have been demonstrated in invertebrates as well as vertebrates. Lutz and Robinson (2013) examined spatial learning and the expression of IEGs in the honeybee. The "task" the investigators studied was the short learning flights performed by bees when initiating foraging behaviors. These flights are known as orientation flights, and during these flights, bees acquire information about landmarks and learn to associate position of the sun, time of day, and directionality. The IEG the investigators chose was *EGR*, the bee homolog of the vertebrate IEG, *EGR-1* (*NGFI-A*). In vertebrates, the gene *NGFI-A* is an activity-induced IEG associated with learning and novelty detection. In addition, as discussed in a previous section, *NGFI-A* mediates the influence of the quality of maternal care in rodents on the stress response of the offspring. The DNA-binding domain of *EGR* shares 89% amino acid sequence identity with the mouse.

Orientation flights were associated with significant upregulation of *EGR* in the mushroom bodies of the bee brain. Orienting bees showed a doubling of *EGR* mRNA (compared to controls) 30 min after flight with return to baseline levels by 60 min. The orientation flight can be broken down into its component experiences: exercise, motor learning, exposure to visual cues, and exposure to visual novelty. Through a series of discriminating experiments, the investigators found that both flight and visual novelty were necessary to induce *EGR* upregulation. The investigators did not examine possible downstream targets of honey bee *EGR*, but other experimental observations suggest that bee *EGR* mediated rapid remodeling of dendritic spines in mushroom body neurons after an orientation flight.

The regulation of IEG induction is clearly an important research area and experimental observations have documented the important role of chromatin (epigenetic) modifications in the regulatory mechanisms. Crosio *et al.* (2003) treated mice with agonists of dopamine receptors, muscarinic acetylcholine receptors, and ionotropic glutamate receptors and then looked at chromatin modifications and *FOS* expression in hippocampal neurons. All three agonists induced chromatin histone (H3) modifications: serine 10 phosphorylation together with acetylation of lysine 9 and lysine 14, in the dentate gyrus and areas CA1, CA2, and CA3 of the hippocampus. H3 serine 10 phosphorylation was detected within 15 min, peaked in 1 h, and returned to basal levels within 3 h. The extent of chromatin modifications was substantially greater in the dentate gyrus than in areas CA1, CA2, or CA3. The same neurons demonstrating chromatin modifications also showed activation of the MAPK/ERK (mitogen-activated protein kinase/extracellular signal-regulated kinase) signaling pathway as well as expression of the IEG *FOS*. The time course of *FOS* expression was the same as that of chromatin H3 serine phosphorylation. Induction of chromatin modifications (specifically H3 serine 10 phosphorylation) appeared to precede induction of *FOS* expression. Histone modifications (H3 serine 10 phosphorylation and lysine 14 acetylation) induced by the glutamate receptor agonist were in the chromatin associated with the gene *FOS*. These experimental observations were consistent with the following sequence. Agonist–receptor binding activates the MAPK/ERK signaling pathway inducing modifications (acetylation/phosphorylation) in the chromatin associated with the IEG *FOS*, resulting in transcriptional activation of *FOS*.

Laboratory rodents have been subjected to a number of learning and memory paradigms: MWM, radial maze, and the forced swim test. Rats subjected to these tests displayed chromatin histone modifications (H3 phosphorylation serine 10 and acetylation lysine 14) in hippocampal dentate gyrus granule cells (Reul *et al.*, 2009). The chromatin modifications were transient and lasted for up to 4 h. When animals are challenged by one of these learning tests, the flow of sensory information activates dentate gyrus neurons. Due to high tonic inhibitory control exerted by GABAergic interneurons, only a few dentate granule cells are actually activated (Reul *et al.*, 2009). The observed histone (H3) phosphorylation/acetylation changes were found to co-localize with the FOS protein in the same dentate neurons and subsequently, these histone modifications were documented to occur in the promoter region of the *FOS* gene. Reul *et al.* (2009) reported that the histone (H3) phospho-acetylation changes and *FOS* gene upregulation observed in the dentate gyrus of rodents subjected to forced swimming and other test situations required the action of both glutamate and corticosterone, acting through their respective receptors. Other studies indicated that the MAPK/ERK signaling pathway acted proximal (upstream) of the chromatin modifications. The construct of Reul *et al.* (2009) is described as follows. The hippocampal dentate gyrus plays a critical role in learning and memory acquisition. Activation of granule neurons in this brain region by neurotransmitters such as glutamate stimulates the MAPK/ERK signaling pathway resulting in histone modifications in gene promoters with up regulation of genes such as *FOS*. Importantly, the establishment of these chromatin epigenetic marks and the induction of genes such as *FOS* required the involvement of activated GRs.

O'Donnell *et al.* (2012) reviewed the sequence of steps responsible for regulating the expression of one specific IEG, *FOS*. Whether this sequence of steps can be generalized for all IEGs is not known. One of the principal cellular signaling pathways activating *FOS* is the MAPK/ERK pathway. This signaling cascade can be activated by growth factors such as BDNF and by neuronal activity through binding of specific neurotransmitters (e.g., glutamate, acetylcholine) to their respective receptors (Korb and Finkbeiner, 2011). Under quiescent conditions, the *FOS* promoter is in a poised state with SRF occupying the serum response element (SRE) and ELK-1 (transcription activator and one of the ternary complex factor, TCF proteins) occupying the *sis*-inducible element (SIE)-binding site. Activation of the MAPK/ERK pathway is associated with increased acetylation of the nucleosomes adjacent to the *FOS* transcription start site. These epigenetic marks alter chromatin structure allowing transcription factors such as NF-1 (nuclear factor 1) access to specific binding sites in the *FOS* gene promoter.

NF1 recruits other chromatin remodeling complexes to the promoter region such as PARP-1 (poly(ADP-ribose) polymerase 1). The sum effect of these chromatin modifications and associated recruitment/binding of transcription factors to the "opened" promoter region is a rapid increase in *FOS* transcriptional activity from basal levels. The chromatin modifications and binding of transcription factors to the *FOS* promoter can have a ripple effect and lead to upregulation of nearby genes, even those as far away as 100 kb.

Whether the mechanisms regulating the transcriptional activity of *FOS* as described by O'Donnell *et al.* (2012) are tissue specific is not clear since Okuno (2011) described a somewhat different sequence of events for neuronal *FOS*. Okuno (2011) noted the importance of transcriptional regulators such as CREB (cAMP response element-binding protein) and SRF. CREB and SRF bind to the cAMP response element (CRE) and serum response element (SRE), respectively. Point mutations in the CRE or SRE of the *FOS* promoter greatly reduce expression of this gene in the brain. Okuno (2011) does not describe histone modifications or the recruitment of NF-1 or PARP-1, whereas O'Donnell *et al.* (2012) do not implicate CREB in their description of regulatory events. However, histone modifications appear to be a critical feature for IEG induction (Crosio *et al.*, 2003; Reul *et al.*, 2009). Crosio *et al.* (2003) observed that the extent of chromatin modifications associated with the induction of *FOS* varied between subregions of the hippocampus, suggesting that there is indeed a spatial specificity to the mechanisms regulating transcriptional activity of IEGs such as *FOS*.

The transcriptional response of IEGs to growth factors or neuronal activity is usually transient. For example, the upregulation of *FOS* to activation by MAPK/ERK signaling reaches a peak within 30–60 min and the return to basal levels of transcriptional activity occurs within 60–90 min after termination of the activation stimulus. This rapid and transient kinetic pattern raises a number of as yet unanswered questions (O'Donnell *et al.*, 2012). How are the transcriptional activation signals extinguished and how are the epigenetic marks reset to the pre-stimulation state? The rapid decline in IEG transcriptional activity after cessation of the stimulus requires that the IEG mRNA transcript stability be low and that degradation mechanisms be available to clear the mRNA. A number of negative feedback regulators have been described which can extinguish the ERK/MAPK signal, but these negative feedback regulators do not explain how chromatin epigenetic marks are reset or how bound regulators are released.

How do experiences which are generally transient result in stable lasting phenotypic changes? If IEGs mediate the transduction of experience-induced neuronal activity into a persistent stable phenotypic change, how does this occur given the rapid activation and deactivation kinetics of IEG expression?

There is considerable evidence that epigenetic mechanisms underlie cognitive functions such as learning and memory formation and consolidation (Day and Sweatt, 2011). For example, contextual fear conditioning which "occurs" within the hippocampus is associated with the establishment of histone protein marks in the CA1 region of the hippocampus, and these marks are not observed in control animals exposed to the same environmental context but in the absence of fear conditioning. A number of experimental observations indicate that specific types of behavioral experiences are associated with specific sets of histone modifications. These histone marks appear to be necessary for memory formation and consolidation. There is evidence that changes in DNA methylation also are an essential component for learning and memory functions. As summarized by Day and Sweatt (2011), epigenetic mechanisms in mature neurons regulate patterns of gene expression in a functionally specific manner and within this context, changes in epigenetic marks can be considered as one of the final steps in a sequence of events that results in transcriptional activation of learning- and memory-related genes. An important upstream event of these epigenetic chromatin modifications is the induction of IEGs. The studies of Ma *et al.* (2009) showed that the IEG, *Gadd45b* mediated the epigenetic reprogramming observed in experienced-induced neurogenesis, while the experiments of Weaver *et al.* (2007) documented the role of the IEG, *NGFI-A* in mediating the epigenetic reprogramming of the *GR* promoter region.

Given all these observations, a reasonable construct is that experience-induced behavioral phenotypes result from the activation of central neurons (region-specific depending upon the behavioral function involved) with stimulation of various intracellular signaling pathways (e.g., MAPK/ERK pathway), and subsequent induction of IEGs which then mediate epigenetic reprogramming in the promoter sequences of various target genes. The net effect of this sequence is a particular pattern of gene expression which manifests as a specific behavioral phenotype.

Are there biological mechanisms other than IEG induction that mediate the transduction of experiences into epigenetic reprogramming with changes to gene expression patterns? Most probably yes.

McClelland *et al.* (2011) reviewed the links between chronic early life stresses and subsequent impairments in hippocampal functions in adulthood. Adverse early life experiences in humans included extreme poverty, loss of parent(s), and social isolation and abuse. An example of social isolation was discussed in a previous section — the institutionalization of orphans in Bucharest. A rodent model was developed in which mothers and offspring were maintained in an impoverished cage environment and this manipulation engendered chronic stress in both mothers and pups. This arrangement was continued for postnatal days 2–9 during which hippocampal function develops. At the end of this period, mothers and offspring were returned to normal cage conditions; by adulthood, rodent offspring showed normal neuroendocrine functioning. Initially the chronic early-life stress offspring, as young adults, performed well in the MWM test of spatial learning and memory. Long-term potentiation, a measure of synaptic strength plasticity, was normal in areas CA1 and CA3 of the hippocampus. However, when these same offspring were tested at age 7–10 months (middle-age), they performed poorly, compared to controls, in both the MWM test as well as in the novel object recognition test. In addition, long-term potentiation was attenuated in areas CA1 and CA3 in these middle-aged, early-life stressed rodents. Dendritic atrophy with loss of dendritic spines and synapses was observed in CA1 pyramidal neurons in association with the impairments in spatial and recognition memory. In essence, a single week of chronic stress in early postnatal life surfaced in later life as a disturbance in memory functions and synaptic plasticity due in part to loss of dendrites, dendritic spines, and synapses. These observations are a striking example of both the plasticity of neurodevelopmental programs and the long latent period that can occur between the inciting adverse early-life experience and the expressed behavioral phenotype.

What are the links between this early life chronic stress experience and later life cognitive impairments (McClelland *et al.*, 2011)? Middle-aged rodents stressed early in life show increased amounts of CRH (corticotropin releasing hormone), both protein and mRNA, in hippocampal pyramidal cells. In hippocampal neuronal cultures, CRH interacts with CRH receptor (CRHR-1) to induce atrophy of dendrites and reduced dendrite spine number. CRH is part of the stress response; CRH is released from cells within the hypothalamus, hippocampus, and amygdala. Hypothalamic CRH stimulates the pituitary to release adrenocorticotropic hormone (ACTH) which stimulates the release of glucocorticoids from the adrenal gland. Within the brain, glucocorticoids interact with GRs to complete a feedback loop. The promoter of the *CRH* gene has a number of binding regions: CRE, a glucocorticoid response element (GRE), and the intron between exons 1 and 2 which contains a RE-1/NRSE sequence. This sequence binds REST/NRSF, a repressor of gene expression. REST/NRSF functions by recruiting chromatin modifying enzyme complexes to the *CRH* promoter region (Ooi and Wood, 2007). Enriched early life experiences (augmented sensory input from mother), in contrast to early-life stress, were associated in the middle-aged adult with enhanced learning and memory and attenuation of the stress response. These middle-aged animals displayed a reduced *CRH* expression which appeared to be the result of repression by REST/NRSF-induced epigenetic reprogramming. In addition, the extent of CpG methylation in the *CRH* promoter was noted to be lower in the adult animals exposed to an enriched early-life experience. Hence, early-life experiences through an epigenetic mechanism can lead to long-lasting changes in the expression of CRH. Up- or downregulation of CRH is associated with respectively impaired or enhanced learning and memory functions and synaptic plasticity in late adult life. McClelland *et al.* (2011) did not describe any involvement of IEGs in the epigenetic reprogramming of the *CRH* gene.

Food intake

Phenotypic effects of changes in the composition of food ingested

The intake of nutrients is clearly essential for the maintenance of the living state, but alterations in the type of food ingested can have profound biological effects. I will refer to non-physiological deviations in the type of food ingested as 'nutritional stress.' An important variable is the period of time during the animal's life cycle that the nutritional stress occurs. During critical periods of development, nutritional stresses can permanently induce changes in the physiology and metabolism of the animal which can endure long after the inciting nutritional stress has been withdrawn. This process has been termed 'metabolic programming' (Patel and Srinivasan, 2002) and is a feature of a plastic developmental program. Nutritional stress represents an important environmental cue. Unlike social experiences which are composed of a number of sensory components that are inputted into the animal through sensory pathways and activate central neurons, food directly enters the animal through the gastrointestinal tract. However, as will be discussed, the eating habits of animals, particularly humans, are not always regulated by physiological processes and eating patterns can be governed by central neuronal reward circuits which are involved in the non-physiological process of addiction. Hence, in the human, food intake as an environmental cue involves both the metabolic effects of the food itself (type, quantity, life cycle stage of animal at which nutritional stress occurs) as well as the biological effects of any superimposed addiction processes.

Perhaps the best example of metabolic programming is that of caste development in the honey bee induced by dietary alterations. Nurse worker bees differentially feed genetically identical female larvae to create mainly workers and when a few queens are required (Singer, 2013). Royal jelly is fed to larvae destined to become queens and a less sophisticated food is fed to the other larvae. Kucharski *et al.* (2008) investigated the genetic basis of this morphological and reproductive/behavioral plasticity in response to the type of food ingested. These investigators found that the underlying mechanism appeared to be differential DNA methylation in response to a dietary cue. Using RNA interference technology to silence the gene encoding DNA cytosine-5-methyltransferase 3 (*Dmnt3*) in newly hatched larvae, Kucharski *et al.* (2008) observed that the majority of methylation-silenced larvae developed into queens (72%). They looked specifically at the methylation of CpG nucleotides within the coding exons of a specific gene, *DCTN4*, coding for the p62 subunit of dynactin. This gene is involved in growth and feeding-related processes. In control animals, CpG methylation in the coding exons was reduced in queens (heads and whole bodies) compared to workers and similarly was less in the heads of queens induced by RNA interference silencing of *Dmnt3* in newly hatched larvae compared to newly hatched larvae injected with a control gene. In this example of metabolic programming, a transient dietary cue induced a permanent morphological and reproductive/behavioural transformation through epigenetic reprogramming.

Interestingly in social wasps, a nutritional cue (differential feeding) during the larval stage, by itself, is insufficient to explain the divergence of worker and queen castes. Suryanarayanan *et al.* (2011) found that in *Polistes* wasps, a mechanical signal (antennal drumming, AD) produced by the worker wasps was also necessary. These investigators proposed that both different frequencies of the mechanical signal AD as well as differential feeding were necessary to account for caste biasing, i.e., whether larvae developed into workers or queens.

In the laboratory rat, Patel and Srinivasan (2002) demonstrated quite dramatically the long-lasting consequences of metabolic programming. Rat pups were fed a high-carbohydrate (HC) milk formula instead of the usual fat-rich rat milk between postnatal days 4 to 24, the end of the weaning period. The HC formula supplied 56% of calories as carbohydrate, whereas only 8% of calories in rat milk are derived from carbohydrate. On day 24, the rats receiving the HC milk formula were weaned onto laboratory chow. The control group (MF) was composed of pups nursed by their own mothers. Pups fed the HC diet developed during the 20-day period a number of metabolic and morphologic changes compared to MF pups: increased blood insulin levels, an increase in the number of smaller sized pancreatic islets, exaggerated insulin secretory response to a glucose challenge, and increased pancreatic islet mRNA levels of preproinsulin and somatostatin transcription factor 1 (PDX-1). Even after withdrawal of the HC milk diet on postnatal day 24, the metabolic changes persisted. Adult rats suckled on the HC diet displayed normoglycemia but persistent chronic hyperinsulinemia as well as an abnormal glucose tolerance test. These rats had a greater mass of insulin-producing pancreatic tissue than controls. Adult rats showed upregulation of the insulin gene (preproinsulin) as well as weight gain and the development of obesity. Interestingly, female rats fed the HC milk diet during their suckling period transmitted their metabolic characteristics to their offspring without dietary manipulation in the offspring themselves. Second-generation rats (offspring of mothers that received HC during their suckling period) as adults displayed elevated blood insulin levels, exaggerated insulin secretory response to glucose, and increased preproinsulin and PDX-1 mRNA levels in islet tissue. Is this an example of inheritance of a phenotype across generations through the transgenerational passage of epigenetic marks?

Patel and Srinivasan (2002) proposed a scheme to account for the direct effects of the HC diet. Briefly, the HC diet induced an increase in a number of factors (increased expression of specific transcription factors, elevated glucagon-like peptide 1 levels, and P13 kinase activation) which are upstream of PDX-1. PDX-1 is an important transactivator of the insulin gene and its expression is increased by the aforementioned upstream factors. Although not discussed by Patel and Srinivasan, the induction of long-lasting metabolic alterations by a brief period of nutritional manipulation most likely involved epigenetic reprogramming. In summary, the ingestion of a HC diet during the suckling period of a rat pup permanently reprogrammed the metabolism of that rat, as an adult, and this metabolic reprogramming was transferrable to the next generation.

When female laboratory rodents are fed a high-fat diet (MHF) during pregnancy and lactation, gross histologic alterations are induced in the liver of adult offspring, specifically changes of (non-alcoholic) fatty liver. Dudley *et al.* (2011) studied the mechanisms underlying this metabolic programming by introducing MHF to pregnant rats throughout pregnancy and lactation and then examining liver tissue of male offspring at postnatal day 2 and postnatal day 27. Weaning occurred at postnatal day 22 and from postnatal days 22 to 27, MHF-fed male offspring were given a standard control diet. A control group (CTRL) of pregnant rats was fed a standard chow diet during pregnancy and lactation. At postnatal day 2, the livers of MHF-fed male offspring showed reduced cell proliferation and a significant inhibition of transition from cell-cycle phase G0/G1 to S phase compared to the livers of CTRL male offspring. By postnatal day 27, these differences in hepatic cell-cycle dynamics between MHF and CTRL offspring were no longer evident. However, even though hepatic cell-cycle dynamics had normalized in the MHF offspring, these offspring still had smaller livers, by weight, at postnatal day 27 than CTRL offspring. However, other body organs such as the brain showed equal weight at postnatal day 27 in MHF and CTRL offspring.

At postnatal day 2, hepatic cell cycle-related gene expression analysis revealed that seven genes were differentially expressed between the two groups of offspring. Five genes encoding cell cycling-promoting proteins were downregulated in MHF compared to CTRL offspring, whereas the gene *CDKN1a*, which encodes the cell-cycle inhibitory protein p21, was upregulated in MHF and not in CTRL offspring. This differential hepatic gene expression pattern was not present in livers from MHF and CTRL offspring examined at postnatal day 27. The investigators measured in the liver the methylation level across the *CDKN1a* CpG island (promoter region). The average extent of methylation across the *CDKN1a* CpG island was lower in MHF neonatal livers at postnatal day 2 than in the livers of the CTRL group. In addition, several site-specific CpG nucleotides showed even greater hypomethylation in MHF than in CTRL livers. There was a significant negative correlation between the extent of methylation (both the average level and the level at these site-specific CpG nucleotides) and the expression of *CDKN1a*. By postnatal day 27, the differences in *CDKN1a* CpG methylation in the liver between MHF and CTRL offspring had disappeared as had the differences in gene expression.

These experimental observations showed that the offspring of mothers fed MHF during pregnancy and lactation showed epigenetic reprogramming of cell cycle-related genes such as *CDKN1a* compared to the offspring of mothers fed a standard diet. Although this epigenetic reprogramming of *CDKN1a* was transient and reversed by postnatal day 27, MHF-fed offspring still showed a persistently smaller liver than that of CTRL-fed offspring. Epigenetic-mediated alterations in the expression pattern of cell cycle-related genes were probably one of the mechanisms responsible for the development of non-alcoholic fatty liver disease in the adult offspring of MHF-fed mothers.

Dietary manipulations during pregnancy can induce not only metabolic programming but also developmental programming of the cardiovascular system in the offspring. A low-protein maternal diet can reset arterial blood pressure levels and cause vascular dysfunction in offspring (Nuyt and Szyf, 2007). The renin–angiotensin system is a major player in this cardiovascular reprogramming. There are experimental observations in the rodent consistent with the construct that fetal "adversity" due to a low-protein maternal diet can result in epigenetic reprogramming (altered DNA methylation marks in the promoter region) of the angiotensin type 1b receptor gene (*Agtr1b*) (Nuyt and Szyf, 2007). The mechanistic links between maternal nutritional restriction and altered epigenetic marks in the fetal *Agtr1b* gene are unknown.

Although there are considerable data supporting the construct that reprogramming of fetal and newborn epigenetic gene expression is one of the final steps in the sequence between maternal nutritional cues and offspring phenotypic changes, the pitfalls and challenges inherent in this research have been well outlined (Niculescu, 2012). In humans, there are data strongly supporting an association between intrauterine fetal adversity and a higher risk for various "chronic degenerative diseases" in that individual's adult life (Couzin-Frankel, 2013). However, teasing out the mechanisms underlying this association represents a daunting task. For example, the cohort who developed *in utero* during the famine in the Netherlands in 1944–1945 had a higher incidence of obesity and insulin resistance as adults compared to individuals not exposed to that nutritional deprivation while *in utero*. Different DNA methylation patterns have been detected for the gene *IGF2* in older adults exposed to the famine while *in utero* compared to older adults not exposed to the famine while *in utero* (Couzin-Frankel, 2013). However, mechanistically linking *in-utero* adversity with differential DNA methylation patterns in later life is problematical. For example, how does one date when the DNA methylation pattern was established? Since DNA methylation patterns change with age as discussed in previous sections, these two observations, i.e., *in-utero* adversity and later-life DNA methylation status, may not actually be causally linked. However, despite these experimental difficulties and limitations, epigenetic reprogramming appears to be the most plausible mechanism to account for the transduction of a time-limited environmental cue into a long-lasting phenotypic change.

This small sample of studies demonstrates that altering the type of food an animal consumes at critical times in that animal's life cycle can induce long-lasting changes in metabolic phenotypes. The terminal step(s) in this process appears to be epigenetic reprogramming of different target genes determined by the specific nutritional stress involved with concomitant alteration in the expression of those genes. However, there is no information as to what pathway(s) mechanistically link the nutritional stress to the altered metabolic phenotype. However, as indicated at the beginning of this section, the ingestion of food can be driven by non-physiological factors. Non-homeostatic eating has been termed 'hedonic eating' and involves cognitive, reward, and emotional factors (Berthoud, 2011). The distinction between hedonic eating and food addiction is more difficult to define (Blumenthal and Gold, 2010; Corwin and Grigson, 2009) although binge-type food consumption in humans is considered a form of food addiction. Two physiological observations support the concept of food addiction. First, the consumption of highly palatable foods and the ingestion of drugs of abuse activate common pathways within the limbic system that mediate motivational (reward) behaviors (Lutter and Nestler, 2009). Secondly, all rewarding drugs or rewarding activities (such as eating very palatable foods) increase dopaminergic transmission from the ventral tegmental area of the midbrain to the nucleus accumbens and other regions comprising the limbic system (Robison and Nestler, 2011).

Before discussing non-homeostatic attributes of eating, an interesting question concerns how an individual actually makes a conscious decision to perform a motor act such as eating. Soon *et al.* (2008) in a very elegant study looked at brain activity using functional MRI (fMRI) in subjects asked to make a conscious decision involving a choice as to which of two buttons to push. The experimental protocol was as follows. While brain activity was measured using fMRI, subjects were shown a stream of letters on a screen. At some point when they felt the urge to do so, subjects were to freely decide between one of two buttons, operated by the left and right index fingers, and press it immediately. The timing of the conscious decision to push one of the buttons was determined by having the subject recall the letter that was on the screen when the motor decision was made. This trial was repeated several times. Neural activity was measured using local patterns recorded from various brain regions which were decoded with statistical pattern recognition techniques.

As expected, two brain regions, the primary motor cortex (right and left) and the supplementary motor cortex encoded the actual motor act itself with local patterns in these areas occurring after the conscious decision to push one of the buttons was made. Two other regions, however, showed characteristic local patterns preceding the conscious decision to push one of the buttons. The lateral and medial frontopolar cortex showed local pattern activity that was already present about 7 seconds before the subject's motor decision. A second region located in the parietal cortex also showed local pattern activity about 7 seconds prior to the subject's decision to push one of the buttons. In summary, two specific regions in the frontal and parietal cortex displayed neural activity reflecting an impending motor response and this neural activity preceded the time of the decision entering consciousness by as much as 7 seconds. Most likely during this preparatory pre-conscious decision phase, information that will shape the upcoming motor decision is gathered an integrated. This information would include previously learned associations between environmental cues (see next sections) and the motor activity. This type of paradigm would apply to feeding behaviors wherein environmental cues would influence and shape the upcoming motor activity of eating.

Hedonic aspects of food intake

Schiltz *et al.* (2007) examined the effects of food cues on the pattern of brain activation using the expression of IEGs as a marker of activity. Two different environmental contexts (A and B) were designed which differed in terms of olfactory, somatosensory, and visual cues. The behavioral paradigm was as follows. In context A, half of the rats were given access to water and the other half to chocolate Ensure. Rats spent about 90 min daily in context A. The same rats also spent an approximately equal amount of time in context B each day. In context B, the rats that received water in context A were given chocolate Ensure and the rats that received chocolate Ensure in context A were given water in context B. This training/conditioning period lasted 15 days. All the rats were housed in a home cage when not in either context A or B. Three days after training, the rats were food deprived for about 12–13 h and then all the rats were reintroduced into context A for 45 min. Half of these rats would have received water in context A and chocolate Ensure in context B, and the other half would have received chocolate Ensure in context A and water in context B. During this test period, locomotor activity was measured. The rats were sacrificed and brain tissue was analyzed for gene expression patterns (mRNA levels). The investigators measured the tissue expression of five IEGs: *Homer-1a, Arc, NGFI-A* (*zif268*), *NGFI-B,* and *c-Fos*. Homer-1a and Arc proteins are involved in the trafficking of neurotransmitter receptors, whereas the proteins NGFI-A, NGFI-B, and c-Fos are transcription factors with differing promoter specificities. The investigators also measured the tissue expression of the proenkephalin gene (*PENK*).

During the test period, rats that had received chocolate Ensure in context A (Ensure group) showed much more locomotor activity than rats that had received water in context A (water group). The group that had received chocolate Ensure in context A also showed increased levels of corticosterone following the test period compared to the group that had received water in this context. The expression of all five IEGs was significantly greater in the Ensure group than the water group, and the regional distribution pattern of IEG expression differed between Ensure and water groups. Appreciable expression of *PENK* was limited to the striatum and was greater in the Ensure group than the water group. In summary, contextual cues that are associated with previous access to palatable foods (chocolate Ensure) elicited conditioned increases in locomotion, plasma corticosterone levels, widespread forebrain expression of IEGs, and increased striatal expression of *PENK*, a gene which encodes endogenous opioid enkephalin peptides. Rats exposed to contextual cues previously associated with a palatable food (Ensure) demonstrated a regional pattern of IEG expression consistent with widespread neuronal connectivity involving higher-order executive functions, energy balance sensing, and motor arousal. In addition, these rats also showed increased expression of endogenous opioids. Rats exposed to the same contextual cues when these cues were previously associated with a neutral "food" (water) did not show the following: motor arousal, a similar regional pattern of IEG expression, a robust corticosterone response, or a similar increase in the expression of endogenous opioid peptides (enkephalins). The observations of Schiltz *et al.* (2007) demonstrate that contextual cues predictive of desirable food availability can elicit widespread neuronal activation associated with cognitive and reward seeking behaviors as well as increased motor activity.

The results of Schiltz *et al.* (2007) are consistent with observations previously reported by Petrovich *et al.* (2005). Petrovich *et al.* (2005) used a different behavioral paradigm. Rats in a hungry state were trained in a discrimination task in which presentations of a specific tone (CS+) were paired with food delivery, whereas presentations of another auditory stimulus, white noise, (CS−) were not paired with food access. After completion of the conditioning process, mice were given free access to food in their home cage for 14 days. On day 14, the rats were subjected to a test situation. In the immediate pre-test period, the rats were allowed to eat to satiety and were then tested for food consumption during a fixed time interval during which the rats received either CS+ or CS− cues. After the test period, neural circuitry in these rats was assessed by two techniques. Injections of FluoroGold (at completion of the conditioning period) into the lateral hypothalamus were used to define neuronal connections histologically between the lateral hypothalamus and other brain regions involved in the control of feeding (e.g., nuclei of the amygdala). The tissue expressions (mRNA staining) for two IEGs, *Arc* and *Homer-1a* were used as a measure of neuronal activation.

During the test period, rats consumed significantly more food pellets in the presence of CS+ compared to tests with CS−. In CS+ tested rats, 19.8% of cells of the basolateral and basomedial amygdala that projected to the lateral hypothalamus showed expression of one or other of the IEGs in comparison to 8% of cells in CS− tested rats. For prefrontal cortical regions that projected to the lateral hypothalamus, the results for the percentage of cells expressing one or other of the IEGs was 19% (CS+ test situation) versus 7.3% (CS− test situation). For cells in the nucleus accumbens that projected to the lateral hypothalamus, the percentage of cells expressing one or other of the IEGs was similar in both test situations: 9.2% CS+ versus 8.2% CS−. These observations demonstrate that an auditory cue paired with food delivery can stimulate food intake even in satiated rodents and can activate neuronal circuits which regulate feeding behaviors.

In summary, if an animal learns to associate an environmental cue with a food attribute (e.g., a specific environmental context paired with a particularly palatable food or an auditory stimulus paired with availability of food), then the presentation of that cue becomes sufficient to enhance the activation of neuronal pathways that control food consumption. In essence, the food intake of that animal will be determined not only by homeostatic need, but also by learned associations between environmental cues and certain food characteristics such as a pleasurable taste or ready accessibility. These environmental cues become the driver of hedonic feeding behaviors.

The characteristics of an environmental cue can be quite nuanced and the physiological state of the animal can modify the effect of an environmental cue. These attributes are illustrated in the following studies.

Plassmann *et al.* (2008) looked at the effect of price on the perceived taste pleasantness of wine by a group of young adults. The experimental protocol involved subjects tasting wine or a neutral substance while undergoing fMRI. Two of the wines were each presented twice: once as a wine of modest price and once as an expensive wine. During the tasting sessions, brain imaging via fMRI showed increased activity in the medial orbitofrontal cortex and ventromedial prefrontal cortex. These brain areas are part of the limbic system and are involved in reward-seeking behaviors including hedonic eating behaviors (Berthoud, 2011; Robison and Nestler, 2011). The subjects' ratings of wine pleasantness showed a definite correlation with price; wines were rated as having a more pleasant taste when presented as having a higher price. In addition, fMRI showed greater prefrontal cortical activity in association with tasting an expensive versus a cheaper "version" of the same wine. For each subject, the investigators tabulated the change in reported taste pleasantness between the high- and low-priced wines and the change in cortical activity by fMRI between the two wine-tasting experiences. There was a direct positive correlation between change in brain activity and change in liking between high-priced and low-priced wine. In addition, there was a direct positive correlation between activity levels in the prefrontal cortex and measures of wine taste pleasantness.

This interesting study showed that the price assigned to a wine modulated the perceived pleasantness on tasting that wine. A wine, when given a higher price, elicited a more "pleasant" taste sensation compared to that same wine when given a lower price, and the experience of tasting the higher-priced "version" was associated with a greater level of prefrontal cortical neural activity compared to the experience of tasting the lower-priced "version." The investigators proposed that a change in the price of the wine modulated the activity of the medial oribitofrontal cortex (part of the limbic system) which altered the perceived taste sensation for that wine. In this case, the environmental cue is information about the retail price of a wine to be sampled. The price is not an intrinsic property of the wine yet this information led to a change in the taste sensation that individual experienced when sampling that wine. This informational environmental cue inputs into the limbic neural system which processes memory, emotional, and reward information (Berthoud, 2011) and modulates, in this example, a taste sensation by associating an informational cue with a taste expectation. This experimental result can probably be generalized to account for the observation that humans appear to assign a higher quality to goods that are more expensive compared to comparable goods that are less costly.

Tindell *et al.* (2009) studied the relationship between the physiological state of an animal and its feeding behaviors. Rats were fitted with electrodes to measure neuronal activity in the ventral pallidum (VP). This brain area has connections with other mesocorticolimbic regions and is involved in regulating motivational and hedonic behaviors. Rats were trained to learn associations between distinct auditory cues and oral infusions given at the offset of the cue. One cue predicted a sucrose solution, a second cue predicted a concentrated salt solution, and a third cue was predictive of no subsequent oral infusion. Tests were carried out over two days. On the test day, the three distinct cues were presented in a random order. Test day 1 was done with the rats in a state of normal homeostasis, while test day 2 was done after the rats had been rendered salt depleted. Each test day began with an extinction phase in which the cues were presented without being followed by any oral solutions.

During the extinction phase (presentation of cue alone), in normal homeostasis, the cue predictive of sucrose activated 29% of tested VP neurons while the cue predictive of salt activated 14%. In the salt-depleted state, 42% of tested neurons responded to the sucrose cue while 31% responded to the salt cue. The significant increase in the number of VP neurons responding to the salt cue during a salt-depleted state is consistent with the incentive (motivational) value of this cue appropriate to the rat's physiological state. In the paired tests, when a cue was followed by an infusion, the sucrose taste activated 49% of tested VP neurons in normal homeostasis and 32% during salt depletion. The salt taste activated 21% of VP neurons in normal homeostasis but 53% during salt depletion. These results revealed that a salt-depleted state increased the number of VP neurons that responded to both a cue predictive of an oral salt infusion and to the oral salt infusion itself. In the case of sucrose, a salt-depleted state induced not much change in the number of VP neurons responding to the cue or the sucrose infusion itself compared to the state of normal homeostasis.

Similar results were obtained when the firing rates of responsive VP neurons were examined. For paired tests (cue followed by infusion) during normal homeostasis, VP neurons showed increased firing associated with presentation of the sucrose cue and a further increase in firing during the sucrose infusion. VP firing was not significantly increased by the salt cue or the salt infusion. During the salt-depleted state, responsive VP neurons increased firing to both the sucrose cue and sucrose solution comparable to that observed during normal homeostasis. Further, during the salt-depleted state, responsive neurons displayed a large increase in firing to the salt cue as well as a large increase in firing to the salt infusion. In fact, neuron firing to the salt solution was greater than that to the sucrose solution.

The investigators also measured the behavioral response of rats to oral infusions of sucrose and salt. The responses were graded into hedonic (liking) and aversive reactions, and a hedonic and aversive score was created. During normal homeostasis sucrose had a high hedonic and low aversive score, while the opposite was true for the concentrated salt solution. During the salt-depleted state, sucrose and salt both had high hedonic scores and low aversive scores.

In summary, a concentrated salt solution represented an aversive taste to the rat. Under normal homeostasis, a cue learned to be predictive of a salt taste activated only a small number of VP neurons and elicited little change in their firing rate. However, if these rats were made salt depleted, the incentive (motivational) value of this salt cue itself was enhanced and presentation of this cue significantly increased both the number of responsive VP neurons as well as their firing rate. Similarly, under normal homeostasis, an oral salt solution activated about the same number of VP neurons as the salt cue alone and the administration of an oral salt solution caused little increase in the firing of these neurons. During salt depletion however, an oral salt solution activated a greater number of neurons and caused a significant increase in the firing of these responsive VP neurons. Hence, a change in a physiological state (induced salt depletion) can change the incentive value of an environmental cue. In this case, the incentive value of a cue predictive of an aversive (highly concentrated salt) taste became enhanced and the aversive character of the salt taste itself was reversed from mainly aversive to mainly hedonic. An alteration in a physiological state can modulate the incentive value of an environmental cue and change an aversive sensory experience to a pleasurable one.

The results reported by Tindell *et al.* (2009) are consistent with the observations reported by Lemos *et al.* (2012) reviewed in a previous section on **social experiences during the life cycle**. In the study by Lemos *et al.* (2012), direct injection of CRF into the nucleus accumbens of rodents led to dopamine release and was associated with a positive preference for the chamber in which the injection was administered. However, if the physiological state of the animal was altered by prior exposure to chronic stress, injection of CRH into the nucleus accumbens under these circumstances was associated with an aversion for the chamber in which the animal received the injection. An alteration in the physiological state of the animal modulated a sensory experience such that a learned place preference reversed to become a place aversion.

There is considerable evidence that foods particularly rich in sugar, fat, or salt can be addictive (Blumenthal and Gold, 2010; Corwin and Grigson, 2009), and the addictive nature of such foods is well known to the food industry. In fact, the sugar, fat, and salt content of foods has been manipulated by the food industry to promote their sale (Moss, 2013). This manipulation of the addictive potential of foods by the large food companies has been compared to the engineering of the addictive potential of cigarettes by the tobacco industry (Moss, 2013). Brownell and Warner (2009) in an in-depth review described the "playbook" used by the tobacco industry to promote its product despite the accumulating evidence that using this product was associated with significant morbidity and mortality. As pointed out by Brownell and Warner (2009), the food industry seems to be adopting the same "playbook" used by the tobacco industry. As expressed by Brownell and Warner (2008); "The experience of tobacco shows how powerful profits can be as a motivator, even at the cost of millions of lives and unspeakable suffering." "Will the food industry adopt a playbook that promotes public health, or will its future come to rival tobacco's past?"

The intersections between the physiological systems regulating homeostatic energy balance and the mechanisms underlying the hedonistic and addictive aspects of appetite and feeding behaviors have been reviewed by Berthoud (2011) and Lutter and Nestler (2009). A system of hormones (e.g., leptin, insulin, gherlin) and neural nutrient-sensing mechanisms convey information concerning overall energy balance to specific regions of the brain. For example, the arcuate nucleus of the hypothalamus contains two populations of neurons considered to be first order with respect to the homeostatic control of feeding in response to nutrient signals (Benoit *et al.*, 2008). One population co-expresses the peptides neuropeptide Y (NPY) and the melanocortin receptor antagonist agouti-related protein (AgRP). These two peptides stimulate food intake (orexigenic). A second population of neurons contains pro-opiomelanocortin (POMC), the precursor of alpha-melanocyte-stimulating hormone and cocaine- and amphetamine-regulated transcript (CART). These two neuropeptides inhibit food intake (anorexigenic). Both of these populations of neurons express receptors for leptin and insulin. Leptin and insulin signaling stimulate the activity of POMC/CART neurons and reciprocally inhibit neurons expressing NPY and AgRP, thus suppressing feeding (Lutter and Nestler, 2009). Gherlin, a gastric-derived peptide, primarily activates NPY/AgRP neurons thus stimulating feeding (Lutter and Nestler, 2009).

The reward circuitry of the brain centers on dopaminergic neurons in the ventral tegmental area (VTA) of the midbrain. Projections from the VTA go to the limbic system: nucleus accumbens, dorsal striatum, amygdala, hippocampus, and parts of the prefrontal cortex (Robison and Nestler, 2011). Virtually all rewarding activities increase dopaminergic transmission from the VTA to the nucleus accumbens as well as other limbic regions. Dopaminergic neurons are thought to fire in unison (Niv, 2013) and release dopamine in an episodic transient pattern in accordance with the principles of reinforcement learning. Recently, Howe *et al.* (2013) reported that dopamine release rather than being transient could also be gradual, increasing in a ramp-like fashion. This more prolonged "ramp-like" dopamine signaling was observed in rats navigating mazes of different sizes and shapes to retrieve rewards and could provide a sustained motivational reward-seeking drive. Hence, activation of midbrain dopaminergic neurons can lead to either transient or a more gradual release of dopamine. The biological significance of these two different patterns of dopamine release has not yet been clarified.

One intersection between the homeostatic and hedonistic "systems" centers on the arcuate nucleus. This nucleus, which contains the neuronal populations expressing POMC/CART and NPY/AgRP peptides that regulate homeostatic food consumption, sends projections to the nucleus accumbens and VTA neurons of the reward system (Lutter and Nestler, 2009).

Robison and Nestler (2011) have reviewed the transcriptional and epigenetic mechanisms involved in reward and motivational neuronal circuits. One transcriptional factor is delta-FosB which is encoded by the gene *FosB*. Delta-FosB is a carboxy-terminal truncation of full-length FOSB and is generated by alternative splicing. This factor is induced primarily in the nucleus accumbens and dorsal striatum by reward-type behaviors such as the consumption of sucrose and high-fat foods. Within these brain regions, delta-FosB is found in medium spiny neurons that express the D1 dopamine receptor. The transcriptional targets of delta-FosB include genes encoding proteins which control dendritic spine morphology and thereby modulate synaptic structural plasticity. Other transcription factors that appear to be important in regulating gene expression within the nucleus accumbens and dorsal striatum include CREB, nuclear factor-kB (NF-κB) and multiple myocyte specific enhancer factor 2 (MEF2) proteins. There is experimental evidence that delta-FosB modulates gene expression patterns through the establishment of histone epigenetic marks (Robison and Nestler, 2011). Changes in DNA CpG methylation in the nucleus accumbens and dorsal striatum have also been associated with drug-induced motivational behaviors.

In summary, nutrient intake (i.e., feeding) is an essential function. A highly complex system monitors the internal milieu and relays information to specific regions of the brain, primarily the hypothalamus and brain stem. Food intake is suppressed or stimulated as a homeostatic response to this information. Feeding behaviors can also be determined by attributes of food that activate the reward system centered in the VTA of the midbrain with projections to the limbic system. Foods rich in sugar, fat, or salt stimulate the "reward" neural circuitry and lead to hedonic and addictive feeding behaviors which are independent of homeostatic requirements. In addition, environmental cues become encoded with learned associations to certain food attributes (e.g., social context in which preferred versus non preferred food is presented) and these cues can increase food consumption beyond homeostatic needs. The physiological state of the animal is another important variable since deviations in a physiological state can convert an aversive experience to a pleasurable one. Hence, the actual food intake of an animal is influenced by multiple inputs: homeostatic requirements, qualities of certain foods that lead to hedonistic and addictive food appetites, and exposure to environmental cues predictive of access to preferred foods.

When hedonistic and addictive feeding behaviors override the homeostatic regulation of food intake, obesity is the result. Although obesity has recently been categorized as a disease by the American Medical Association, Katz (2014) argued that labeling obesity a disease is a misguided decision. The relationships between obesity, food, and addiction are becoming better understood from a neuroscience perspective (Potenza, 2014). In a recent article, Willyard (2014) reviewed the complex interactions between genetics, epigenetics, and environmental influences that underscore the development of obesity and the challenge will be untangling the threads comprising these interactions.

In this chapter, the focus has been on the effects of environmental cues (experiences) in shaping molecular and behavioral phenotypes. In general, environmental cues act by reprogramming epigenetic marks, thereby changing gene expression patterns. Behavioral phenotypes are also genetically determined as exemplified by the process of domestication. In addition, domesticated animals represent a good ageing model since these animals compared to their wild ancestors experience many of the same health issues as humans. The process of domestication and the relevance of the dog as a suitable model of ageing are discussed in **Chapter 3**.

CHAPTER 3

Domestication

The Domestication Process: The Russian Silver Fox Experiment

There exists an extensive literature on the process of domestication and in fact, it has even been suggested that the shaping of human social behaviors represents a form of self-domestication (Kukekova *et al.*, 2012). Debates concerning the origins of animal domestication have focused upon the issue of intentionality; i.e., was domestication the result of deliberate human choices and actions (Trut, 1999)? However, domestication represents a long, complex process that has involved a number of mammalian species. Probably the first domesticated species was the dog, which diverged from the grey wolf about 15,000 to 16,000 years ago, likely as a result of multiple domestication events (Boyko, 2011; Lindblad-Toh *et al.*, 2005). Early phases in the domestication process of the dog are unknown, but the Russian evolutionary biologist D. K. Belyaev proposed that the key factor selected for in domestication was a behavioral one — tameability (Trut, 1999). To test this hypothesis, Belyaev in 1959, set up a selective breeding program based solely on behavioral criteria using farmed silver foxes as the test animal.

Silver foxes have been subject to captive breeding for over a century, but still retain a predominantly fear–aggressive response to humans.

Belyaev chose the silver fox because of the close taxonomic relationship between the fox and the dog and because the cage breeding of the silver fox had already adapted this animal to a new social environment (Trut *et al.*, 2009). Researchers assessed the tameability of fox pups on the basis of their responses to hand feeding and to attempts to touch or pet them. Less than 10% of the tamest animals of every generation were used as parents of the next generation. By the fourth selected generation, pups appeared that responded to humans by dog-like tail wagging. By the sixth generation, some pups sought out human contact by whining, whimpering, and licking in a dog-like manner. Such foxes were classified as the elite of domestication, the most highly tamed. Elite pups constituted 1.8% of the population in the sixth generation, 17.9% in the 10th generation, 49% in the 30th generation, and almost all of the pups by the year 2006 (Trut *et al.*, 2009). A parallel selective breeding was created which developed a strain of foxes that were aggressive and difficult to handle (Kukekova *et al.*, 2011a). Behavioral phenotypes for these two populations of foxes, domesticated and aggressive, were quantitated using a variety of behavioral tests and analyzed by the technique of principal component analysis (Kukekova *et al.*, 2011b). Two principal components were developed: principal component 1 encompassed a suite of behaviors on the tame–aggressive spectrum, whereas principal component 2 reflected behaviors more along the bold–shyness axis. PC1 and PC2 were found to account for 33% and 9% of the total behavioral variation respectively (Kukekova *et al.*, 2011b). Domestic foxes, similar to dogs, were also able to read human social cues such as gestures and glances. These tame and aggressive behavioral phenotypes were genetically determined (Kukekova *et al.*, 2011a).

Although these foxes were selectively bred using tameability as the only criterion, morphological and other phenotypical changes also developed. Many of the domesticated foxes displayed a change in their reproductive behavior. Wild and farmed foxes mate only once a year during the period from January to March. However, a small number of elite (highly domesticated) foxes were observed to mate outside the breeding season and a few domesticated females even mated more than once a year. In addition, some of the domesticated foxes showed morphological features that mirrored those of dogs. For example, curly tails had a frequency of 9.4% in domesticated foxes but only 0.8% in farm-bred foxes. For floppy ears, the respective frequencies were 0.23% and 0.017% (Trut *et al.*, 2009). The linkage between behavioral and morphological phenotypes is one of the fascinating observations to emerge from the silver fox domestication experiment. Even though captive silver foxes were artificially selected on the basis of a behavioral trait only, across the generations, specific morphological changes also appeared as these animals became more domesticated. As pointed out by Trut *et al.* (2009), these morphological features i.e., white spotting of the coat, floppy ears, and curly tails are similar to those observed in other domesticated mammals.

There are clues as to possible mechanisms underlying this behavior–morphologic linkage. In chickens, alleles of the gene *PMEL17* regulate both plumage color as well as social behaviors (Wiener and Wilkinson, 2011). In the silver fox, behavioral traits that encompassed the range from aggressive to tame appear to be associated with a broad QTL on chromosome 12 (Kukekova *et al.*, 2012; Wiener and Wilkinson, 2011). Lindberg *et al.* (2005) examined gene expression differences in three brain regions (amygdala, frontal lobe, and hypothalamus) between wild and farmed foxes (domesticated and non-domesticated). They used a cross-species hybridization technique of fox mRNA to human microarrays, a technique that is not very specific and generally detects only large gene expression differences. The three brain regions of wild foxes contained many genes with expression differences when compared to the brains of domesticated or non-domesticated farmed foxes. However, the brains of domesticated foxes compared to the brains of non-domesticated farmed foxes showed only a limited number of genes with expression differences. These results suggest that the transformation of the silver fox from a wild to a farmed (captive) state was associated with a much greater divergence in brain gene expression than the change from a non-domesticated to a domesticated condition. Kukekova *et al.* (2011a) measured differential gene expression in the prefrontal cortex of a tame and an aggressive fox. Only a relatively small number of genes (355) had expression levels differing by more than twofold between the two animals. Twenty-seven genes had a 10-fold difference in expression between the tame and aggressive foxes. One gene, *HTR2C*, was specifically commented on by the investigators. This gene had more than ninefold greater expression in the tame fox. This finding in the fox is similar to that which has been observed in the rat; a higher expression of *HTR2C* is found in tame compared to aggressive rats. This gene plays an important role in dopaminergic and serotonergic signaling in the frontal cortex and hippocampus of rodents.

The activity of the HPA stress axis also differs significantly between domesticated and non-domesticated farmed foxes (Trut *et al.*, 2009).Comparative analysis of the HPA axis was initiated at generation 10 of the domestication experiment and at this stage, tame foxes showed a lower blood level of glucocorticoids in all seasons compared to non-selected foxes. By generation 20, basal cortisol levels in tame foxes were almost twofold lower than in non-selected foxes. In generation 45, basal and stress-induced blood cortisol levels in tame foxes were three- and fivefold lower than in non-selected foxes. These data suggest that selection for the behavioral trait of tameability also selects for genes regulating the overall stress response.

In summary, selection of foxes for breeding based solely upon the behavioral trait of tameability resulted in the domestication of this animal after a number of generations. This domestication process led to a behavioral phenotype in which the animal sought out human contact and was able to "read" human social cues. The specific genetic mechanisms responsible for this behavioral change have not been clearly defined but an alteration in the expression of a number of genes within certain brain regions has been documented. Surprisingly, the genetic mechanisms underlying the behavioral changes also "select" genes which control a number of other body systems such as the stress response, reproductive behaviors, and the development of certain morphological characteristics.

The Dog

The dog is probably the most intensely studied domestic animal and current genomic techniques are being utilized to better understand the dog's genetic background. The grey wolf is considered the ancestor of the domestic dog with the predominant input being from wolf populations in the Middle East and to a lesser extent, Europe. However, the geographic origin of the dog and the time at which domestication took place are still controversial. More recent studies suggest that the grey wolf population that gave rise to the modern dog is now extinct and that domestication actually occurred in Europe between 18,800 and 32,100 years ago (Pennisi, 2013b).

Vonholdt *et al.* (2010) analyzed almost 50,000 SNPs in 912 dogs from 85 breeds as well as studied 225 grey wolves. Using the SNPs they catalogued, Vonholdt *et al.* (2010) looked for adaptive substitutions that would point to genomic regions involved in the domestication process rather than breed formation. For a subset of 43, 452 SNPs, a fixation index (Fst), and the extent of cross-population extended haplotype homozygosity were calculated between populations of non-admixed wolves and modern dogs. Genomic regions were ranked according to these calculations.

Fst is a measure of how populations differ genetically. It has a value between 0 and 1. High values would indicate a low level of shared alleles, i.e., significant differentiation between the sampled populations (e.g., wolves and modern dogs), whereas low Fst values indicate that members of the two populations have many shared alleles. The analysis is usually done by pair-wise SNP comparison between the two subpopulations (Norrgard and Schultz, 2008). Extended haplotype homozygosity is a measure of the length of homozygosity around a core SNP; i.e., the probability that two extended haplotypes around a given locus are the same given that they have the same allele at that locus (Zhong *et al.*, 2011) and can be used to measure linkage disequilibrium.

Genomic regions with the five highest rankings in descending order were located on chromosomes 5, 30, 13, 17, and 1. Two of the three top regions identified by this method to be associated with domestication were located near genes which in other mammals (human and mouse) have been functionally linked to memory formation and certain social behaviors. In the domesticated silver fox, the QTL on (fox) chromosome 12 associated with variation in tameness and aggressive behavior included a region homologous to the region on (canine) chromosome 5 associated with domestication in the dog (Kukekova *et al.*, 2012).

The genomic structure of the modern dog has been shaped by several key population bottleneck events (Lindblad-Toh *et al.*, 2005). The first population bottleneck was associated with the domestication process which took place approximately 15,000 years ago. A second population bottleneck occurred more recently during Victorian times when intensive selection and breeding were used to create the multiple current modern breeds. Both of these bottlenecks, one old and one recent, left their distinctive signatures on the canine genome — a reduction in genetic variation. However, the more recent bottleneck was tighter and associated with a greater reduction in genetic diversity as manifested by long regions of linkage disequilibrium (LD; non-random association of genes at different loci). A third factor that has modified the dog genome is a relaxation of the strength of natural selection acting on modern purebred dogs (Boyko, 2011).

Dog domestication occurred before the advent of agriculture and the role that dog domestication played in the transformation of humans into agriculturists is still debated (Boyko, 2011). In the Victorian era, hundreds of modern breeds were created through a process of controlled breeding. This artificially orchestrated selection and breeding program resulted in the sharing of identical causal genetic variants across different breeds

(Boyko, 2011; Vonholdt *et al.*, 2010). For example, at least 19 distinct breeds with foreshortened limbs (e.g., dachshunds, Pekingese, and basset hounds) share the same retrotransposed version of the gene on chromosome 18 encoding FGF4. Likewise, short-snouted breeds share a haplotype on chromosome 1 and floppy-eared breeds share a haplotype on chromosome 10. As already noted, the evolutionary background of dogs characterized by domestication in the remote past and more recent intense breed creation has resulted in a genetic architecture "simpler" than that of humans. In dogs, most of the variation in phenotypic features such as body size, coat color, relative leg length, and body proportions can be accounted for in each case by a few QTLs with large effects (Boyko, 2011; Vonholdt *et al.*, 2010). In contrast, the top 180 QTLs associated with human height account in total for only 10% of the variation in this body phenotype (Boyko, 2011).

In summary, by virtue of its domestication and a stringent breeding program, the dog possesses a genomic landscape characterized by reduced variation, and as a result, the dog is a wonderful model for examining the genetic underpinnings of phenotypic variation. The mapping of complex traits and disease loci is technically easier in dogs than in humans (Boyko, 2011; Lindblad-Toh *et al.*, 2005; Wayne and Ostrander, 2007). The modern dog population consists of over 300 partially inbred genetic isolates called breeds, and gene flow between breeds is restricted by the pedigree barrier. Most modern dog breeds are relatively young and many have been derived from a small number of founders. In addition, the domesticated dog is characterized by more than 360 genetic "diseases," many of which have close human analogs (Ostrander and Kruglyak, 2000). Hence, the advantageous features of the dog's genomic structure means that this animal represents a good starting point for identifying potential "disease" causing genes (Boyko, 2011).

Mapping morphologic traits

To map QTLs and determine the genomic location of candidate genes potentially responsible for specific phenotypes and disease processes, researchers use SNPs with known genomic loci (Gross, 2010; Lindblad-Toh *et al.*, 2005). The number of evenly spaced SNPs required for mapping depends on the degree of LD present. Within a given breed, dogs display reduced genetic diversity as manifested by the presence of long lengths of breed specific LD across the entire genome; LD is approximately 50-fold greater in length in the dog compared to the human (Lindblad-Toh *et al.*,

2005). The implication of this observation is that QTL mapping can be performed in the dog with far fewer evenly spaced SNPs than in the human. Since humans and dogs share many similar "diseases," causative genes will be easier to localize in association studies done in dogs than in humans.

Sutter *et al.* (2004) examined for breed specific LD in five breeds: Akita, Bernese Mountain Dog, Golden Retriever, Labrador Retriever, and Pekingese. SNPs and indels (insertions and deletions) were mapped and used to determine the extent of LD. One locus was chosen from each of five chromosomes (1, 2, 3, 34, and 37). At each locus, sequence reads were clustered into five regions: a central region and two regions on each side of the center in each direction along the chromosome. This scheme was used to calculate the extent of LD at these loci. There was considerable variation of LD across the five loci and across the five breeds. However, the extent of LD in each of the five dog breeds exceeded the extent of LD reported for human populations, even human populations that are comparatively isolated. A parameter D′ (a measure of the independence of different gene loci) can be used to summarize the extent of LD. The genomic length at which D′ falls to half its maximum value for the different breeds was: Golden Retriever, 370 kb; Labrador Retriever, 680 kb; Akita, 2.4 Mb; Bernese Mountain Dog, 3.0 Mb; and Pekingese 3.3 Mb. For human populations, the value is in the range of tens of kb. Sutter *et al.* (2004) also looked at haplotype diversity. The increase in LD resulting from population bottlenecks is generally associated with a decrease in nucleotide diversity (Ardlie *et al.*, 2002). For the five breeds studied by Sutter *et al.* (2004), between 2 and 3.4 haplotypes explained 80% of chromosomal sequences in each breed. The breed with the fewest number of haplotypes was the Pekingese and this breed has a history of very tight population bottlenecks. Sutter *et al.* (2004) also looked at haplotype sharing between breeds and observed that over the 10 breed pairings, an average of 63% of the chromosomes within each breed in a given pair carried haplotypes shared between both breeds.

Lindblad-Toh *et al.* (2005) also measured LD and haplotype diversity in different dog breeds. Within breeds, a plot of LD as a function of genomic distance across 15-Mb regions showed a biphasic pattern with a sharp initial drop by 90 kb, followed by a long extended shoulder that declines to background by 5–15 Mb. Across breeds (and in humans), LD falls rapidly to background levels by 100 kb, corresponding to the first component only of the within breed pattern. No extended shoulder is present. The first component (and the "curve" for the overall, across breeds, dog population) probably reflects short-range LD based upon ancestral dog haplotype blocks which emerged at the time of the domestication process. The second component of LD in the within breed pattern reflects long-range LD that emerged when modern breeds were created within the past several hundred years. Lindblad-Toh *et al.* (2005) assessed haplotype diversity using a 10-kb window on chromosome 6. Across 24 breeds, there were a total of seven haplotypes in this window. All but three were observed in multiple breeds at varying frequencies. In four specific breeds (English Springer Spaniel, Irish Wolfhound, Pug, and Rottweiler) there were only 2 to 5 haplotypes with one or two major haplotypes accounting for the majority of the chromosomes.

Boyko *et al.* (2010) have elaborated upon many of the observations of Lindblad-Toh *et al.* (2005) and Sutter *et al.* (2004). Specifically, these investigators looked at LD, haplotype diversity, and runs of homozygosity (ROH) using genomic data from 59 dog breeds and a population of village dogs and wolves. ROH are regions of the genome where the copies inherited from the two parents are identical because the parents have inherited them from a common ancestor at some time in the past. Within each breed, LD extended over a long distance (>1 Mb), whereas across all dogs, LD decayed very rapidly comparable to that observed in village dogs but greater than that of wolves. ROH were much longer within breeds than in wolves or village dogs. Haplotype diversity, measured as the number of haplotypes within each sampled genomic window, was greater in the wolf and village dog populations compared to any of the examined dog breeds. Hence, consistent with the studies of Lindblad-Toh *et al.* (2005) and Sutter *et al.* (2004), these results show that within a given dog breed there is considerable genetic uniformity (small number of haplotypes, long runs of homozygosity and extensive linkage disequilibrium) compared to across dog breeds, wolves, or village dogs.

Within breeds, there is morphological uniformity, but across breeds, there is extensive morphological divergence. In fact, variation in body size across dog breeds is greater than in any other land species (Boyko *et al.*, 2010). However, groups of breeds do share certain morphological features such as short-leggedness (chondrodysplasia) which is a defining characteristic of approximately 19 breeds. These chondrodysplastic breeds have a shared allele, an inserted retrogene of *FGF4*, on chromosome 18. Boyko *et al.* (2010) mapped morphologic traits (20 different tape measurements, long bone and skull/tooth dimensions, size variation, and ear floppiness) across 80 breeds. These investigators observed that for most of the morphological traits examined, variation across breeds could be accounted for by a few QTLs of large effect. This simplified genetic architecture underlying extensive trait variation has resulted from the domestication process coupled with the creation of relatively closed genetic breed populations.

Mapping behavioral traits

Dog behaviors can be conceptualized on a number of levels. On one level, the domestication process itself which took place about 15,000 years ago was associated with the development of a suite of behaviors. On another level, modern breeds have been "created" to enhance certain more specific behaviors such as herding, pointing, tracking, hunting, and retrieving to name a few. In addition, as demonstrated by the Russian Silver Fox experiment, selection and breeding based solely on a behavioral trait (most likely the initiating mechanism underlying domestication) also results in the emergence of morphological changes; i.e., the domestication process is associated with mechanistic linkages between behavioral and morphologic phenotypes. Finally, as with all animals, there is individual variation in behavioral traits between individual dogs within the same breed.

Li *et al.* (2013) compared genomic divergence between gray wolves, Chinese Native Dogs (CNs), and German Shepherds (GSs). Wolves represent the ancestor of the dog while the CNs were used as a surrogate for ancient-type village dogs which predated modern breeds. These test populations allowed the investigators to examine for genomic changes that could be attributed to the process of domestication and breed creation. The final populations consisted of 20 CNs, 14 wild gray wolves, and eight GSs. When LD was quantified based upon SNP mapping, a long LD decay pattern was observed in GSs as expected, while a shorter LD decay pattern was present in CNs. CNs had longer LD than wolves. This LD pattern is quite consistent with the evolutionary history of the dog involving a remote population bottleneck at the time of domestication and more recent population bottlenecks at the time of breed formation (discussed in a previous section). Pair-wise SNPs were compared in terms of Fst values between the two populations, wolves, and CNs and a number of differences were noted. Genomic differentiation between wolves and CNs was greatest in the brain compared to nine other tissues and in particular the prefrontal cortex. Genomic regions which showed differentiation contained a number of interesting genes. One gene, *CCRN4L* is involved in regulating circadian rhythms. Another differentiated region between wolves and CNs contained a serotonin receptor (*HTR4*). As already noted in a previous section, a gene *HTR2C*, which is involved in dopaminergic and serotonergic signaling, showed a significant expression difference in the prefrontal cortex when tame and aggressive foxes were compared (Kukekova *et al.*, 2011a). A comparison between CNs and GSs identified over 600 candidate genes that were differentiated but unlike the situation between CNs and wolves, these differentiated candidate genes did not have a preferential brain expression pattern. These data plus the studies of Kukekova *et al.* (2011a,b; 2012), Lindberg *et al.* (2005), and Vonholdt *et al.* (2010) all support the proposal that domestication was a behavioral-driven process and that the suite of behaviors associated with domestication are neurodevelopmentally determined.

The social cognitive skills of dogs have also been acquired as part of the domestication process. Hare *et al.* (2002) in a series of experiments examined the use of human directed social cues in dogs compared to chimpanzees, dogs compared to wolves, and in puppies of various ages. Generally, the experimental design involved the capacity of the animals to use human-directed cues to find hidden food. Dogs outperformed chimpanzees and wolves in these experiments. However, in a separate set of experiments based on memory-based food finding without involvement of human cues, wolves and dogs performed equally well. Puppies were able to use human social cues quite skillfully, which function was independent of age or rearing history. These observations of Hare *et al.* (2002) are most consistent with the proposal that the dogs' social communicative skills with humans emerged as part of the domestication process and are also neurodevelopmentally determined.

In summary, dogs as a species are characterized by a suite of behaviors including social cognitive skills toward humans, which have evolved as part of the domestication process. As such, these behaviors are neurodevelopmentally determined. Based upon current evidence, the involved genomic regions have been localized to specific chromosomes and to specific brain regions. The few candidate genes that have been identified have been noted to play a role in dopaminergic and serotonergic signaling (neural) circuits.

Dogs also display breed-specific behaviors such as herding, retrieving, pointing, and tracking for example. These behaviors are engrafted onto the domestication associated behaviors already discussed. Candidate genes associated with specific behavioral phenotypes in the dog are the same genes that have been linked to novelty seeking behaviors in humans and to attention deficit hyperactivity disorder in children. One such gene is the dopamine D4 receptor gene (*DRD4*).

The canine *DRD4* gene has polymorphisms involving exon 3 and intron 2 (Hejjas *et al.*, 2007; 2009). These polymorphisms involve the insertion of a variable number of tandem repeat (VNTR) sequences; seven alleles have been identified for exon 3 (alleles 1 to 7 of increasing length due to more VNTRs) and two alleles for intron 2, a long version (Q, three VNTRs) and a short version (P, two VNTRs). Hejjas *et al.* (2007) measured *DRD4* exon 3 polymorphisms in German Shepherds and could only detect two of the known seven alleles: allele numbers 2 and 3a. Hejjas *et al.* (2009) genotyped five dog breeds and European gray wolves for *DRD4* intron 2 polymorphisms. All five dog breeds possessed both P and Q alleles, whereas in the gray wolf, the Q allele could not be detected. Hejjas *et al.* (2009) behaviorally phenotyped 96 unrelated German Shepherds using an applied greeting test that measured social impulsivity — the friendliness of the dog towards a human stranger. A low score indicated a dog that was unfriendly to the greeting human, whereas a high score characterized a dog that was not only friendly but followed the greeting human when that human moved away. Dogs having two copies of the intron 2, long Q allele, or the exon 3, number 3a allele had significantly higher social impulsivity scores than dogs possessing one or two copies of the intron 2, short P allele, or the exon 3, number 2 allele. The effects of the exon 3 and intron 2 polymorphisms were additive. Dogs homozygous for both the long Q allele of intron 2 and the number 3a allele for exon 3 had the highest social impulsivity score, whereas dogs homozygous for the short alleles, P for intron 2 and number 2 for exon 3 had the lowest score. Dogs possessing one copy of either short allele had intermediate scores. These results in the German Shepherd are consistent with the observation that humans with a high number of repeats of a certain sequence in *DRD4* exon 3 are generally highly "novelty seeking" (Takeuchi *et al.*, 2009).

Takeuchi *et al.* (2009) performed genotyping and behavioral phenotyping in 77 unrelated Shiba Inu dogs. They examined various polymorphisms on the exons of nine neurotransmitter-related genes. A questionnaire, completed by the dogs' owners, was used to measure behavioral traits. Two factors distilled from the questionnaire data, aggression to strangers and reactivity (excitability, excessive barking, timid with strangers), were used in assessing for associations between genetic variants and behaviors. Only factor 1, aggression to strangers, showed a significant association with variants of three of the genes examined: tyrosine hydroxylase (*TH*), *DRD4*, and *SLC1A2* (solute carrier family 1 {neuronal/epithelial high affinity glutamate transporter} member 2). The strongest association was with gene

SLC1A2 and a thymine-to-cytosine substitution at nucleotide 471 (c. 471 T > C polymorphism). Dogs whose genotype was CC were much less aggressive than dogs whose genotype was TC. The TT genotype could not be tested because of the small sample size. The *SLC1A2* gene product removes the excitatory neurotransmitter glutamate from the extracellular space and helps maintain low extracellular glutamate concentrations. Takeuchi *et al.* (2009) also reported a significant association between VNTR polymorphisms in exon 3 of *DRD4* and a cytosine-to-thymine substitution at the 264th nucleotide (c.264 C > T polymorphism) of TH. It is interesting that this substitution at position 264 in TH is silent, i.e., does not result in an amino acid substitution (Takeuchi *et al.*, 2005).

Domesticated dogs and humans share the same environmental space and Wang *et al.* (2013) have reported that dogs and humans display an overlapping set of genes for digestion, metabolism, neurological processes, and cancer that have undergone positive selection over the domestication period, i.e., a case of parallel or convergent evolution. Three important characteristics of the modern purebred dog make this animal a wonderful model for studying the genomic underpinnings of phenotypic traits: the dog's unusual genomic architecture by virtue of its evolutionary background which simplifies genomic mapping of morphological and behavioral traits, the circumstance that modern dogs and humans occupy the same environmental space, and the dog's social cognitive skills which enable the dog to relate to humans on a very sophisticated level.

The behavioral characteristics of the dog and the unique bond between the dog and its companion human have been well chronicled by Horowitz (2009).

Since modern dogs and humans share a common protected environment, it is not surprising that dogs manifest many of the same "diseases" as that of humans. This topic is discussed in the next section.

Diseases in dogs

Modern dog breeds were created about two hundred years ago through a process of artificial selection and breeding. In addition, a strict pedigree system was established which discouraged mating across breeds. Hence, dog breeds have remained relatively genetically pure. As noted by Parker *et al.* (2004) on the basis of genotype data alone, 99% of individual dogs could be assigned to the correct breed. In some breeds, behavioral traits were enhanced and this is most evident in the so-called working breeds.

In many breeds, specific morphologic traits were developed, dictated to some extent by the wishes of their human companions. Even a cursory examination of the various breeds highlights that many morphologic features are clearly maladaptive. Dogs live in a shared space with humans and definitely in the Western world have access to human-type medical care. Similar to the situation in humans (Singer, 2013), the strength of purifying natural selection has been relaxed in dogs. Considering all these attributes and circumstances, one might expect that the dog was more likely to accumulate deleterious mutations than its wild ancestor the wolf. Cruz *et al.* (2008) and Toll-Riera *et al.* (2011) have examined this question.

Toll-Riera *et al.* (2011) assessed lineage-specific variation in natural selection in six mammalian species: human, macaque, mouse, rat, dog, and cow. The dog breed(s) used was not given. The original data set used comprised 10,267 orthologous genes, and from this set, 2,929 orthologous gene sequence alignments were chosen. For each orthologous gene, the number of non-synonymous substitutions per non-synonymous site (dN, amino acid altering) and number of synonymous substitutions per synonymous site (dS, silent, i.e., not amino acid altering) were measured. The ratio dN/dS can be used to assess changes in the strength and type of selection. In the complete absence of selection, one would expect a ratio of 1. Most genes have a dN/dS much lower than 1 indicating a strong effect of negative or purifying selection. A dN/dS value greater than 1 is indicative of positive selection. If a species experiences an increase in the ratio dN/dS, this could be due to the effect of positive selection or to a reduction (relaxation) in purifying selection. Conversely, if a species experiences a decrease in the dN/dS ratio, then this situation indicates an increase in purifying selection. Toll-Riera *et al.* (2011) calculated mean and median dN/dS ratios for the full set of 2,929 orthologous genes in the six mammalian species. dN/dS ratios were not uniform across these species; mean and median ratios were larger in primate lineages (including humans) than in the other species. For example, in humans, the mean ratio was 0.163 whereas in dogs the mean ratio was 0.115. The investigators attributed the higher ratio observed in primates to a weakening of purifying selection in this lineage. At the level of the individual gene, after correcting for species-specific differences, about 25% of the 2,929 orthologous sequences showed species-specific deviation in the ratio (dN/dS). In about half of the cases, the deviation was an increase in the value of the ratio and in the other half, the value of the ratio decreased. Hence, for this set of orthologous mammalian genes, there were both species-specific and across these species, gene-specific deviations in the type and strength of selection pressures. However, at the species level, the dog did not show any relaxation of selection pressures (i.e., a weaker purifying selection) than the five other mammalian species examined.

Cruz *et al.* (2008) noted that dogs appeared to have accumulated non-synonymous mutations in mitochondrial DNA at a greater rate than wolves, and that this observation implied a weakening of the strength of purifying selection in dogs. To examine for this possibility, Cruz *et al.* (2008) compared nuclear genomic sequences in the wolf, Boxer, and cat. The dog–wolf–cat alignment comprised about 247 kb of coding DNA and this alignment was used to compare dN/dS ratios. This ratio was 57.3% higher in the dog compared to the wolf. The actual number of non-synonymous substitutions was comparable in both lineages, but the wolf had a higher number of synonymous substitutions which resulted in a lower value for the dN/dS ratio in the wolf compared to the dog. The functional impacts of the non-synonymous substitutions were estimated by classifying them as radical or conservative using standardized criteria. There were 27 radical and 17 conservative changes in the wolf, and 22 radical and 27 conservative changes in the dog. Cruz *et al.* (2008) concluded that the higher dN/dS ratio for protein-coding DNA sequences in the dog compared to the wolf reflected a decrease in the strength of purifying selection in the dog, i.e., a relaxation of negative selection and as a result, the dog had accumulated a number of deleterious mutations.

The experimental designs of Cruz *et al.* (2008) and Toll-Riera *et al.* (2011) are not comparable. Toll-*Riera et al.* (2011) measured the strength of different selection pressures on a number of orthologous genes in six mammalian species. They observed that the strength of selection (both positive and purifying) on orthologous genes showed specific variations across the different species. However, their data did not indicate that purifying selection was weaker in the dog compared to the other mammalian species examined. Cruz *et al.* (2008) in their experiments specifically compared the dog (Boxer breed) with the wolf and noted that the dog showed a reduction in the strength of purifying selection compared to the wolf and that this observation probably accounted for the apparent accumulation of deleterious mutations in the dog.

The modern dog lives in a protected environment that it shares with its human companion. As such, one would expect under these circumstances the strength of purifying selection to be weaker. In support of this construct is the documentation that a total of 624 "disorders" have been catalogued in the dog — website: Online Mendelian Inheritance in Animals (OMIA). Two hundred and forty-two of these disorders have a Mendelian inheritance and in 167 of these, the key mutation is known. Fewer disorders have been defined in other domestic animals: 444 in total in cattle and 318 in total in cats. Of the 624 disorders in the dog, 332 are considered to be potential models for human "diseases." The reasons that livestock have a lower frequency of genetic diseases than companion animals were discussed by Patterson (2000). The breeding of livestock is driven by the goal of developing relatively uniform populations of animals that were healthy, both in terms of general reproductive fitness and in traits that were economically important such as milk, meat, and wool production. In contrast, the goals of dog breeding are much more diverse and loosely defined. These goals are driven by changing fashions in physical appearance based upon performance at "shows" and in some breeds by the desire for certain behavioral traits. Hence, the selection and breeding process in livestock is designed to eliminate deleterious mutations while such is not the case with respect to selection and breeding in modern dogs.

Fleming *et al.* (2011) used the Veterinary Medical Database (VMDB) to examine mortality patterns in North American dogs. The VMDB was initiated in 1964 and contains medical records submitted by 27 participating veterinary teaching hospitals in North America. The database scrutinized for this study included records between the years 1984 to 2004. Causes of death were classified on the basis of both the organ system involved and the causative pathological process. From the database, the records of 72,376 dogs were available for analysis. The relative frequencies of death within these two categories were tabulated across 82 breeds. In 69 of the breeds, neoplasia was the leading pathological process causative of death; the relative frequency of deaths due to neoplasia varied from a high of 0.546 in the Bernese Mountain Dog to a low of 0.060 in the Miniature Dachshund. Hence, the proportionate death rate due to cancer varied across breeds. For all adult dogs (age one or greater), the percentages of deaths involving different organ systems were: neurologic 12.5, musculoskeletal 11, gastrointestinal 11, and cardiovascular 9. The percentage of deaths due to specific pathological processes in all dogs were: neoplasia 31, traumatic 10, infectious 6, and metabolic 5. Degenerative and metabolic diseases increased slightly with age, whereas neoplastic diseases showed a marked increase in relative frequency with age, about a fourfold increase. When the influence of size (body mass in kg) was considered across breeds, larger breeds had a much higher frequency of death due to musculoskeletal and neoplastic diseases than smaller-sized breeds: neoplastic > musculoskeletal. The data of Fleming *et al.* (2011) highlight some breed specific differences. For example, the Bernese Mountain Dog has a relative mortality of 0.546 from cancer, whereas equal-sized breeds such as the Standard Poodle and Old English Sheep dog, have lower relative cancer mortalities of 0.271 and 0.360, respectively. There were also significant breed differences with respect to the organ systems characterizing the mortality pattern. In Dachshunds, a high percentage of deaths involved the neurological system, while in Fox Terriers, a high percentage of deaths involved the cardiovascular system.

As already noted, the dog is emerging as an important model for human "disease" disorders. Two of these disorders, cancer and diabetes, are examined in more detail in the following sections.

Cancer in dogs

The overall mortality statistics for cancer are similar in humans and dogs. In adult dogs, cancer accounts for about 31% of deaths (Fleming et al., 2011), whereas in humans living in Canada, cancer accounts for 29.8% of deaths (Statistics Canada, 2012). However, the types of cancer differ between these two species (Dobson, 2013). Breast cancer is common in female humans and in female dogs, while prostate cancer is common in male humans but uncommon in male dogs. Lung and colorectal malignancies are the leading causes of cancer deaths in humans (excluding breast malignancies in females; Canadian Cancer Society, 2011), but these malignancies are quite uncommon in dogs. Tissue sarcomas are common in dogs but in contrast are relatively rare in humans. Another feature shared by both species is that cancer deaths increase in number with increasing age (Fleming *et al.*, 2011; Cancer Research UK, 2008–2010). In dogs and humans, cancer is predominantly a "disease" of ageing.

Dobson (2013) described features of some of the more common malignancies found in dogs. Histiocytic sarcoma is relatively common in the Bernese Mountain Dog and the Flat-Coated Retriever and somewhat less frequent in Rottweilers and Golden Retrievers. Genomic studies in Bernese Mountain Dogs and Flat-Coated Retrievers demonstrated deletions in several tumor suppressor genes (*CDKN2A/B, RB1, and PTEN*). (Tumor suppressor genes regulate cellular functions such as division, DNA repair, and apoptosis). Osteosarcoma is the most common bone malignancy in dogs and occurs primarily in large and giant breeds such as the Rottweiler,

Great Dane, Irish Wolfhound, Greyhound, and Saint Bernard. In a study of osteosarcomas in Rottweilers, about 50% showed deletions in at least one of the *WT1*, *TP53*, *CDKN2A*, *PTEN*, or *RB1* tumor suppressor genes. Other tumors have a reported predilection for certain breeds: visceral hemangiosarcoma in the German Shepherd, cutaneous mast cell tumors and lymphomas in the Boxer and Bull Dog breeds, and cutaneous melanomas in breeds with heavily pigmented skin such as Schnauzers and Scottish Terriers. In these cancers, alterations in some of the tumor suppressor genes have been detected. The observation that specific types of cancer have a predilection for certain dog breeds (Dobson, 2013; Shearin and Ostrander, 2010) makes the dog a valuable model for studying the genetic underpinnings of cancer.

The study of the genetic architecture of canine cancers is still in its infancy particularly when compared to the vast literature on genomic changes in human cancers. Recently, Alexandrov *et al.* (2013) reported on the mutational processes present in 30 different human cancers. Mutations (DNA structural variants) can be categorized as being inherited, *de novo* or somatic (Poduri *et al.*, 2013). An inherited mutation is present in all the cells including germ cells of one or both parent(s). This mutation is transmitted to the offspring and will be present in all the cells of the offspring including germ cells. A *de novo* mutation can occur during gametogenesis in one of the parents. This mutation will not be present in the somatic cells of the parent but will be present in all the cells of the offspring including germ cells. Somatic mutations can occur after zygote formation during mitotic cell divisions (through errors in DNA replication or chromosome segregation) that generate the embryo. If the mutation occurs at an early stage after zygote formation, most or all of the individual's tissues will contain the mutation although only a portion of the cells in each tissue will harbor the mutation. If the mutation occurs at a later stage after zygote formation, then only certain tissues in the individual will contain the mutation and only some of the cells in those tissues. Hence, somatic mutations (DNA structural variants) can result in a state of mosaicism which is much more common than previously suspected (Lupski, 2013). In fact, it is possible that within a specific tissue, each cell will have its own "personal" genome. The notion that somatic mosaicism is quite common is supported by experimental observations reported in several recent publications (Abyzov *et al.*, 2012; Jacobs *et al.*, 2012; Laurie *et al.*, 2012; Macosco and McCarroll, 2012; Macosko and McCarroll, 2013; McConnell *et al.*, 2013; O'Huallachain *et al.*, 2012). This topic and these reports are discussed in **Chapter 5**. The biological consequences of genomic mosaicism have been discussed by Lupski (2013) and Poduri *et al.* (2013).

Mosaicism can also occur in the germline (Zlotogora, 1998) as a result of different mechanisms. One mechanism involves the accumulation of DNA structural variants during the successive mitotic cell divisions occurring in the somatic stem cells which ultimately form germ cells. This mechanism was experimentally demonstrated by Liang *et al.* (2008). These investigators used well characterized mouse embryonic stem cells which were capable of differentiating into germ cells. Embryonic stem cells in culture undergo 30–40 mitotic divisions before resuming their developmental route into the germline and during these successive mitotic divisions, DNA structural variants (mutations) such as copy number variants (CNVs) can occur. For example, an embryonic stem cell clone designated AB2.2, which can be transmitted into the mouse germline, contained two CNVs denoted CNV-2 and CNV-18. Each of four sub-clones derived from this clone contained these two CNVs as expected. However, these four sub-clones also contained five additional CNVs that occurred during their mitotic divisions: CNV-12, CNV-22, CNV-32, CNV-38, and CNV-48. The distribution of the CNVs among these four sub-clones were: clone 1 had CNVs 2 and 18; clone 2 had CNVs 2, 12, and 18; clone 3 had CNVs 2, 18, and 22; and clone 4 had CNVs 2, 18, 32, 36, and 48. Hence, the four sub-clones derived from the parental clone displayed a mosaic genomic pattern. If the mutational events occur at a very early developmental stage after zygote formation prior to the commitment of stem cells to germline differentiation then that individual will manifest both somatic and germline mosaicism for those mutations (Lupski, 2013; Zlotogora, 1998). In this case, the mutations have occurred in stem cells that differentiate into both germ cells and other somatic cells.

A second mechanism that leads to germline mosaicism is the process of meiotic recombination which is a normal feature occurring during the generation of mature gametes. Meiotic recombination generates genetic variation through the exchange of genetic material between homologous chromosomes. However, the presence of "unstable" genomic architectures (so-called hotspots), such as low copy repeats, can lead to processes such as non-allelic homologous recombination (NAHR) occurring during meiosis. NAHR can result in the generation of deletions, duplications, inversions, and translocations (Osborne, 2008; Turner *et al.*, 2008). Such DNA structural variants are associated with significant alterations in gene function. Since germline mosaicism due to meiotic recombination occurs during formation of mature gametes, it will not be accompanied by somatic mosaicism.

Germline mosaicism has been implicated in the transmission of "genetic" disorders from an apparently "unaffected" parent to an offspring (Lupski, 2013; Osborne, 2008; Turner *et al.*, 2008; Zlotogora, 1998).

In summary, germline mosaicism is the rule since meiosis involves the process of meiotic recombination which is designed to exchange genetic information between homologous chromosomes. However, the presence of "hotspots" which are sites of genomic instability create conditions such that meiotic recombination also generates extensive DNA structural variants that can have significant functional consequences. In addition, errors during the cell mitotic divisions that precede meiosis are another source of genomic structural variants some of which will be functionally deleterious.

The mutations found in a cancer genome are primarily somatic mutations and can occur at any stage in the history of that cellular lineage from fertilized egg to the sequenced cancer cell (Alexandrov *et al.*, 2013). These mutations may arise as a consequence of errors occurring during DNA replication, exposure to exogenous or endogenous mutagens, enzymatic modification of DNA, or defective DNA repair. Exogenous mutagens include tobacco smoke in the case of lung cancers and ultraviolet light in the case of skin cancers.

Different mutational processes can generate combinations of mutation types termed "signatures." The mutations catalogued by Alexandrov *et al.* included base substitutions, small insertions, and small and large deletions from 7,042 primary cancers of 30 different classes. In all cases, normal DNA from the same individuals was sequenced to establish the somatic origin of the mutations. The investigators defined mutational signatures primarily on the basis of nucleotide base substitutions and included information on the DNA sequence context of each mutation — nature of the bases immediately 5′ and 3′ to each mutated base. Twenty-one distinct mutational signatures were defined. In most cancer classes, at least two mutational signatures were observed with a maximum of six signatures in cancers of the liver, uterus, and stomach. Signatures 1A and 1B are probably initiated by the spontaneous deamination of 5-methyl-cytosine to thymine and were observed in 25 out of 30 cancer classes. Signatures 1A/B showed strong positive correlations with age in the majority of cancer types and hence, these mutational signatures can be considered a function of the ageing process. Signatures 4, 7, and 16 included mutations in DNA repair mechanisms. In addition, signature 4 was primarily observed in cancers of lung, head and neck, and liver, most of which are known to be caused by tobacco smoking. Signature 4, then, is partially a reflection of tobacco-induced mutations.

Signature 7 was mainly found in malignant melanoma and is primarily the result of ultraviolet radiation (UVR)-induced mutations. Signature 6 contained a large number of small insertions and deletions which resulted in defective DNA mismatch repair and was seen in colorectal, uterine, liver, kidney, prostate, esophageal, and pancreatic cancers. Large deletions were seen in signature 3 which occurred in breast, ovarian, and pancreatic cancers. Signature 3 was associated with mutations in the *BRCA1* and *BRCA2* genes which are involved in DNA double-strand break repair.

These mutational signatures underscored the diversity and complexity of mutational processes underlying carcinogenesis in humans. For some of the signatures, the mechanistic origin of the mutations was partially understood, e.g., tobacco carcinogens for signature 4 and UVR for signature 7. However, for the majority of signatures the mechanistic basis of the mutations was unknown. One avenue of research that could help clarify the mechanistic origin of human cancer mutational patterns is a better understanding of mutational patterns in canine cancers. Given, the unusual genomic architecture of the dog by virtue of its evolutionary background and the apparent predilection of certain cancers for specific breeds, unraveling the genomic mutational patterns underlying various cancers will probably be easier in the dog than in the human.

Diabetes mellitus in dogs

In North American adult dogs, metabolic "diseases" which would include diabetes mellitus account for about 5% of deaths (Fleming *et al.*, 2011). Unlike the situation in humans in which most cases of diabetes are type 2, in dogs diabetes mellitus is primarily the result of deficient endogenous insulin production (Catchpole *et al.*, 2013). However, like humans, diabetes is a heterogeneous disorder and is observed predominantly in middle-aged and older dogs. There is also a breed-specificity to the incidence of diabetes. In the UK, the odds ratio of developing diabetes was 35.8 in the Samoyed, 10.39 in the Tibetan Terrier, and 9.76 in the Cairn Terrier, but only 0.23 in the German Shepherd and 0.07 in the Boxer. Less than 10% of diabetic dogs in the UK Canine Diabetes Register were diagnosed under seven years of age. Diabetic Samoyeds commonly express one of two DLA genotypes (analogous to the HLA genotype), i.e., heterozygous DLA DRB1 009/DRB1 015 or homozygous DLA DRB1 015. In contrast, the highest risk haplotype DLA DRB1 009 is rarely observed in the Boxer and German Shepherd breeds which have a low incidence of diabetes. Variable VNTRs have been detected in the 5′ end of intron 2 of the canine insulin gene. The allele of the insulin gene containing a double VNTR was more common in diabetes susceptible breeds than in diabetes-resistant breeds. These observations support a genetic mechanism underlying diabetes mellitus in the dog, but the results are much too preliminary to define specific causative biological processes at this stage.

Summary

The modern dog is emerging as an important model for the study of the ageing process with particular relevance to ageing in humans. As a result of the domestication process and the selective breeding of modern dogs, this species displays extensive diversity of morphological and behavioral phenotypes across breeds but relative homogeneity within breeds. The dog (Boxer) genome has now been sequenced. The genomic architecture of the dog is characterized by long runs of linkage disequilibrium and long blocks of haplotype homozygosity which facilitate the genetic mapping of phenotypes and "diseases" as compared to the human. Modern dog breeds are relatively genetically distinct. In addition, the dog shares our environmental space, generally lives in our homes, and has well-developed social cognitive skills which leads to a special "bond" between the human and their companion dog. The dog is also subject to a number of inherited "diseases," many of which are similar to those observed in the human. In the previous sections, the discussion was restricted to two "diseases": cancer and diabetes mellitus. Cancer as a cause of death has about the same frequency in the dog and the human although the types of malignancies differ between the two species and also between different breeds of dogs. However, in both species, cancer is a "disease" of ageing. Diabetes in the dog is a heterogeneous disorder largely due to deficient insulin production, and it varies in incidence between different breeds and appears to have a genetic basis.

The dog lives in the human social environment and has learned to be very much in "tune" with human behaviors. This animal represents a remarkable natural animal model for dissecting out the relative roles of environmental cues (experiences) and genetic factors underlying the ageing process and so-called disease phenotypes.

Social Context and Culture

Overview

The social structure in which an animal lives can have a significant influence on that animal's ageing pattern. This observation is particularly true for humans and domesticated animals. One of the messages of evolutionary medicine is that biological evolution is much slower than cultural change and hence mismatches arise between our bodies and modern environments (Stearns *et al.*, 2010 b). The myriad of biological processes within an organism are subject to unavoidable trade-offs and constraints such that organisms can be considered bundles of compromises shaped by natural selection to maximize reproductive success (Stearns *et al.*, 2010b). For the human, the effects of culture and environmental changes are profound and modern environments are far different than the ones in which humans evolved.

Genomic tools can be used to examine relationships among populations living in different geographical areas and to determine population ancestries. Li *et al.* (2008) used 650,000 SNPs from the Human Genome Diversity Panel which represented approximately 1,064 individuals from 51 populations from Sub-Saharan Africa, North Africa, Europe, the Middle East,

South/Central Asia, East Asia, Oceania, and the Americans. Interestingly, Li *et al.* (2008) found that most of the genetic diversity in these populations was accounted for by variation between individuals (89%) as opposed to variation across geographical regions (9%). The relationships and ancestries revealed by the genomic data were consistent with a single origin for modern humans in Sub-Saharan Africa followed by population expansion through the successive migration of a small fraction of individuals out of the previous location, starting from this single origin.

Humans are the products of their environment and climate is believed to be the engine driving human evolution (Gibbons and Pennisi, 2013). Since the human lineage originated in East Africa, scientists are now trying to reconstruct the nature of the climate at important times of evolutionary change by extracting sediment cores in dried out lake beds in East Africa (Kenya and Ethiopia) next to key fossil sites at the lake margins. The goal of this research is to examine the influence of climate variability in human evolution by obtaining a record of climate indicators from the very places where our ancestors lived and died (Gibbons and Pennisi, 2013).

A number of years ago, Richard Potts proposed that evolution of the *Homo* species was driven by climate variability and not by any single type of habitat or environmental trend (Potts, 1996). His proposal has been presented in more detail in recent reports (Gibbons and Pennisi, 2013; Smithsonian Institution, 2014).

Human and chimpanzee lineages are believed to have diverged from a common ancestor about 6–8 million years ago. At approximately the same time, the African climate was becoming more arid, which was associated with changes in the makeup of plant communities. Grasses and sedges became more prominent than trees and shrubs. Hence, the human lineage was "born" at the time when grasses were on the rise and ancestral humans responded by broadening their diets to include grasses and trees and shrubs. Ancient human remains have been discovered in a variety of different habitats: wooded tree areas as well as savanna grasslands. Overall, the human fossil record and the environmental record indicate that humans evolved during a highly variable climatic period. This climate variability favored adaptations that helped ancestral human species cope with a number of different environments. (These large-scale environmental fluctuations occurred on time scales of tens of thousands of years).

The ancient human *Ardipithecus ramidus* (4.4 million years old) lived in two different Ethiopian sites: wooded areas and wooded grasslands in which grazing animals predominated. In addition, the foot structure of

A. ramidus allowed for both terrestrial bipedality and arboreal climbing and grasping (Ward *et al.*, 2011). By about 3.2 million years ago, the foot structure had been transformed to one designed for bipedal propulsion (Ward *et al.*, 2011). This type of foot structure is seen in *Australopithecus afarensis* (named 'Lucy'), an ancient human who was adapted to the widespread grasslands. The savanna environment would have favored the more effective bipedal gait made possible by the foot structure characteristic of *A. afarensis*.

The first stone tools date from around 2.6 million years ago and these tools would have allowed ancient humans to obtain food from a variety of sources: the cutting of meat, to access the marrow of bones, the grinding of various plants, and the digging for tubers. The ability of humans to adapt to a variety of climates and habitats allowed for migration out of Africa by about 1.8 million years ago. These early migrations were to China, Java, and Georgia. In these locations, ancient humans encountered distinctly different environments, vegetation, animals, and climates. Some human species were able to survive during these times of migration and environmental instabilities, whereas other species not able to cope became extinct. More recently, over the last 100,000 years, behaviorally modern humans have colonized most of the globe.

Over the past 800,000 years, the human brain size increased dramatically, which favored increased problem solving and hence allowed humans to produce solutions to the challenges created by new and diverse environmental and climate changes. New more sophisticated tools were developed and trading of materials between groups allowed for the sharing of resources. By 130,000 years ago, humans were exchanging goods over distances as large as 300 km. Communication with symbols may have originated about 250,000 years ago.

Following the end of the last Ice Age 14,000 years ago, there was a major warming trend that brought global temperatures to their current levels. Between 10,000 and 12,000 years ago, in the Fertile Crescent and a little later elsewhere, there was a transition from hunter-gatherer to agricultural societies (Voight *et al.*, 2006). The initial spread of agriculture into Europe involved the active dispersal and migration of people rather than the cultural dispersion of agricultural knowledge and technologies without a demographic expansion of people practicing farming (Pinhasi and von Cramon-Taubadel, 2009).

Skoglund *et al.* (2014) explored population details occurring at the time Neolithic farmers were spreading into Europe. These investigators used mitochondrial DNA from six Neolithic hunter-gatherers and four Neolithic farmers (all about 7,000 years old) to examine relationships between these two groups as well as relationships between these two groups and modern Europeans. Principal component analysis based on the genetic landscape of Western Eurasian populations segregated the Neolithic farmers with present-day Europeans, while Neolithic hunter-gatherers formed a separate group. The Neolithic hunter-gatherer group had much less genetic diversity than the farmer group probably for several reasons. The population size of the hunter-gatherers was low for various demographic and climatic reasons (small populations are associated with low genetic diversity) and gene flow (admixture) from the hunter-gatherers into the farmer group increased the genetic diversity of the farmers. There was no significant evidence of admixture of incoming farmers into the hunter-gatherers. In summary, as the Neolithic farmers radiated into Europe, the subsequent demographic mix resulted from a greater degree of hunter-gatherer admixture into farmers with little introgression from farmers into hunter-gatherers. Neolithic farmers, the ancestors of modern Europeans, became more genetically diverse as a result of this admixture.

The development of agricultural practices including animal domestication made it possible for humans to live in greater population densities.

Another influence in the evolution of modern humans has been the interbreeding between different lineages. There are genomic signatures of archaic DNA from Neanderthals and Denisovans in present day humans (Stewart and Stringer, 2012; Stringer, 2012). Whether this archaic DNA is tied to any visible traits is unknown.

Two significant observations emerge from this brief overview of human evolution. The first is that adaptations that have shaped the human lineage were driven by the occurrence of fluctuations or instabilities in local climates and environments and not by the selective pressures of specific habitats or ecological niches. Humans who were able to adapt to these climatic and environmental fluctuations survived, while species that lacked adaptability became extinct. Secondly, agricultural practices and cultural evolution allowed groups of humans to share resources and to acquire knowledge and skills from each other. Through these mechanisms, humans were able to develop reliable food sources and make themselves less susceptible to fluctuations in climatic and environmental conditions. In our modern urban societies, humans "control" the microclimate within their houses, have ready access to food and medical care, and are able to travel great distances in relatively short times.

The early human lineage evolved under a certain set of environmental pressures. Concomitant cultural evolution with the emergence of gene–culture co-evolution allowed humans to expand in terms of numbers and to radiate throughout the globe. Modern humans inhabit a wide variety of environmental settings, but cultural practices are not uniform across the whole human population. There are geographical regions in which humans have a culture that can be characterized as that of an advanced technological society. There are other geographical regions in which the culture is much less advanced and features poor public safety and restricted access to food, sanitation, and adequate housing. An intriguing question is whether the evolutionary path of the human species will differ between these regions.

The evolution of species is driven by positive selection. Traits that are adaptive for that animal's environmental pressures and that are heritable will be selected and increase in frequency throughout the population. If the animal has adapted to its environment and if that environment remains relatively constant, then negative or purifying selection assumes a very important role in maintaining that animal's fitness. Deleterious mutations that reduce fitness within the animal's stable environment will be eliminated. Positive selection of adaptive traits is a slow process and hence will not be an effective mechanism for adaptation when environmental pressures change relatively rapidly. If a given species cannot adapt quickly enough for the changing environmental pressures, the species will become extinct. Cultural evolution, a process unique for humans, has allowed humans to adapt to rapidly changing environmental pressures and gene–culture co-evolution is now considered the dominant mode of evolution in our species. Examples of gene–culture co-evolution are discussed in the next sections.

Genomic studies in humans have documented the presence of both positive and negative selection (Fay *et al.*, 2001; Khurana *et al.*, 2013). Positive selection of adaptive genomic variants (mutations) is the cornerstone of Darwinian evolution, but the majority of genomic variants are either neutral or slightly deleterious. Negative (purifying) selection prevents deleterious variants from reaching significant frequencies (Fay *et al.*, 2001). These investigators used the ratio of non-synonymous SNPs (nsSNPs) to synonymous SNPs (sSNPs) to measure the frequency of deleterious variants. They calculated that about 20% of nsSNPs were neutral and 80% deleterious — one third slightly deleterious and two thirds strongly deleterious. They examined for positive selection by comparing divergence data for sSNPs and nsSNPs using 182 orthologous human and old-world monkey genes. They estimated that on average, one advantageous substitution occurred in humans every 200 years since humans and old-world monkeys diverged 30 million years ago.

Khurana *et al.* (2013) used sequencing data from 1,092 humans to look at patterns of selection for genetic variants: SNPs, small indels, and larger structural variants (SVs). These investigators used as a metric for measuring negative selection the fraction of rare variants (number of rare variants divided by the total number of variants) in categories of genes. A higher fraction of rare variants (allele frequency less than 0.5%) suggests stronger negative selection. The strength of negative selection was examined in both coding and non-coding genomic regions. Different gene categories showed differential selection consistent with their known phenotypic consequences. Among protein-coding genes, those that are cancer causal are under the strongest selection. The proteins that are encoded by "cancer" genes normally regulate cell proliferation, cell differentiation, and cell apoptosis (Futreal *et al.*, 2004). Mutations in genes that mediate DNA repair processes can also cause cancer. There was tissue specificity to the strength of negative selection for protein-coding genes with the strongest constraint observed in the ovary and brain and the weakest constraint in adipose tissue. Non-coding genomic regions were also under negative selection and certain transcription factor-binding sites were among the loci under the strongest negative selection. Many so-called hub genes, genes that are central to gene networks, were under negative selection and there was a correlation between the strength of selection and the centrality of the gene. Although negative selection was widespread throughout the genome, some sites within these constrained regions actually experienced positive selection. For example, loss of function mutations in the gene *ABCA12* result in a severe skin disorder, whereas an SNP in the second intron of this gene showed extreme differentiation between Europe/East Asia and Africa — allele frequency of over 90% in Europe and East Asia but only 13% in Africa. This polymorph shows evidence of positive selection in non-African populations possibly reflecting adaptations of the skin to reduced levels of sunlight outside Africa.

Thus, the human genome shows evidence of both positive and negative selection and as expected, the strength of negative selection is strongest at those genomic regions where structural variants would have the most deleterious functional consequences.

There is evidence that humans have also experienced a relaxation of negative selection at certain genomic loci. Somel *et al.* (2013b) tested for human specific relaxation of negative selection in 54 unrelated individuals of diverse ancestry. Their database consisted of a genome-wide cataloguing of over 64,000 nsSNPs and over 54,000 sSNPs in protein-coding sites of these individuals. They used three criteria for identifying the occurrence of human-specific relaxation of negative selection on a group of protein coding genes: (1) the gene group should have low levels of divergence across mammals as measured by the ratio dns/ds but high levels of diversity for nsSNPs in humans. Mammals on average show low values of the ratio dns/ds consistent with the action of negative selection to prevent nsSNPs from reaching high frequency; (2) the gene group should have higher nsSNP diversity than the genome average in the population of humans being studied; and (3) human diversity for nsSNPs should be greater than that observed for chimpanzees. Using these criteria, the investigators detected relaxation of negative selection for two groups of genes: olfactory receptor genes and genes associated with the proteosome complex. Relaxed selection associated with olfactory receptor genes has been previously documented, but the finding of relaxed selection with respect to genes associated with the proteosome complex was a new observation. The proteosome complex is involved in the degradation of damaged or misfolded proteins so marked for destruction by the attachment of ubiquitin molecules. The proteosome contains a catalytic core consisting of seven alpha- and seven beta-subunits and two sets of regulatory particles. All mammalian proteosome gene sets have significantly lower values of dns/ds compared with the rest of the genome, indicating the effect of strong negative selection. However, the investigators noted that in the human, proteosome genes bearing nsSNPs were restricted to those encoding beta-subunits only. The genes encoding alpha-subunits contained only sSNPs. Thus, the core human proteosome showed a clustering of nsSNPs of high frequency. In contrast, chimpanzee proteosome genes did not show a higher frequency of nsSNPs.

In summary, Somel *et al.* (2013b) found that in the human, olfactory receptor genes and genes associated with the core component of the proteosome showed evidence of a relaxation of negative selection. The reduced strength of negative selection on olfactory receptor genes is thought to be the result of a transition from nocturnal to diurnal activity in the human lineage. The possible reasons for the relaxed selection associated with protesome genes are unknown. However, the finding of proteosome nsSNPs in the Denisovan genome suggests that this relaxation in negative selection appeared about 800,000 years ago.

In modern human populations, survival through the peak reproductive ages is generally assured, so reproductive success has become essentially the only metric of Darwinian fitness (Stearns *et al.*, 2010a). Based on data from the Framingham Heart Study, Byars *et al.* (2010) calculated the actual lifetime reproductive success (number of live births per individual over the reproductive period; LRS) in women categorized by their year of birth; the time frame covered was from 1890 to 1950. For women born between 1890 and 1930, LRS was relatively stable or actually increasing. However, for women born since 1930, there has been a steady decline in LRS which by 1950 was significantly lower than previous eras. This data appear to indicate that there has been a reduction in fitness since the 1930s, suggesting the accumulation of deleterious mutations as a result of weaker negative selection. In the section on **diseases in dogs** in the previous chapter, genomic evidence was presented that the dog, our companion animal, in comparison to its ancestor the wolf, has experienced a relaxation of negative selection in various coding genes.

Are Humans Still Evolving?

The answer to this question is yes. Current genomic technologies have provided evidence for continuing human evolution over the past 10,000–20,000 years. There is evidence of recent positive selection for specific genes as well as evidence of recent positive selection for specific genomic loci. Before considering these two data sets, it is useful to look at the basis of the genomic evidence.

The genomic evidence for recent positive selection is based on a number of considerations. In response to strong positive selection, a selected allele will rise rapidly in frequency (Bersaglieri *et al.*, 2004). This rapid rise in allele frequency will reduce the time in which genetic recombination can reshuffle the different alleles linked to this selected allele (genetic hitchhiking), and hence, long-range associations with nearby polymorphisms will be maintained — the long-range haplotype. This reduction in variation among nucleotides in the neighborhood of the positively selected allele is referred to as a selective sweep. A selective sweep is defined as a positively selected haplotype consisting of the positively selected allele and its linked ("hitchhiking") neighboring alleles. A selective sweep is considered complete if the favored allele reaches a frequency of 100% in the population (gene fixation). The sweep is considered partial if the frequency of the favored allele is less than 100%. In summary, the genomic signal of strong positive selection for a given allele is that the allele has increased in frequency so quickly that it ends up in an unusually long haplotype of low genomic diversity. Genomic regions not carrying the favored allele have levels of haplotype diversity and LD typical of the genome as a whole (Voight *et al.*, 2006). It is important to realize that long-range haplotype homozygosity can occur under conditions other than strong positive selection. For example, population bottlenecks that characterize the evolutionary history of the dog are believed to be responsible for the extended haplotype homozygosity in this animal. However, natural selection will act on only certain genomic regions, while complex demographic events such as population bottlenecks will affect the whole genome.

As discussed by Scheinfeldt and Tishkoff (2013), characterizing an adaptive variation requires four steps: identifying a candidate adaptive locus, identifying possible functional alleles (e.g., SNPs) within this locus, characterizing the phenotype associated with the candidate adaptive allele(s), and determining the relationship between the phenotype of the candidate allele(s) and reproductive fitness. In most experimental studies, all four steps are not achieved; often, the relationship between the phenotypic expression of the favored allele and reproductive fitness is not clear (Scheinfeldt and Tishkoff, 2013).

Lactase persistence

Infants almost universally produce the enzyme lactase and can digest the lactose in their mother's milk. However, humans and most mammals downregulate the lactase gene shortly after the weaning period. In some humans, lactase production persists and globally about 35% of the human population can digest lactose (Curry, 2013; Leonardi *et al.*, 2012). The emergence of lactase persistence in the adults of certain human populations occurred only after the onset of dairying, and the data indicate that dairying is always adopted before the acquisition of the ability to digest milk; the lactase persistence-associated alleles are positively selected in dairying populations (Itan *et al.*, 2009; Leonardi *et al.*, 2012).

Milk became available for human consumption only after animal domestication (cattle, sheep, and goats) started in the Fertile Crescent and Turkey about 11,000 years ago. Milk production predated by several thousand years the evolution of the allele(s) associated with lactase persistence, so ancient dairy farmers converted milk into cheese and yogurt and in this way reduced the lactose concentration of milk products. The practice of dairying expanded in concert with the Neolithic culture spreading to Greece and the Balkan region about 8,400 years ago. About 7,500–6,500, years ago, lactase persistence, the ability of adults to digest milk, emerged in Central Europe, probably Hungary (Curry, 2013). Today, lactase persistence is prevalent in Northern Europe, West Africa, the Middle East, and South Asia (Curry, 2013). In Southern Europe, lactase persistence is relatively uncommon — less than 40% in Greece and Turkey but more than 90% in Britain and Scandinavia. Lactase persistence is considered the best known example of gene–culture co-evolution.

The protein lactase is encoded by a single gene, *LCT*, located on chromosome 2. Upstream of the *LCT* transcription start site, a SNP (13,910 C/T) was detected in a *cis*-acting regulatory element located in an intron of a neighboring gene (*MCM6*). The T variant (13,910 T) of this SNP was found to be a much stronger enhancer of *LCT* transcriptional activity than the C variant (13,910 C) (Leonardi *et al.*, 2012). The 13,910 T allele has not been detected in early Neolithic Europeans (Leonardi *et al.*, 2012). Itan *et al.* (2009) proposed the following sequence of events. The co-evolution of lactase persistence and dairy farming originated in a region of Central Europe and the Northern Balkans about 6,200–8,600 years ago. After the arrival of the Neolithic culture in Southeastern Europe and the increasing importance of cattle herding and dairying, natural selection started to act on a few lactase persistence individuals. After an initial slow increase in the frequency of lactase persistence in these populations, around 7,500 years ago, the prevalence of lactase persistence rose more rapidly in a gene–culture co-evolutionary process coupled with a population expansion. This dramatic rise in the lactase persistence frequency over a relatively short time interval cannot be explained by genetic drift alone and indicated that lactase persistence was under very strong positive selection. However, as discussed by Leonardi *et al.* (2012), although it is well accepted that lactase persistence was driven by strong positive selection, it is not clear what was the actual "fitness" advantage underlying the acquisition of lactase persistence. Bersaglieri *et al.* (2004) suggested that lactase persistence provided a selective advantage of added nutrition by allowing humans to consume fresh milk, while Curry (2013) estimated that people possessing the 13,910 T allele would have produced about 19% more fertile offspring than those who lacked this variant.

Although the 13,910 T polymorphism appears to account for lactase persistence in Europe, it cannot account for lactase persistence in people residing in the Near East, Middle East, and in Africa. This observation means that other alleles must exist to explain lactase persistence in these areas and in fact, other alleles have been described (Curry, 2013; Leonardi *et al.*, 2012).

Bersaglieri *et al.* (2004) reported genomic evidence of strong positive selection for lactase persistence. Two polymorphisms upstream from the *LCT* gene and tightly associated with lactase persistence (13,910 T and 22,018 A) were assayed. The genomic signatures of positive selection used by these investigators included differences in allele frequencies and measures of haplotype homozygosity. DNA samples from European-Americans, African-Americans, and East Asian populations were used. The frequency of these two alleles varied considerably across the different populations: 77% in European-Americans, 13% in African-Americans, and 0% in East Asians. The recent origin of these alleles is supported by the observation that the haplotype containing these alleles had very extended homozygosity. This long stretch of homozygosity associated with this haplotype was not due to a low recombination rate since other haplotypes in this genomic region showed shorter extents of homozygosity. Based on the genomic data, the investigators estimated that these lactase persistence alleles began to increase in frequency sometime between 2,200 and 20,600 years ago. The additional nutrition provided by dairy products presumably conveyed a fitness advantage and hence supplied the stimulus for the strong positive selection for the alleles associated with lactase persistence into adulthood in populations that practiced dairy farming.

Lactase persistence represents one of the best-known examples of gene-culture co-evolution. This example underscores the importance of culture, in this case the practice of dairy farming, in the evolution of humans. Culture operates at the level of the individual as well as at the level of the larger social environment. At the individual level, culture includes socially disseminated information that can direct individual behaviors, while at the level of large social groupings, culture involves the transmission of certain practices, customs, and skills such as dairy farming. Dairy farming allowed humans to exploit a food source and this cultural activity was the driver for the rapid increase in the prevalence of genetic variants (alleles) that enabled humans to digest lactose into adulthood. Natural selection acts on phenotypes or traits. If a trait confers a fitness advantage and that trait is heritable, then natural selection can act on that trait and the trait can be passed down to descendant generations. Culture works at the intersection between a trait and its association with fitness. Here, the cultural practice of dairy farming allowed humans to have a ready and reliable source of milk products. However, most humans were lactose intolerant and had to use milk products low in lactose, i.e., cheese and yogurt. For some reasons, individuals with lactase persistence and the ability to consume fresh milk had a fitness advantage over individuals without lactase persistence. The association of dairy farming and hence a supply of fresh milk coupled with the presence of alleles conferring lactase persistence in a subset of the population set up a gene–culture co-evolutionary process. This process resulted in a rapid increase in the frequency of alleles linked to lactase persistence in this population. Lactase persistence in the setting of dairy farming gave humans the capacity to consume fresh milk. Why the ability to consume fresh milk should confer such a strong selective advantage for that population remains unclear (Itan *et al.*, 2009). This example of gene–culture co-evolution demonstrates the power of a cultural practice (dairy farming) to drive the positive selection of a trait (lactase persistence) given that the trait is heritable (alleles 13,910 T, 22,018 A, plus probably other as yet unknown alleles).

Brain development

There is evidence that genomic loci related to brain development have shown recent evolutionary change. Lahn and his colleagues (Evans *et al.*, 2005; Mekel-Bobrov *et al.*, 2005) have reported genomic evidence of recent strong positive selection for two genes: microcephalin (*MCPH1*) and *ASPM*. These two genes represent two of six known loci which regulate brain development and size. Mutations in these loci result in a condition of primary microcephaly in which the brain is severely reduced in size with otherwise normal neuroarchitecture.

Microcephalin is believed to control the proliferation and differentiation of neuroblasts. This gene has 14 exons and in exon 8 at position 37995, a SNP was detected in which a G was replaced by a C. Evans *et al.* (2005) sequenced the core region containing this SNP in 89 individuals from a variety of global populations. Eighty-six haplotypes were identified and using the 37995 G>C SNP as a diagnostic site, the haplotypes were divided into two groups: those that contained the C allele and those that contained the ancestral G allele. The first group was designated as haplogroup D (derived allele). Haplogroup D comprised 43 of the 86 haplotypes and one of these haplotypes, number 49, had a frequency of 33% — 59 out of 178 chromosomes. The remaining 85 haplotypes had frequencies of 0.6–6.2%. Positive selection of an allele is characterized by a significant increase in the frequency of the allele as well as extended LD or extended haplotype homozygosity around the favored allele. Haplogroup D chromosomes displayed extensive LD across the entire core region of the *MCPH1* gene. The data were consistent with a selective sweep involving the *MCPH1* allele containing a C nucleotide at position 37995 (haplotype 49) over a short time period during which this haplotype increased significantly in frequency. During the sweep, minor variants of haplotype 49 also emerged at a much lower frequency. Together, haplotype 49 and its minor variants comprise haplogroup D. Evans *et al.* (2005) estimated the age of haplogroup D as about 37,000 years (14,000–60,000).

In a companion study, Mekel-Bobrov *et al.* (2005) presented genomic evidence of a recent selective sweep for the gene *ASPM*, another gene involved in the proliferation and differentiation of neural stem cells. These investigators sequenced the entire genomic region of *ASPM* in samples of 90 individuals from diverse backgrounds. One hundred and six haplotypes were identified and one of these haplotypes, number 63, had an unusually high frequency compared to the other haplotypes. Haplotype 63 contained two SNPs in exon 18: one substitution involved an A>G at position 44871 and the second substitution involved a C>A at position 45162. The high frequency of haplotype 63 as well as its greater prevalence in Europeans and Middle Easteners compared to other populations were very suggestive of this haplotype having undergone a selective sweep. The investigators grouped together all the haplotypes containing the G allele at position 44871 and referred to this grouping as haplogroup D with the predominant haplotype in this group being number 63. Haplogroup non-D contained the A allele. Haplogroup D chromosomes showed extended LD across the entire genomic region of *ASPM*, while haplogroup non-D did not show any unusual LD across the core region of *ASPM*. These observations were consistent with a strong selective sweep involving one of the alleles of the *ASPM* gene — the variant containing the G nucleotide at position 44871. The age of haplogroup D was estimated at 5,800 years (500–14,100).

Interpreting the observations reported in these two papers is not straightforward given the lack of any specific phenotypes linked to the favored alleles of the *MCPH1* and *ASPM* genes involved in the selective sweeps (Balter, 2005). This and other concerns regarding these two studies were brought up by Timpson *et al.* (2007) and Yu *et al.* (2007). These concerns were subsequently addressed by Mekel-Bobrov and Lahn (2007). As these investigators point out, studies of genomic positive selection are rarely based on phenotypic data. Although a selective event is driven by a phenotypic trait, the phenotypic traits driving the positive selection of the genes *MCPH1* and *ASPM* are clearly unknown. However, it is reasonable to suggest that expansion of the human brain was necessary to make it possible for humans to cope with the demands of complex social living arrangements. This proposal is at the core of the "social brain hypothesis" (Singer, 2013). In this context, the evolving complex cultural practices of humans, which would have conferred survival and reproductive benefits, required a larger brain capable of more sophisticated cognitive functions. Under this construct, the complex social living arrangements of ancestral humans with its attendant fitness benefits coupled with the presence of alleles favoring the expansion of brain size set in place a gene–culture co-evolutionary process that resulted in a strong selective sweep involving alleles of the genes *MCPH1* and *ASPM*. This is certainly a reasonable proposal, but still speculative. In addition, the expansion of brain size with a capacity for more complex cognitive functions would have required the involvement of many genomic loci, both coding and regulatory. The genes *MCPH1* and *ASPM* could be considered just markers of these many other genetic elements.

The "social brain hypothesis" posits that the capacity for sophisticated cognition made possible complex social interactions between individuals, and hence, the adaptive advantages of this capacity drove evolution of the modern human brain. However, evolution of the modern human brain has also been linked to the development of bipedalism (Falk *et al.*, 2012). Falk surveyed ontogenic closure times of the frontal bone metopic suture in non-human primates and in early hominins. The Taung specimen represents an endocast of a hominin who lived about 2.5 million years ago. This individual was about 3.8 years at the time of death. Taung has been classified as belonging to the group *Australopithecus africanus* and had a cranial capacity similar to that of other fossil finds of *Australopithecus africanus* (Falk, 2004). In non-human primates, the metopic suture closes shortly after birth, whereas in modern humans, this suture closes later after eruption of the first deciduous molar. The Taung specimen showed a partially fused suture. Based on a comparative analysis of variation in time at which the metopic suture fuses in chimpanzees, humans, and fossil hominins, Falk *et al.* (2012) made the following proposals concerning relationships between the evolution of bipedalism and the evolution of brain size and organization.

The evolution of bipedalism led to widespread skeletal changes including changes in the capacity of the birth canal which put constraints on the size and shape of the mature fetus in terms of successful delivery. Patency of the metopic suture and anterior fontanelle facilitated delivery by allowing molding of the fetal cranium during the birthing process. Hence, delayed fusion of this suture until a later postnatal time would have represented an adaptive advantage in terms of favoring successful births. Humans, unlike other primates, show high brain growth rates throughout the early postnatal period which is made possible by delayed fusion of the metopic suture. The Taung specimen had only a partially fused suture indicating that rapid postnatal brain growth could have evolved as early as in *Australopithecus africanus*. In addition, Falk *et al.* (2012) hypothesized that delayed fusion of the metopic suture was associated not only with extensive postnatal brain growth, but also with anatomical reorganization particularly in the frontal neocortex of early hominins. At least 10 key genes mediate neurocranial suture fusion and one of these genes *RUNX2* affects metopic suture fusion. There is genomic evidence for positive selection of the *RUNX2* gene (Falk *et al.*, 2012).

In summary, the data reviewed by Falk and co-workers indicate that bipedalism, morphological re-configuration of the birth canal, and brain size and organization are all intertwined.

There have been a number of studies designed to identify specific genetic changes that have evolved in humans which made possible our cognitive capacities and language functions (Lieberman, 2013; Somel *et al.*, 2013a). These studies generally involve a comparison of the human and non-human primate brain. Although many genetic differences between the human brain and other primate brains have been identified, relating these to differences in cognitive functions has not been straightforward. One gene that has garnered a lot of attention is *FOXP2* which encodes a transcription factor forkhead- box protein P2 (Lieberman, 2013). This gene is involved in vocalization and language acquisition, and recently, it was found that this transcription factor regulates a gene which encodes the protein sushi repeat-containing protein X-linked 2 (*SRPX2*). *SRPX2* promotes mammalian vocalization by controlling the formation of synapses in the mammalian cerebral cortex (Lieberman, 2013). The human protein product of *FOXP2* differs from that of other primates by two amino acid substitutions which predate the modern human–Neanderthal divergence (Somel *et al.*, 2013a). However, it appears that positive selection for the gene *FOXP2* was directed not at these amino acid substitutions but rather at regulatory elements in the proximity of the gene (Somel *et al.*, 2013a). Hence, even though the gene *FOXP2* differs between humans and other primates, the phenotypic expression of this difference has not been defined.

The other important question related to the evolution of the human brain is the nature of the driving force. Did the evolution of human-specific culture drive brain evolution or did evolution of the human brain predate and make possible the evolution of human-specific culture? This is an important question for which at present there is no agreed-upon answer.

Skin pigmentation

The human melanocortin-1 receptor (MC1R) is found in skin and hair follicles and when bound by alpha-melanocyte- stimulating hormone, a signaling pathway is activated that stimulates synthesis of brown and black melanins. In the absence of MC1R signaling, the synthesis of yellow and red melanin is the default pathway. Hence, MC1R can be considered a regulator of the amount and type of pigment produced (Martinez-Cadenas *et al.*, 2013). A number of variants of the gene encoding *MC1R* have been described and Martinez-Cadenas *et al.* (2013) examined the patterns of selection on these variants.

One study population consisted of 1,217 Spanish individuals and genomic sequencing of *MC1R* in this population detected 38 SNPs, one new deletion, and one new insertion. The most frequent variant was the SNP V60L (rs 1805005-T; where T is the derived allele and G is the ancestral allele). The coding region of *MC1R* has a high content of CpG dinucleotides (CpG islands) and has a high mutation rate which is responsible for the nucleotide diversity of this coding region. Most of the variants identified had a very low frequency consistent with the action of purifying selection. The exception was V60L which had a high frequency of about 0.15. The variant V60L is associated with partial loss of *MC1R* function and results in a "red hair, fair skin" phenotype. However, the frequency of the V60L allele was not homogeneously distributed within the Spanish population; Northern Spanish populations had a frequency of about 0.17, whereas Southern Spanish populations had a significantly lower V60L allele frequency of 0.09. The spatial segregation of V60L between northern and southern populations appeared to reflect the geographical UV-B irradiation pattern; highest UV-B irradiation levels were found in the southern-most parts of Spain with progressively lower levels of irradiation occurring as one traveled to northern parts of Spain.

The high frequency of V60L and its geographical segregation suggested the effects of positive selection. To examine for positive selection acting on the V60L allele, Martinez-Cadenas and co-workers measured the extent of haplotype homozygosity on each side of the V60L SNP. The results were consistent with the signature of positive selection. Interestingly, sequencing the *MC1R* genomic region in individuals with malignant melanoma also showed that the V60L variant was the most frequent allele in these subjects compared to controls.

In summary, human *MC1R* showed the effects of two selective forces. Most of the variants were acted upon by negative (purifying) selection and were maintained as the ancestral alleles. The exception was the allele V60L. This derived allele had a much higher frequency than other alleles and had been subject to positive selection in humans residing in geographical areas that were characterized by low levels of UV-B irradiation (scarce sunlight). In these areas, individuals harboring the V60L allele had reduced skin pigmentation, a phenotype which conferred upon them a fitness advantage. The nature of this fitness advantage is best understood within the context of the evolution of skin pigmentation in humans.

The most important variable explaining the geographic distribution of levels of skin pigmentation in humans is the pattern of UVR exposure (Parra, 2007). Melanin is a mixture of biopolymers synthesized in melanocytes and functions as a natural sunblock. It is effective in protecting the skin from the harmful effects of UVR which include damage to DNA and destruction of folate. These harmful effects of UVR have been implicated in the genesis of skin cancer and as already noted the V60L variant of *MC1R* is considered a risk allele for melanoma. The main source of vitamin D for humans is cutaneous synthesis mediated by exposure to sunlight. The geographical distribution of pigmentation patterns in humans is believed to be the result of a balance between natural selection favoring darker pigmentation as protection against sunburn and folate destruction in regions where UVR exposure is high, and selection for lighter pigmentation in regions far from the equator in order to facilitate cutaneous vitamin D synthesis (Parra, 2007). This hypothesis would explain the fitness advantage of darker pigmentation in areas of high ultraviolet exposure and lighter pigmentation in areas of weaker ultraviolet exposure, and hence, would account for the positive selection of the V60L allele of *MC1R* in northern areas of Spain in which sun exposure is low. This allele is one of the genomic mechanisms resulting in lighter skin pigmentation. An interesting byproduct of positive selection on the V60L allele is that this variant also turns out to be a melanoma-risk allele. Clearly, the positive fitness advantages of this allele in terms of conferring reduced skin pigmentation in individuals harboring this allele must outweigh the negative fitness consequences of its cancer-promoting properties.

Greaves (2014) has proposed a different selective pressure as responsible for variations in human skin pigmentation. It is generally agreed that variations in skin pigmentation are related to variations in exposure to solar UVR. UVR is mutagenic to skin and is considered the predominant causal factor in the genesis of skin cancer. The first ancient African humans to lose body hair, a feature of other primates, probably had white or at least pale skin. A selective sweep involving a variant of the gene *MC1R* which favored dark skin occurred in these early human ancestors. What was the environmental pressure that brought about selection for dark skin? Greaves pointed out that albinism in black ethic groups in Sub-Saharan Africa is quite common and that the presence of white skin in these individuals chronically exposed to high levels of UVR is a significant risk factor for the development of skin cancer. If untreated, albinos with skin cancer generally die during their reproductive years. Greaves proposed that early humans in Sub-Saharan Africa being pale-skinned would have similarly suffered a high incidence of lethal skin cancers during their reproductively active years. The detrimental impact of pale skin on reproductive fitness in this environment characterized by high UVR levels would have served as the driver for the selective sweep of a *MC1R* gene variant promoting eumelanin synthesis and dark skin.

Regardless of which specific selective pressures were actually operative, variations in human skin pigmentation represent an example of gene–culture co-evolution. The positive selection of alleles favoring different levels of skin pigmentation was driven by variations in the levels of chronic UVR exposure experienced by different human populations. As humans migrated out of Africa and into more northern climates, alleles favoring changes in skin pigmentation appropriate for the new level of UVR exposure would have been positively selected. The selection of these alleles would confer a fitness advantage to humans in these new environmental settings compared to the ancestral alleles derived from their previous environment.

As noted by Tang *et al.* (2007), other groups of genes have been shown to display strong signatures of positive selection. These groups include drug-metabolizing genes *CYP3A4* and *CYP3A5*, other skin pigmentation genes *MYO5A* and *SLC24A5*, immune system transcription factor interleukin 4 (*IRF4*), and sepsis-resistant gene caspase 12 (*CASP12*).

Many animals share information, but human culture is unique. Although there are many definitions of culture, it can be defined as the sum total of information, practices, and behaviors acquired by individuals from others through a variety of social learning processes including teaching and imitation. The transmission fidelity of culture across generations is sufficiently high for this process to be considered heritable. Genes and culture can be viewed as two interacting inheritance systems with very different response times; cultural evolution occurs much more rapidly than "genetic" evolution. The linkage between these two systems constitutes the basis of gene–culture co-evolution (Richerson *et al.*, 2010). Lactase persistence is probably the best-understood and -characterized example of gene–culture co-evolution. In this example, the practice of dairy farming by a population in which some individuals had a genetic variant for lactase persistence, a variant that conferred a fitness advantage within the setting of that cultural practice, resulted in strong positive selection for that genetic variant. The details of gene–culture co-evolution have not been well defined. Simplistically, two scenarios can be considered. Does a cultural practice create a setting in which a genomic variant becomes adaptive and then subsequently spreads throughout that population, or does the existence of a genomic variant with certain pre-adaptive characteristics in a population precede and make possible the evolution of culture within that population? In other words, is culture the leading or the lagging variable in the gene–culture co-evolutionary process? In the case of lactase persistence, cultural evolution appears to have preceded and driven genetic evolution. In the case of the evolution of the human brain, the sequence of events is not as clear. As reviewed in a previous section, several genes involved in brain development and size have shown recent strong positive selection. However, the actual selective pressures driving human brain development are not known. It is not clear whether the evolution of a large brain capable of complex cognitive functioning preceded and made possible the evolution of culture or whether the evolution of culture was the primary driver of brain development and size.

This "chicken-and-egg" argument with respect to genes and culture is probably not an appropriate way to address this issue. As will be discussed in **Chapter 5**, culture and genes actually form a "loop-like" interaction. Cultural practices can modify the local environment and thereby modify selection pressures acting within that environment. Modification of selection pressures will alter the action of natural selection and change the course of human evolution within that local environment. In essence, culture becomes a driver of evolution and evolution in turn is a driver of cultural practices. Therefore, in the process of gene–culture co-evolution, culture is both the leading and lagging variable.

Species other than humans adapt to new or changing environments through the process of biological evolution. However, humans can adapt to new or changing environments much more rapidly through the process of cultural evolution, a process unique to our species. Cultural evolution and genetic evolution have become interwoven in a gene–culture co-evolutionary process which is considered the dominant mode of human evolution (Richerson *et al.*, 2010). It is also clear that the complexity of human culture requires considerable human cognitive sophistication and by inference a brain with a large capacity (Richerson *et al.*, 2010).

Darwinian fitness involves both survival and reproductive success. Clearly, these two components are linked since an animal must survive to reproduce. Modern human cultures have generally eliminated the survival component so that fitness is now measured in terms of reproductive success only (Stearns *et al.*, 2010a). Culture can act on selection patterns by shaping the behaviors of individual humans. Behaviors such as food addiction, consumption of tobacco or alcohol, or a low level of physical activity can have negative effects on reproductive fitness. The negative effects of tobacco consumption on reproductive fitness can be transmitted across at least one generation since maternal smoking can epigenetically change placental gene expression and thereby modify the fetal intrauterine environment (Singer, 2013). This mechanism probably accounts for the low birth weight of infants born to mothers who smoke and the subsequent reproductive fitness of those infants will probably be reduced. This example involving maternal smoking could be included under the rubric of gene–culture co-evolution.

Genomic loci

In addition to specific candidate genes, a number of genomic regions have undergone recent positive selection consistent with ongoing human evolution. Voight *et al.* (2006) analyzed genome-wide SNP data from individuals of diverse ancestry: East Asian (Japanese and Chinese), Northern and Western European, and Nigerian. The signature of strong positive selection for a given allele used by these investigators was an allele with a relatively high frequency located within an unusually long haplotype of low diversity. Alleles not subject to positive selection would have frequencies and surrounding haplotype diversity more characteristic of the genome as a whole. All SNPs with a minor allele frequency greater than 5% were analyzed for the extent of haplotype homozygosity. The strength of evidence for positive selection was assessed for each SNP. Alleles showing a signature of positive selection tended to segregate into clusters. Fourteen genomic regions containing clusters of alleles showing evidence of recent positive selection were identified. The ages at which these regions experienced positive selection were between 6,600 and 10,800 years. The timing of these occurrences of positive selection fell within the agricultural phase of cultural evolution. Voight *et al.* (2006) assessed the types of genes located within these regions. Examples of gene categories, based on biological functions, included in these regions were: chemosensory perception and olfaction, gametogenesis, spermatogenesis and fertilization, metabolism of lipids and carbohydrates, skin pigmentation, skeletal development, and four genes in the CYP450 cluster involved in drug metabolism. Several of these gene categories in particular such as those involved in skin pigmentation, carbohydrate and lipid metabolism, and drug detoxification would appear to reflect adaptations to recent stages of cultural evolution.

As an aside, it is interesting that there have been different interpretations reported with respect to the pattern of selection acting on human olfactory receptor genes. Gilad *et al.* (2003) reported that in the human, more than 60% of olfactory receptor genes bear one or more DNA sequence variants likely resulting in a loss of function (pseudogenes). Humans do not rely on their sense of smell compared to other primates, and Gilad *et al.* (2003) interpreted this extensive array of deleterious mutations in human olfactory receptor genes as a manifestation of relaxed negative selection. In a previous paper, Gilad *et al.* (2000) had suggested that the diversity of sequence variants in the human repertoire of olfactory receptor genes was consistent with the action of weak positive selection. Clark *et al.* (2003) compared human and chimpanzee lineages and concluded that in the human lineage, olfactory receptor genes were under positive selection. These investigators implied that the olfactory receptor genes they studied were intact genes and not pseudogenes. Voight *et al.* (2006) in the report discussed in the preceding paragraph, interpreted their data as similar to the results of Clark *et al.* (2003) and consistent with a pattern of positive selection associated with human olfactory receptor genes. Finally, Somel *et al.* (2013b) considered the high rate of pseudogene formation in the human olfactory receptor gene family as consistent with relaxed negative selection. Hence, there does not appear to be a consensus concerning the nature of the selection pattern that has been acting on human olfactory receptor genes. The two prevailing views are that these genes have experienced either positive selection or relaxed negative selection. The reasons underlying these alternative interpretations are not clear but may reflect whether the genes studied by the different investigators were either the set of functionally intact olfactory receptor genes or the many functionally inactivated genes (pseudogenes).

Several other sets of investigators have reported evidence for sites of recent positive selection in the human genome (Hawks *et al.*, 2007; Sabeti *et al.*, 2007).

Sabeti *et al.* (2007) used data published as part of the International Haplotype Map Project to assess patterns of selection. This data contained sequencing information for over three million SNPs genotyped in three populations: European, African, and Asian. The criteria used to identify recent positive selection included the identification of high-frequency alleles carried on unusually long haplotypes. Twenty-two candidate genomic regions containing clusters of many SNPs with a long haplotype signal were identified. By applying additional criteria with respect to the strength of positive selection to all of the SNPs in each candidate region, the investigators were able to reduce this number significantly. For example, there were 9,166 SNPs in total in all the 22 candidate regions that displayed a long haplotype signal. The application of these additional criteria reduced this number to 41 SNPs in all 22 regions combined. Eight of these 41 SNPs encoded non-synonymous substitutions in coding sites, while the other 33 SNPs were in non-coding sites such as transcription factor motifs, introns, and untranslated regions. This methodology allowed the investigators to make predictions as to which genetic alleles were likely targets of positive selection. Some of the candidate genes having alleles identified by these investigators as having experienced strong positive selection were: *LARGE* which encodes a glycosylase which modifies alpha-dystroglycan, the cellular receptor of arenaviruses such as the one responsible for Lassa fever; *DMD* which encodes an adaptor protein that binds to alpha-dystroglycan and is important for its function; and *EDAR* and *EDAR2* genes which are in the pathway involved in the development of hair, teeth, and exocrine glands. *LARGE* and *DMD* have been selected in the Nigerian population where Lassa fever is endemic while *EDAR* and *EDA2R* have been selected in the Asian population. The biology underlying the positive selection of these alleles is currently not understood. For example, *EDAR* and *EDA2R* have been linked to thicker hair in Asian populations.

Hawks *et al.* (2007) examined the ages of selective sweeps in modern human populations. Their data base was the HapMap of 3.9 million SNPs genotyped in Europeans, African, and Asian populations. The detection of positive selection was based on the frequency of an allele and the finding of extended haplotype homozygosity surrounding the given allele. These investigators only examined for partial sweeps since alleles with a frequency greater than 78% were excluded. The methodology used detected clusters of SNPs with the signature of positive selection given the extensive LD surrounding these alleles. Each cluster was considered to represent a single selection event. The rate at which LD decayed along the genome from a high-frequency allele was used to estimate the age at which selection events occurred. For the European population, the modal age estimate was 5,200 years while for the African population the estimated modal age estimate was 8,000 years. These modal ages indicated significant acceleration of positive selection events over the past 80,000 years of human evolution.

Williamson *et al.* (2007) used a somewhat different methodological analysis of genome-wide SNPs to look for recent complete selection sweeps; i.e., the beneficial mutation had attained a frequency of 100% in the tested population, in African-American, European-American, and Asian-American populations. Table 1 of their report listed 101 genomic regions with strong evidence for recent selective sweeps. The table included the chromosomal location of these regions and the most likely candidate gene or genes associated with the sweep positions. One of the regions with the strongest evidence is that associated with the gene *DTNA* in the Asian-American and European-American populations. This gene encodes the dystrobrevin protein, a component of the dystrophin protein complex. This complex functions as an important structural component of muscle tissue. However, which specific allele of this gene and which "muscle" phenotype had been the targets of positive selection were unknown. Several other categories of genes that have shown recent positive selection included genes that encoded keratin associated proteins and genes that encoded heat-shock proteins.

Williamson *et al.* (2007) also made several other interesting observations. First, they noted that since the pattern of selective sweeps varied between subpopulations, adaptations to specific local environmental pressures must have been an important process in recent human evolution. Secondly, they reported evidence for a selective sweep in a genomic region on chromosome 6 in the vicinity of the *HFE* gene in Asian-American and European-American populations.

The protein encoded by *HFE* is involved in regulating the interaction between the transferrin receptor and transferrin (Rochette *et al.*, 1999). This interaction helps to determine gastrointestinal absorption of iron. *HFE* is characterized by several high-frequency SNPs such as C282Y (G>A substitution at position 845) and H63D (transversion C>G at position 187). Approximately >70% of individuals with hereditary hemochromatosis (an iron-overload disorder) are homozygous for the derived allele; 282Y (Rochette *et al.*, 1999; Yen *et al.*, 2006). The occurrence of the C282Y polymorphism is very high in Northern Europeans with a frequency in Ireland, for example, of 12.3% for the derived allele (Beutler, 2004). In a sample of almost 100,000 North Americans, the frequency of the derived allele 282Y was 10% for Caucasians but only 0.12% for those of Asian ancestry (Adams *et al.*, 2005). Williamson *et al.* (2007) determined that the genomic position of the selective sweep they identified was probably at a cluster of histone genes about 150 kb away from *HFE*. These investigators raised the possibility that the high frequency of the *HFE* C282Y polymorphism in certain populations was not the result of this allele being the direct target of positive selection, but rather this derived allele had attained a high frequency through an association with a nearby beneficial allele (genetic hitchhiking). A similar suggestion was made by Toomajian *et al.* (2003). As an aside, the observation by Williamson and co-workers is somewhat perplexing. They identified a selective sweep in the vicinity of the *HFE* gene and implied that the European-American and Chinese-American populations in their study had high frequencies of the 282Y allele. However, it has been documented that in Asian populations the 282Y allele is very uncommon.

Rochette *et al.* (1999) examined the frequency of *HFE* polymorphisms in the populations of Myanmar and Sri Lanka — geographical areas with a heterogeneous population distribution. In the populations of Myanmar and Sri Lanka, the frequency of the ancestral allele C282 was 100% and 99.2%, respectively. The derived allele 282Y had a frequency of 0% in Myanmar and 0.8% in Sri Lanka. For unaffected European French, the allele frequencies were 98% C282 and 2% 282Y. The frequencies for the H63D allele were different. For Myanmar and Sri Lankan populations, the ancestral allele H63 had frequencies of 97% and 89.2%, respectively. The derived allele 63D had frequencies of 3% and 10.8% in these two populations. In unaffected European French, the ancestral (H63) and derived (63D) alleles had frequencies 79.5% and 20.5%, respectively. In summary, this study and published data indicated that the derived C282Y allele, 282Y, is limited to Northwestern European ancestry, whereas the derived H63D allele, 63D, is found at frequencies >5% in Europe, countries bordering the Mediterranean, in the Middle East, and in the Indian subcontinent. The investigators examined the haplotype pattern of chromosome 6 in the vicinity of the *HFE* gene in both Myanmar and Sri Lankan populations and compared these patterns with those observed in Europeans. The data indicated that in Sri Lanka both the C282Y and H63D polymorphisms arose independently from the same polymorphisms that occurred in Europe.

The various studies designed to define the pattern of selection acting on the *HFE* gene illustrate the difficulties associated with interpretation of the results. For example, there does not appear to be a consensus as to whether the C282Y and H63D polymorphisms themselves have been the target of positive selection in certain populations or whether these variants have been carried along by their tight linkage to a nearby positively selected allele (Beutler, 2004; Toomajian *et al.*, 2003; Williamson *et al.*, 2007;). The *HFE* gene is located on the short arm of chromosome 6 embedded among the HLA class I set of genes (Toomajian *et al.*, 2003). The *HFE* polymorphisms could very well be hitchhikers in a selective sweep involving these *HLA* genes.

If the *HFE* polymorphisms were the target of a selective sweep, then what was the phenotype driving this sweep? This question has been considered by Beutler (2004). Individuals heterozygous for the C282Y allele show an increase in gastrointestinal iron absorption although the magnitude of this increase is very small. What might be the adaptive advantage of the heterozygous state for the C282Y polymorphism? The protein encoded by the wild-type *HFE* gene facilitates the interaction between transferrin iron and the transferrin membrane receptor. The 282Y allele of *HFE* encodes a protein in which a cysteine has been replaced by a tyrosine. This mutant protein is unable to interact with the transferrin receptor (Yen *et al.*, 2006). Both Beutler (2004) and Rochette *et al.* (1999) have suggested that the presence of this mutant protein might protect against various infectious microorganisms although Beutler and Rochette *et al.* suggested different mechanisms to explain such a possible effect.

Finally, paleogenomics, the study of ancient DNA, has been used to detect recent selective sweeps (Shapiro and Hofreiter, 2014). Meyer *et al.* (2012) reported the genome sequence of an ancient Denisovan individual who lived approximately 74,000 to 82,000 years ago. A comparison of the Denisovan genome to that of the modern human showed that since the divergence of *Denisova* and the modern human lineage, many of the genomic loci that were subject to positive selection in modern humans were involved with brain function or nervous system development.

Many researchers believe that positive selection has been more important in human evolution than the random accumulation of mutations (Gibbons, 2010) and gene–culture co-evolution is probably the dominant process driving recent human evolution. It is also becoming clear that selective sweeps involving a single mutation are probably rare and that positive selection acts more commonly on traits that are polygenic-based or acts from standing variation (Gibbons, 2010).

In summary, there is considerable genomic evidence for positive selection within recent human evolution. However, on the whole, the target gene (or more specifically the target allele) of the selective sweep is not known and in addition, in most cases, the adaptive phenotypic trait driving the selective sweep has not been identified. Many of these selective sweeps are probably part of the spectrum of gene–culture co-evolution, but this is speculative on my part. In addition, as noted in the previous paragraph, selective sweeps probably act on clusters of genes that contribute to a given trait.

Contemporary Evolution

Stearns and colleagues (Byars *et al.*, 2010; Stearns *et al.*, 2010a) examined for natural selection in contemporary human populations. Their approach is based on identifying candidate phenotypic traits and physiologic parameters that natural selection could act upon to increase fitness. These investigators pointed out that there are a number of multigenerational data bases containing clinical, demographic, and epidemiological information (such as the Framingham Heart Study) that can be used to measure selection on the basis of phenotypic traits. In modern societies, most individuals live from birth through the reproductive period, hence survival is not a good metric of fitness. A better measure is completed family size — lifetime reproductive success. Phenotypic approaches to studying natural selection require three components; differences in fitness associated with different traits, heritability of the trait under consideration, and within-population variation in the trait. Stearns *et al.* (2010) summarized the results of 14 studies that were consistent with selection in contemporary human populations. Some of the findings from these studies were as follows. Men and women were under selection for earlier age at birth of their first child and women were under selection for later age at last birth. These trends indicated a broadening of the reproductive time window. Women were under selection for an increase in weight but a decrease in height, while men were under selection for an increase in height. These studies used recent databases so the selection trends identified by these studies were quite early in their course.

Byars *et al.* (2010) used data from the Framingham Heart Study to measure natural selection in a contemporary population of women. The Framingham Heart Study was established in 1948 in Framingham, Massachusetts by the National Heart, Lung, and Blood Institute and Boston University and was designed to identify risk factors for cardiovascular disorders. It is a long-running multigenerational study. Relative lifetime reproductive success (LRS) was used as a measure of fitness. This parameter was calculated by dividing each woman's number of births by the mean for her birth cohort. Interestingly, mean LRS for different birth cohorts has actually declined after about 1930, indicating reduced fitness in more recent female populations. The phenotypical and physiological traits extracted from the Framingham Heart Study database included total cholesterol, height, weight, arterial blood pressure, blood glucose concentration, age at first birth, and age at menopause. The investigators calculated selection gradients for each trait which measured the extent and direction of selection, i.e., the relationship between an increase or decrease in the level of the trait and the resulting change in level of fitness (relative LRS). Data analysis suggested that natural selection in women was acting to decrease height, systolic blood pressure, total cholesterol and age at first birth, and to increase weight and age of menopause. The driving force behind these directional changes was fertility (LRS); these directional changes led to an increase in LRS, a measure of Darwinian fitness. Blood glucose concentration did not appear to have an association with LRS. Byars *et al.* (2010) projected that over the next 10 generations, women in this population would show on average a 3.6% decline in total cholesterol, a 1.3% decline in height, a 1.9% decline in systolic blood pressure, a 1.7% decline in age at first birth, a 1.4% increase in weight, and a 1.6% increase in age of menopause. Traits such as age of first birth and age of menopause have obvious links to LRS since these traits define the reproductive time window. Traits such as weight, height, systolic blood pressure, and total cholesterol do not have obvious direct links to LRS and presumably work indirectly through links to other traits that effect LRS.

Interestingly, a recent report by Biro *et al.* (2013) reported observations consistent with the selection gradients calculated by Byars *et al.* (2010). Biro *et al.* (2013) followed a cohort of young girls, enrolled at age 6–8, to assess age of onset of breast development and its relationship to various anthropometric characteristics. A major finding was that in non-Hispanic white girls, the age of onset of breast development was earlier than in two previous studies conducted 10 and 20 years earlier. A higher BMI (body mass index) was the strongest predictor of earlier age of breast development. Hence, the data of Biro and co-workers showed that over the past 10–20 years for non-Hispanic white girls, there has been a decrease in the age of onset of breast development associated with an increase in BMI. The earlier age of breast development would correlate with an earlier age of menarche and an earlier age for the onset of the reproductive period. These temporal trends reported by Biro and co-workers are in the same direction as the temporal selection trends calculated by Byars *et al.* (2010). The observations of Biro *et al.* would suggest that an earlier age of breast development and a higher BMI would increase LRS.

Are humans still evolving? The studies reviewed in this section clearly indicate that the answer is yes. However, there are many unknowns which cloud the interpretation of the experimental observations. With regard to genomic evidence of selective sweeps, as already pointed out, in most cases the target allele of the sweep and the phenotype driving the sweep are generally unknown. Sometimes, it is difficult to distinguish whether a particular allele has been the target of positive selection or a relaxation of negative selection. As discussed with respect to the alleles of the *HFE* gene, the high frequency of specific alleles in certain populations may be due to genetic hitchhiking rather than the result of a selective sweep directly involving these alleles. The "phenotypic trait approach" of Stearns and co-workers starts with known phenotypic and physiologic traits and measures the action of natural selection on these traits. However, this approach does not give information as to what genomic mechanisms might underlie identified directional selection changes, and in some cases, the relationship of the trait to fitness is not clear. The effect of culture on human evolution is very profound and a factor unique to our species. The evolutionary changes projected by Byars *et al.* (2010) are predicated on the assumption that conditions that were operative over the past 100–150 years will continue to act over the next 100 years and more.

One question posed by studies of human evolution is what makes humans human. One method used to answer this question is to examine the evolved biological differences between the human lineage and our closest relative the chimpanzee. One would think that these biological differences would be distinct given the obvious differences in anatomy and behaviors between the human and chimpanzee. Such, however, is not the case. As discussed in the **Introduction**, evolution is a tinkerer. Evolution is a conservative process that builds upon existing successful biological designs rather than creating new ones. Varki *et al.* (2008) reviewed the known evolved biological differences between the human and chimpanzee. They looked at differences in protein-coding genes, DNA structural variants, and in noncoding (presumably) regulatory genomic regions. They noted that only a few studies have looked for epigenetic differences. One clear difference between the human and chimpanzee is culture which is a unique human attribute and can directly transfer information and behaviors across generations. Culture allows humans to adapt to rapidly changing environmental conditions — a process unique to humans. Culture can also drive human evolution through the process of gene–culture co-evolution. One quote from the paper by Varki *et al.* (2008) sums up the situation: "With the exception of the *FOXP2* gene, in which mutations cause a defined phenotype in humans, and some human-specific consequences of the *CMAH* gene mutation, much of the discussion about genes involved in human uniqueness has been somewhat speculative." Just as it has been difficult to define the genomic basis of phenotypic differences between the human and chimpanzee, it will be even more difficult to define the relationships between genomic differences and phenotypic variation across humans.

Society, Culture, and Ageing

With the evolution of human social groupings, culture became a more dominant evolutionary force and a modifier of ageing patterns. The origins of human society have been the subject of much debate and recently Shultz *et al.* (2011) presented a new perspective on the possible sequence of events. These investigators applied a phylogenetic approach and mapped the composition of foraging groups for 217 primate species onto a primate consensus tree. The phylogenetic component of their model indicated that the evolution of social organization among closely related primate species was more similar than would be expected by chance. The model suggested by their approach consisted of a series of transitions. The ancestral primate state was one of solitary foraging. About 52 million years ago, loose social aggregates composed of multiple males and multiple females appeared. From these unstable forager groups, a transition to stable social groups based on either kinship or reproductive ties (e.g., pair-living, single-male harem systems) appeared about 16 million years ago. The catalyst for the evolution of primate group living from the ancestral solitary state was probably the shift from a predominantly nocturnal to a predominantly diurnal pattern of foraging. A diurnal pattern of activity would have exposed primates to an increased predator risk which could be counter balanced by the anti-predator benefits of group living. A somewhat "deeper" question is whether these early social groups were composed of biologically related individuals (kin selection) or whether they were formed by the process of group selection (Wilson, 2012). This question is still being debated.

There are several important implications of the model of Shultz and co-workers (Silk, 2011). First, the evolution of primate social living was not a linear progression from solitary to small groups and then to larger groups. Hence, social complexity is not a simple function of group size. Secondly, the shift from unstable social aggregates to more stable pair bonds must have occurred after the divergence of chimpanzees and humans from their common ancestor about six million years ago. Thirdly, the data of Shultz and co-workers are not consistent with the socio-ecological model which posits that the evolution of group living was driven by individual responses to and competition for limited food resources.

Domestication of dogs occurred approximately 15,000–20,000 years ago and agricultural practices emerged and replaced foraging about 10,000–12,000 years ago. The emergence of agriculture resulted in a "rapid" increase in the global population. Just prior to the introduction of agricultural practices, the world's population is estimated to have been about six million, rising to a staggering seven billion in 2011 — a roughly 1,200-fold increase in 11,000 years (Bocquet-Appel, 2011). This shift from foraging to agriculturally based food-producing social groupings is known as the Neolithic Revolution. The rapid rise in population was due to an increase in maternal fertility as a result of the availability of high-caloric agriculturally based foods compared to the low-caloric food available to mobile foraging groups (Bocquet-Appel, 2011). The increased birth rate was closely followed in time by an increased mortality rate producing the historical growth rate of 0.2–0.1% per year characteristic of pre-industrialized farming populations. The factors determining the mortality pattern in these pre-industrialized societies included lack of clean drinking water, absence of sewage disposal systems, and candidate infectious diseases generally associated with diarrhea.

The sedentary farming villages that developed with the emergence of agriculture were capable of food production and hence were not constrained by the limits of nature which was the case for pre-agricultural forager villages. This freedom from nature-set limits allowed for demographic growth and made possible the so-called urban revolution (the origin of cities) which took place around 6,000–7,000 years ago, initially in an area currently occupied by Kuwait, Syria, and Iran (Lawler, 2012). The catalyst for this urban revolution was growing contact among different groups of humans and from this heterogeneous mix of peoples developed our complex modern society (Lawler, 2012).

The trend toward urbanization has become even stronger in our modern technologically based world. There is a strong correlation between urbanization and economic development, and by 2008, about half of the world's population was living in cities (Glaeser, 2011). Although the economic benefits of cities are apparent, the density of urban groupings also has significant downsides. The (negative) costs of city living, particularly in megacities of the developing world, include insufficient infrastructure with respect to clean water and sewage systems, crime and violence, traffic congestion with long commutes, and barriers to connect with other people (Glaeser, 2011). From the perspective of the individual, both the positive benefits and the negative costs associated with the urban environment would be translated into repetitive and cumulative personal experiences that through epigenetic mechanisms could modulate that individual's ageing pattern.

There are data that give glimpses into aspects of ageing and mortality patterns associated with different stages of the evolution of human social groupings. Since the rate of ageing is difficult to assess and is not available for many species including humans, longevity is often used as a proxy for ageing (de Magalhaes *et al.*, 2007).

Caspari and Lee (2004) looked at longevity during different stages of the evolution of the human lineage. These investigators used dentition samples from four hominid groups: later australopithecines, early and middle Pleistocene *Homo* and Neanderthals from Europe and Western Asia, and post-Neanderthal early Upper Paleolithic Europeans. The Pleistocene and Paleolithic Periods cover essentially the same time frames (1.8 million to about 11,500 years ago) but are defined by different criteria. The Pleistocene Period is based on geological phenomena, whereas the Paleolithic Period is based upon human evolution. The Paleolithic Period begins with the first evidence for stone tool use by the human lineage and the end of this period is defined by the emergence of agriculture and the Neolithic Revolution. Ageing of individuals was done on the basis of the fossil dentition pattern. Eruption of the third molar (M3) was taken as occurring at the time of reproductive maturation set as 15 years of age. Older adults were defined as twice the age of reproductive maturation or 30 years of age. Based on whether M3 had erupted and if so the extent of its wear, individual fossil samples were classified as coming from either a young (Y) or old (O) individual. For each of the ancient human groups the ratio of old to young (OY) individuals was calculated based on the dental criteria.

The total number of dentition samples was 768, distributed as follows: Australopithecines, 353; early *Homo*, 208; Neanderthals, 113; and early Upper Paleolithic, 74. The OY ratio for these four groups were: 0.12, 0.25, 0.39, and 2.08, respectively. Although the differences among the ratios for all four groups were significant, the most dramatic change in OY ratio was observed with the post-Neanderthal early Upper Paleolithic Europeans.

Caspari and Lee drew several conclusions from their data. First, the increasing OY ratios across these four groups indicated a trend for increased survival of older adults through the course of human evolution. Secondly, the greatest change in longevity occurred with the early modern humans of the Upper Paleolithic Period. For the first time, this group showed a greater number of older adults than younger adults in the death distribution. The investigators postulated that this dramatic increase in survivorship of older adults was linked to the evolution of social groupings and culture. The survival of older adults made possible the evolution of connections between generations since an adult of 30 years of age or older could theoretically be a grandparent. In other words, the Upper Paleolithic Period may have been the first time that there was a strong likelihood that social groups could contain more than one generation. The presence of grandparents would have made possible the intergenerational transfer of resources from older to younger individuals within the social group. Such an intergenerational transfer would have provided an additional layer of support to the care of and provisioning of offspring and grand offspring.

The observations of Caspari and Lee have implications for the development of culture. Culture, defined as the transmission of practices, behaviors, and information across generations, appears to be unique to the human lineage although it is still an evolved phenomenon. However, one must distinguish between culture and sociality although these terms are often used interchangeably. Many animal species are social but have not developed a culture equivalent to that of humans. Culture was made possible by the evolution of social groupings and the increase in longevity of the human lineage. These two factors led to the emergence of social groups containing more than one generation. Such multigenerational groups would have facilitated and promoted the transmission of practices, behaviors, and information between generations, i.e., the evolution of culture. Cultural evolution, in addition to requiring the establishment of multigenerational social groups, also required that these social groupings be of a reasonable large size and that its members be capable of a set of cognitive skills that included accurate imitation (Richerson, 2013). Humans have the cognitive capacity for high-fidelity imitation which allows them to learn complex skills and ideas from others. (The social brain hypothesis posits that the requirement of sophisticated cognitive functions for cultural development underlies the evolution of the human brain.)

The social networks must be large enough such that individual humans can acquire knowledge from a diverse set of mentors thus promoting the spread of the best ideas and practices throughout the social grouping. Small groups will generally not have individuals with a wide variety of different skill sets and knowledge, and experimental observations indicate that the evolution of complex ideas and practices does not occur when the social network is too small and thereby lacks diverse human resources (Richerson, 2013). In summary, the soil for the evolution of culture was the emergence of multigenerational social groups. However, the development of social groups alone was not sufficient. The evolution of culture also required that social groupings were of sufficient size and that humans possessed a sophisticated repertoire of cognitive capacities that included the ability to accurately imitate and the communication skills to facilitate the transfer of information and practices across generations, i.e., culture represents a non-genetic form of an inheritance system.

Investigators have begun to examine cultural evolution in non-human primates and this information should give insights into the origins of human culture. Chimpanzees live in discrete communities which are characterized by specific behavioral features transmitted from one generation to the next (Haslam, 2014). Langergraber *et al.* (2014) estimated the ages of eight East African chimpanzee communities using an analysis of haplotypes on the Y chromosome. Chimpanzees are male philopatric (males tend to stay in their birth territory) and chimpanzee communities do not intermingle. The Y chromosome is passed from fathers to sons and by using Y chromosome haplotype analysis, Langergraber and co-workers were able to estimate when different communities diverged from a common ancestor. The eight communities that were studied showed ages varying from 125 to 2,625 years. Assuming new cultural variants arise over time through innovation, one would have expected the number of cultural variants expressed in a community to be greater in older communities compared to younger ones. Such was not the case. Three communities of ages 445 to 2,625 years showed a similar number of cultural variants. This observation implies that at least for chimpanzee communities, a certain level of cultural complexity is reached and then remains stable. By contrast, humans exhibit what is referred to as cumulative culture. Dean *et al.* (2014) defined cumulative culture as the modification, over multiple transmission episodes, of cultural traits (behavioral patterns transmitted through social learning) resulting in an increase in the complexity and efficiency of those traits over subsequent generations.

According to Dean and co-workers, although other animals are capable of social learning and non-human primates do exhibit culturally specific behaviors, only humans are capable of cumulative culture, i.e., successive generations building on what went before. Human cumulative culture appears to be rooted in a number of factors. Humans possess a suite of cognitive capacities that include teaching, high fidelity imitation, complex language and communication skills, and prosocial behaviors. In addition, humans form large social groups which facilitate cooperation and transmission of knowledge and practices between individuals. Since other animals display innovation and social learning, studying cultures in other animals, particularly non-human primates, will help define why cumulative culture has only evolved in the human species.

Why did a significant increase in longevity occur in the Upper Paleolithic Period and was culture the driving force for this increased longevity or vice versa? Perhaps it is best to consider the observations of Caspari and Lee as another example of gene–culture co-evolution.

The dramatic increase in longevity in the Late Paleolithic Era suggests that underlying changes in the rate of ageing were taking place. Clearly, assessing the characteristics of ageing in an ancient human population is difficult, but there are observations and studies that give tantalizing insights into the lifespan and ageing characteristics of selected ancient groups.

In 1991, a frozen Tyrolean Iceman estimated to be 5,300 years old was discovered. This ancient human has been subjected to morphological, radiological, and genomic analyses. Computer tomography imaging revealed widespread vascular calcifications in both carotid arteries at the sella turcica, left carotid artery in the neck, distal aorta, and right iliac artery (Murphy Jr *et al.*, 2003). The spine showed evidence of degenerative arthritis in the cervical and lumbar regions and the right hip showed osteoarthritis. It has been estimated that this ancient human was between 40–50 years of age at the time of death which was believed to be due to an arrow wound to the chest (Murphy Jr *et al.*, 2003). In 2012, Keller *et al.* reported the results of whole-genome sequencing using DNA extracted from a left iliac bone biopsy. These investigators identified 2,200,000 SNPs of which 1,700,000 had been reported in the current SNP database. Three of the SNPs in the Iceman's genome have been associated (on the basis of genome wide association studies in modern humans) with an increased risk for atherosclerosis. In addition, the Iceman was probably lactose intolerant since he did not have the allele for lactase persistence. Keller and co-workers postulated that the vascular calcifications in this ancient human had a genetic basis.

In mummified ancient Egyptians, Allam *et al.* (2011) detected evidence of vascular calcifications consistent with atherosclerosis. These investigators used computed tomography (CT) imaging to examine the vascular beds of 52 ancient Egyptian mummies. The mummies lived between approximately 4,000 to 1,650 years ago and since the financial costs of mummification were substantial, the mummies would have had high socioeconomic status. The finding of calcification in an identifiable artery was considered diagnostic of atherosclerosis. Calcification along the expected course of an artery was classified as probable atherosclerosis. Of the 52 mummies, 44 had identifiable vascular or cardiac tissue. The mean age of death of these mummies was 39.3 years and was not different by sex. Twenty or 45% of the 44 mummies had definite or probable atherosclerosis with evidence of calcification in the aorta, peripheral arteries (femoral, popliteal, tibial), carotids, iliacs, and coronary vessels. Mummies with atherosclerosis (probable or definite) were older, mean age of 45.1 years, than the mean age (34.5 years) of mummies with cardiovascular tissue but no atherosclerosis. The prevalence of atherosclerosis did not have a gender predisposition. The prevalence of atherosclerosis was found in mummies from all of the ancient Egyptian eras spanning from about 4,000 to 1,650 years ago. With respect to the lifestyle of these ancient Egyptians, tobacco was not available and without modern transportation, they were probably reasonably active. Egypt had an organized agricultural society and their diet contained meat (beef, goat, sheep, fowl) as well as bread and cake.

Why might ancient humans living between 5,300 and 1,650 years ago in quite different geographical areas have vascular calcifications? Are the SNPs detected in the Iceman's genome, which have been found to be associated with atherosclerosis in recent human genome wide association studies, responsible for the vascular calcifications in this ancient human? I doubt that these three risk-associated alleles in the Iceman's genome are responsible for his vascular calcifications since genome-wide association studies detect only statistical correlations, which are generally weak, not causality between a phenotype and a genomic locus (Chakravarti *et al.*, 2013). Also pre-industrial populations have a low death rate due to degenerative diseases (Gurven and Kaplan, 2007). The finding of vascular calcifications in disparate ancient humans, Egyptians who had lived 4,000 to 1,650 years ago, and the Tyrolean Iceman who lived 5,300 years ago, with different lifestyles and ancestries points to this feature being a variant of the ageing process (Singer, 2013). As discussed by Singer (2013), age-related hemodynamically mediated vascular injury induces a repair process and calcification is a morphological feature of that repair process.

The ability to analyze DNA samples obtained from ancient humans has become much better developed and now an ingenious technique has been used to measure the methylation status of ancient DNA samples (Pennisi, 2014). Recently, Gokhman *et al.* (2014) analyzed the DNA methylation pattern of single Neanderthal and Denisovan genomes. These ancestral humans diverged from the modern day human lineage about 550,000 to 765,000 years ago. About 99% of these two archaic genomes showed no significant methylation differences compared to the present day human. However, a small number of DMRs were identified and these were classified as being Denisovan specific, Neanderthal specific, or present- day human specific. An interesting observation was reported related to so-called disease-related genes. In the human genome, approximately 10.8% of genes are statistically linked to diseases. However, for genes associated with modern human-specific DMRs, the percentage of disease linked genes is about 20%. These DMR-containing genes are enriched for neurologic and psychiatric disorders. I interpret this finding as consistent with the proposal that in the evolution of the present day human, emergence of "new" phenotypic traits was due primarily to alterations in the epigenetic regulation of gene expression and not to variations in protein-coding genes. One could further speculate that these epigenetic changes were due to environmental influences.

Gurven and Kaplan (2007) have reported on the mortality profiles of all extant hunter-gatherers for which sufficient high-quality data are available. They have used this data to provide a lens into the mortality patterns that probably existed in human groups about the time of the Neolithic Revolution. As noted by Gurven and Kaplan, human-specific characteristics such as large brains, relatively long lives, pair bonding, male investment in offspring, long child dependency on parents, and grandparental support of grandchildren appear to have evolved during the pre-agricultural stages of human evolution.

Their database consisted of demographic details collected on 20 populations classified as follows: five were hunter-gatherers (HG), six were forager-horticulturalists (FH), nine were acculturated hunter-gatherers and one was a transitional population (AHG). Acculturated groups have had exposure to modern medicines and other modern amenities. The data on these various populations were collected between 1979 and 2007. A Swedish population between 1751 and 1759 was used as a pre-industrial control group. The percentages of children surviving to age 15 years was: HG, 57%; FH, 64%; and AHG, 67%. For the Swedish group, the value was 60%. For those who reached 15 years of age, 64% of HG, 61% of FH, and 79% of AHG reached the age of 45. For the Swedish "control" group, the value was 71%. Two other parameters are noteworthy: mean life expectancy at birth and mean life expectancy at age 45. The respective values were: HG, 31 years and 20.7 years; FH, 33 years and 19.8 years; and AHG, 37.6 years and 24.6 years. For the Swedish group, the values are 34 years and 20 years.

The data showed a clear mortality pattern for these pre-industrialized populations. Over the first 10 years of life, the mortality rate was high at about 3.5–4% per year. Between ages 10 and 40 years, the mortality rate was lower and slowly increased from about 0.75% per year between ages 10 and 20 years to slightly over 0.9% per year between ages 30 and 40 years. Thereafter, the mortality rate accelerated, going from 1.0% per year between ages 40 and 50 years, 2.2% per year between ages 50 and 60 years, to over 4% per year between ages 60 and 70 years. The pattern was one of a high mortality rate during early life followed by a low mortality rate until age 40 and then a steeply rising mortality rate beyond age 40. The Swedish group showed a similar pattern but wild chimpanzees showed a very different pattern. These primates have a life expectancy at birth of about 13 years. The mortality rate is very high over the first 10 years of age (6% per year) falling to 2.5% per year between ages 10 and 20 years. Between ages 40 and 50 years, the mortality rate is about 5% per year and by 60 years of age, there are essentially no survivors. Captive chimpanzees have a better survival than wild chimpanzees, 20% living to age 45 years, compared to 3% of wild animals, but total lifespan for captive chimpanzees is still about 60 years.

Gurven and Kaplan calculated modal ages of adult death, conditional upon surviving to 15 years of age. Modal age of death refers to the peak in the distribution of deaths. The average modal age of death for the three populations (HG, HF, and AHG) was 72 years with a range of 68–78 years. Wild and captive chimpanzees had modal ages of death of 15 and 42 years, respectively. For the Swedish sample (1751–1759), the modal age of death was 72 years and for the United States population in 2002, it was 85 years.

Causes of death among the study populations were mainly due to illnesses which were subdivided into respiratory, gastrointestinal, fever, and other. These illnesses accounted for 70.1% of deaths for all ages combined. Degenerative diseases accounted for 9.2% of deaths with the percentage being 9.5% in those less than 15 years of age and 28.2% in those 60 years of age and older. The actual degenerative diseases were not well assessed, but in the older ages appeared to be cerebrovascular problems and "old age." Obesity, hypertension, heart attacks, and definite strokes were rare. Violence and accidental deaths were a significant cause of death accounting for 18.8% for all ages combined. When broken into age groups, this category accounted for 30.3% of deaths in the 15–59 year old group, but only 17.2 % of deaths in those over 60 years of age.

Comparisons of mortality rates for several groups were quite revealing. The ratio of mortality rates between wild chimpanzees and HG increased dramatically with age. By age 10 years, wild chimpanzees had a mortality rate about three times that of HG and this value increased to about eightfold by age 45 years. A similar pattern was observed when captive chimpanzees are compared to HG although the results were not as dramatic. At age 45 years, the mortality rate of captive chimpanzees was about four times that of HG. Even though captivity does improve living conditions and leads to reduced mortality during infancy and early childhood, the mortality of captive chimpanzees increased significantly during adult life and exceeded that of HG by many fold.

Gurven and Kaplan also included estimates of life expectancies for Paleolithic populations prior to the evolution of agriculture. The problems with the accuracy of this data were discussed. As already reviewed, the study by Caspari and Lee (2004) suggested that longevity increased significantly in the Upper Paleolithic Period. According to the data reviewed by Gurven and Kaplan, the Paleolithic population had a mean life expectancy at birth of about 20 years and a mean life expectancy at age 45 years of five years.

The findings of Gurven and Kaplan can be summarized as follows. Adult mortality appears to have two stages: a low mortality rate of about 1% per year from about age 10 until age 40 and then an accelerating mortality rate characterized by a doubling in rate approximately every 10 years. About 40% of a birth cohort will survive to age 45 and these individuals will have a life expectancy of another 20 years. Hence, survivorship to grandparental age is quite common among the HG, FH, and AHG populations. The mean modal age of adult death is 72 and for the total sample (HG, FH, and AHG) the causes of death included illnesses 70%, violence and accidents 20%. and degenerative diseases 9%.

Gurven and Kaplan argued that the long post-reproductive lifespan characteristic of many individuals in these pre-industrial societies underscored the benefits of longevity. These benefits derive from the intergenerational transfers of resources and provisions. Older individuals in these societies increase their inclusive fitness by enhancing the fertility of their offspring and the survivorship of their grand offspring through these intergenerational transfers. For these pre-industrial populations, the mean modal age of adult death was 72 and modeling by Gurven and Kaplan suggested that in the case of females this was the age at which the importance of intergenerational transfers, i.e., the need for grandparental investment, was greatly diminished. This was also the age at which the mortality rate for post reproductive females increased steeply.

There has been considerable discussion concerning the contribution of intergenerational transfers to the ageing and mortality pattern in social species such as humans (Lee, 2003). In species in which females have a significant post-reproductive lifespan, these females can contribute to the survival of offspring through transfers of resources and provisions, and direct parental and grandparental care. In such species, natural selection against mortality in post-reproductive females would remain in force as long as these individuals can continue to make significant intergenerational transfers to offspring. With respect to juvenile mortality, if there is continuing parental investment in the maturing juvenile, then the force of selection against juvenile mortality would rise with juvenile age and juvenile mortality would fall; i.e., as the juvenile ages the cumulative investments made by parents and grandparents increase the "net worth" of the juvenile and hence it is advantageous that this juvenile survive. In essence, the force of selection on mortality at a given age can be considered to be a weighted average of several components of selection for Darwinian fitness. One component is related to remaining fertility and another component is related to remaining "transfers" that an individual can give to others. The data of Gurven and Kaplan (2007) are consistent with this model. Hunter-gatherers and forager-horticulturalists show a high infant mortality rate which declines with increasing juvenile age reaching a low value at about age 10. Thereafter, mortality rate remains low until age 40; during this time parental and grandparental intergenerational transfers occur. After age 40, mortality rate accelerates doubling almost every decade. The modal age of death in these populations is 72 which, at least for females, is the age at which grandparental investment into offspring becomes of limited value. At this age, the mortality rate rises steeply.

Female post-reproductive lifespan is not restricted to humans or even mammals (Cohen, 2004; Lee, 2003). However, among mammals, humans and some species of whales have the longest female post-reproductive period (Cohen, 2004; Foote, 2008).

Foote (2008) looked at fertility and mortality patterns of females from three species of whales. The killer whale and short-finned pilot whale are characterized by a long post-reproductive period (about 14 to 20 years), while the long-finned pilot whale has a very short to non-existent post-reproductive lifespan. All three species form stable multigenerational matrilineal social groups and exhibit kin-directed behaviors such as food provisioning. The mortality rates at age 15 years for the long-finned and short-finned pilot whale were similar but their adult mortality rate doubling times (a measure of the rate of ageing) were very different; long-finned pilot whales had a mean doubling time of 9.76 years, and short-finned pilot whales had a mean doubling time of 20.21 years. In the killer whale, the mean adult mortality rate doubling time was 14.20 years. Although both species of pilot whale have similar social group structures, they differ significantly in their life history traits. The short-finned species and the killer whale have a maximum lifespan of 65 and 90 years, respectively and cease reproducing at just over halfway through the maximum lifespan. The long-finned species has a maximum lifespan of about 59 years and females as old as 55 years have been found to be pregnant. The different life history traits of these two species of pilot whales (the female short-finned pilot whale has a slower rate of ageing and a much longer post-reproductive life span than the female long-finned pilot whale) cannot be attributed to differences in intergenerational transfers since these two species live in similar multigenerational matrilineal societies with similar offspring rearing behaviors. Foote does not put forward in this report an explanation to account for the differences between these two species of pilot whale.

Ward *et al.* (2009) looked at reproductive and somatic life spans in two neighboring but discrete populations of killer whales. The data consisted of longitudinal sightings of individual whales made over a 30- year period. The female killer whale, as already noted, has a maximum lifespan of about 90 years. Reproduction ceases at an age (years) in the mid- to late 40s. Hence, the female killer whale has a very long post-reproductive period. Male killer whales have a shorter lifespan seldom exceeding 50 years of age. In contrast, another large mammal, the African elephant, has a maximum lifespan of about 65–70 years and a very short to nonexistent post-reproductive period. Births to African elephants over age 60 years are well documented (Moss, 2001). Ward *et al.* (2009) were not able to find an association between the long post-reproductive life span in the female killer whale and enhanced survival of grand offspring or enhanced reproductive success by daughters.

The observations of Foote (2008) and Ward *et al.* (2009) indicate that the benefits accrued by intergenerational transfers from post-reproductive grandmothers to offspring are not easily identified. Female short- and long-finned pilot whales have very different post-reproductive lifespans but live in equivalent multigenerational matrilineal social groups which are characterized by kin-directed behaviors. The (long) post-reproductive lifespan characteristic of the killer whale does not appear to be associated with enhanced grand-offspring survival or enhanced reproductive success of daughters. Perhaps the focus on post-reproductive lifespan is misplaced. Grandparental investment in offspring is clearly beneficial irrespective of whether the grandmother is post-reproductive or not. If this is the case, why then has it evolved that the females of certain mammalian species survive a considerable length of time following cessation of reproduction? In the human lineage, post-reproductive longevity evolved thousands of years ago, prior to the introduction of agriculture (Gurven and Kaplan, 2007).

The evolution of early reproductive cessation and a prolonged post-reproductive survival is clearly more nuanced than can be accounted for by the "mother" and "grandmother" hypotheses (Ward *et al.*, 2009). Johnstone and Cant (2010) explored the role of age-related kinship relationships in the evolution of an early menopause in species such as the human and killer whale. Most social mammals exhibit male-based dispersal and female based philopatry (remaining at the family home). This is not the case in humans and killer whales. In most human populations, at the local scale, there is a history of female transfers or dispersal. For example, in forager societies, females are transferred to the husband's family at marriage. In the case of killer whales, mating behaviors include philopatry for both males and females with mating outside the local group. When these specific types of mating arrangements were modeled, the kin group dynamics showed an increase in relatedness with increasing female age; i.e., as the female ages more and more of the members of her social grouping have a kin relatedness to her. This increase in kin relatedness as the females aged, favored later-life helping behaviors and at least partially explained the evolution of a long post-reproductive lifespan in humans and killer whales.

Another example in which culture has been an important determinant of behavioral phenotypes involves the acquisition of language skills in non-human primates. Whether animals other than humans have language is a controversial topic (Slobodchikoff, 2012). Part of the problem is how we, as humans, define language. If we define language from an anthropocentric viewpoint, then it is likely that we will not find this human-type of communication behavior in other animals. Our definition of language must be broader (Slobodchikoff, 2012).

Many experiments have used primates and the design of the experiments usually involved a human teaching the primate a human-type language skill (Savage-Rumbaugh *et al.*, 2000). In the experiments reviewed by Savage-Rumbaugh *et al.*, (2000), the language skills of the primate were measured using a lexical keyboard (composed of small two-dimensional printed patterns) linked to a computer. This instrument, which allowed the primate to create lexigrams composed of combinations of symbols, allowed the experimenters to measure language comprehension and literacy rather than vocal language production. Two important changes were made in the experimental protocol. First, the objective of the experiments was to measure language comprehension rather than production of vocal language. Comprehension appeared to be the initial step by which primates acquired language literacy. In addition, the test primates were housed in a cultural environment that was rich in complexity and much more appropriate for the primate. This environment contained a variety of objects, human participants, other primates, visitors, and a forest area which contained other animals. This cultural environment fostered relevant cross-communication between primates and the transfer of skills between these animals through imitation. For example, one of the primates had been trained on the lexical keyboard and was competent with 256 symbols on this keyboard. Two new primates in the group were exposed to the use of this lexical keyboard by this one primate and these two new primates acquired the skills to use the 256 symbols much faster than had the original primate. Hence, the original primate "transferred" this cultural practice of keyboard skills to the two new primates; these two new primates did not require direct instruction by a human. In fact, the two new primates built upon these transferred skills and created more complex lexigrams than had been done by the original primate. Moreover, the original primate had learned to flake stone by instructions from a human. It took this primate almost two years to acquire this tool making skill. One of the new primates was able to observe the original primate performing this tool-making skill and was able to acquire this skill within weeks. By observing a primate perform this skill rather than a human, the new primate learned this skill much faster than had the original primate.

The investigators concluded that culture was an important driver of adaptations. When placed in an appropriate cultural context, primates learned language comprehension and literacy (measured using a lexical keyboard) as well as cultural practices such as tool making much more easily and quickly than when placed in a more artificial cultural context in which language comprehension and literacy as well as tool making were "taught" directly by humans. Primates acquired skills such as language comprehension and literacy "spontaneously" when occupying a cultural context that promoted the relevance of these skills between the primates themselves and also fostered the transfer of practices such as tool making from primate role models.

The observations of Savage-Rumbaugh and co-workers hightlight two important points. First, the right question must be asked. Looking for human language in a non-human will fail. Animals communicate with one another for a variety of reasons but each species will evolve a language system that is adaptive for its own specific ecological niche (Slobodchikoff, 2012). Secondly, once the right question has been framed, the experiment must be designed to make it possible for that question to be answered. Primates showed spontaneous acquisition of language skills and tool-making skills only when occupying a cultural setting that fostered the transfer of cultural practices (lexical keyboard skills and stone flaking skills) between primates themselves.

In summary, human evolution shares a number of features that are common to the evolution of other social mammalian species. What is unique to humans is culture. Human-specific culture is an evolved phenomenon but it is not clear when it diverged from its common ancestral roots. The emergence of human culture required a certain set of conditions: survivorship long enough to allow for the formation of multigenerational social groupings, and a repertoire of cognitive capacities that included the ability to communicate information and ideas to others as well as the ability to accurately imitate the behaviors and practices of others. It is not clear whether the requisite cognitive functions evolved first and made possible the evolution of culture or whether cultural evolution was the driver for the evolution of the sophisticated human brain. Regardless, human evolution is primarily shaped by the process of gene–culture co-evolution which is unique to the human species. Gene–culture co-evolution represents a process based upon two mutually interacting inheritance systems. Gould (1996) pointed out the major differences between these two systems of inheritance. Darwinian evolution is based upon the process of natural selection which is slow, inefficient, and lacks obvious directionality. Cultural change is rapid and leads to the accumulation of favorable innovations and hence has an obvious directionality which is generally referred to as "progress." Although the modal forms of the human body and brain have not changed significantly over the past 100,000 years (stabilizing selection), over the last 15,000 years, humans have introduced agricultural practices, developed permanent cities, and developed our modern technological culture and society (Gould, 1996).

Culture allows for the transgenerational transmission of information, ideas, practices, and behaviors. Culture has made it possible for humans to adapt quickly to fluctuations in environmental conditions and has made it possible for humans to live all over the globe. Genes comprise an inheritance system responsible for individual phenotypes and which responds very slowly to environmental changes. Future human evolution will be dominated by the process of gene–culture co-evolution and at present, there are some troubling cultural trends occurring within our society. Four such cultural trends will be considered.

In a recent report, the U.S. National Research Council (NRC) and the Institute of Medicine (IOM) described the "health disadvantage" of the U.S. compared to other wealthy developed countries (Bayer *et al.*, 2013). The U.S. ranked near the bottom in almost every measure of population health compared to 17 peer countries. These measures included adverse birth outcomes, heart disease, injuries from motor vehicle accidents and violence, sexually acquired diseases, and chronic lung diseases. Whereas previous reports had focused on individual behaviors (smoking, alcohol consumption, obesity) as the root causes of poor population health, the current report explored the relationship between structural and cultural factors and the lower health status of U.S. citizens (health disadvantage). The factors considered were income inequality, poverty, single-parent households, divorce, and incarceration. The authors of the current report also examined the role of individualism in American social ideology, the structure of the welfare state, and patterns of gun ownership. This shift in emphasis from individual health behaviors to cultural and structural factors was partially based on the observation that compared to the U.K., the U.S. as a whole was "sicker." Although there were social heath gradients in both countries, individuals at the top of the social ladder in the U.S., despite their access to advanced medical care had a poorer health status than their British counterparts (Bayer *et al.*, 2013). Life expectancy at birth in the U.S. increased from 68.9 years in 1950 to 79.2 years in 2009, yet the U.S. world ranking in this category decreased from 12*th* to 28*th* place between these time points.

The well-being of urban areas has been examined for neighborhoods in the greater Toronto area, GTA (Urban HEART @ Toronto, 2014). This survey measured 15 indicators across five domains for 140 neighborhoods. The five domains included economic opportunities, social and human development, civic engagement, physical environment and infrastructure, and physical and mental health. The latter domain included premature mortality rates for all causes of death aged less than 75 years in both sexes, prevalence of diabetes, and hospitalization rates for several chronic diseases. The survey showed that residents in neighborhoods with lower incomes and higher rates of unemployment and social assistance had worse health outcomes (domain five). Hence, structural and cultural factors acting selectively across neighborhoods within a single large urban environment can differentially influence ageing patterns and longevity across those neighborhoods.

These observations underscore the significant role of the social and cultural environment in determining ageing patterns and lifespan. The relationship between socioeconomic position and health has been appreciated for more than 30 years since the Whitehall Study and Black Report were released in the U.K. (Bayer *et al.*, 2013). However, now we know that an individual's cultural and social environment engender experiences can influence that individual's behaviors and physiological functioning through epigenetic mechanisms. The process of gene–culture co-evolution is the primary driver of human evolution. An important, but unanswerable question, is how the social and cultural factors identified in the NRC/IOM report will shape the future direction of evolution?

A second example of a "negative" cultural trend involves the wastage of food that occurs within the food supply chain (source to consumer) in wealthy developed countries. This problem was discussed by Gunders in a report issued by the National Resources Defense Council (NRDC) in August, 2012. According to Gunders, within the food supply chain, the consumer is the actor responsible for the largest fraction of losses. The average American consumer wastes 20 times as much food as someone in Southeast Asia. These food losses can be translated into corresponding losses of resources such as land and water that were "used" to produce the food. In addition, this wasted food ends up in landfills where rotting organic matter accounts for 16% of U.S. methane emissions. The breakdown of food losses down the supply chain can be illustrated by using grain products as an example. The overall loss of grain products was listed as 43% which can be broken down as follows: production losses 2%, postharvest, handling and storage losses 2%, processing and packaging losses 10%, distribution and retail losses 2%, and consumer losses 27%.

Gunders acknowledged that the issue of food wastage is a highly complex one, but two basic realities were very important in understanding the problem. For many consumers, food does not represent a large fraction of the household budget, making the financial cost of wasting food too low to outweigh the convenience of "throwing" food away. From an economic perspective, the more consumers waste food, the more those in the food industry are able to sell. These two realities mean that there is not a strong incentive for either consumers or food suppliers to reduce food wastage.

This issue of food wastage in modern societies represents an inversion from the food strategies of early human societies. In hunter-gatherer populations, a reliable food supply was not available. The introduction of agriculture allowed for a more reliable food source as well as the production of foods with a higher caloric content. A reliable supply of high-caloric foods enabled the human population to expand. Social groupings increased in size and urban city life became possible. We have become blasé about our food supply which to most North American consumers is abundant, readily available, and relatively inexpensive. This blasé attitude is translated into poor food choices by consumers, i.e., poor choices with respect to the quantity of food consumed as well as its nutritional value. Forty percent of food in the U.S. goes uneaten; that is more than 20 lbs of food per person per month (Gunders, 2012). This highly inefficient and wasteful food supply system plus the attitudes of consumers and food industry personnel that foster these inefficiencies will unfortunately establish conditions in which metabolic disorders such as obesity and food addictions can develop — phenotypes that have implications for ageing patterns and lifespan.

A third example concerns the relationship between human conflict and climate changes which now becomes more important given the evidence that human-mediated climate changes are accelerating. Human conflict can have significant social and cultural consequences. Hsiang *et al.* (2013) have collected 60 primary studies that gave information on the climate–conflict relationship using the following criteria. The term 'conflict' was used to include a range of outcomes from individual-level violence to country wide political instability and civil war. The effects of climate changes were measured by comparing social outcomes in a given population at points of time when that population was exposed to different climatic conditions. The authors also used paleoclimate reconstructions to look at episodes of inter-group violence and collapse of societies that occurred 5,000 to 3,000 years ago.

The main climatic variables examined were large fluctuations in temperature and precipitation. Such severe fluctuations often coincided with periods of drought. The authors found a strong correlation between conflicts at all scales and levels of social organization (individual level violence to institutional and civil breakdown) with climatic events; i.e., large deviations from normal precipitation and mild temperatures systematically increased the risk of many types of conflict. The data indicated that for each one standard deviation change in climate toward warmer temperatures or more extreme rainfall, the frequency of interpersonal violence increased 4% and the frequency of intergroup conflict increased 14%. Although the correlation between human conflict and climate change was a robust one, the mechanistic links between these variables are not well understood. One hypothesis implicates a role for economic factors; climate change can cause economic productivity to decline which adversely affects labor markets and promotes human conflict. A second hypothesis postulates that climate changes exacerbate social and economic inequalities which then lead to increased human conflict. Severe climate changes could also lead to forced migrations and urbanizations which conditions could increase the occurrence of human conflict. All told, climate changes, depending upon the severity and frequency, could induce significant social and cultural dislocations, e.g., breakdown of civil institutions, violence, migrations (climate refugees) and negative economic consequences. The consequences of these social and cultural dislocations include loss of shelter and an adequate food supply, loss of medical care, and premature death due to violence. These consequences will adversely influence individual ageing patterns and longevity.

A recent paleoclimatic study reported by Pederson *et al.* (2014) supports the correlation between human conflict and climate change documented by Hsiang *et al.* (2013). Pederson and co-workers used a tree-ring reconstruction methodology to calculate environmental water balance and temperature trends in central Mongolia over the time period 900 CE to 2010 CE. They found that between 1180 CE and the early 13th century, dry climatic conditions prevailed which coincided with political instability in Mongolia. The existing political order collapsed and during this period, Genghis Khan emerged and re-established central leadership under his rule. In the 13th century, there was a 15-year period of highly favorable climatic conditions consisting of persistent moisture and warm temperatures. This 15-year period of favorable climatic conditions would have translated into increased food productivity and increased resource availability. Interestingly, this same 15-year period also coincided with the time in which Genghis Khan successfully carried out multiple conquests with the consolidation of a vast empire. The investigators postulated that the abundant resources available because of these favorable climatic conditions helped fuel these conquests. In the early part of the 21st century, a period of drought and higher than average temperatures coincided with political and economic instability. Consistent with the views of Hsiang *et al.*, Pederson and co-workers proposed that future climate change will probably result in more frequent episodes of drought and heat in Mongolia and inner Asia with their social, economic, and political consequences.

A fourth example involves the correlation between climate changes and "mental" health. Willox *et al.* (2013) explored the effects of climate changes on the cultural practices of an Inuit community living in Northern Canada. The Inuit practice a nomadic and subsistence-based lifestyle intimately connected to the natural environment. Land-based cultural activities include hunting, trapping, foraging, fishing, and traveling to family cabins. There are no roads connecting Inuit communities, so travel is by water in the summer months and snowmobile in the winter. Their land-based activities are considered by the Inuit to be integral to their culture, identity, and spirituality. Climate changes being experienced in these Northern regions include increased seasonal temperatures, decreases in sea ice thickness and extent, decreased multi-year pack ice, warming permafrost, rising sea levels, fluctuations in weather patterns, and regional changes in flora and fauna patterns. The land-based cultural activities practiced by the Inuit are very dependent on weather conditions and are becoming imperiled by climate changes that are occurring. Inuit identity, conceptions of the self, and mental wellness are directly linked to the environment and their ability to hunt, trap, fish, forage, and continue to practice cultural traditions related to being "on the land." The land serves as a basis for Inuit well-being and for strong mental health (Willox *et al.*, 2013). The investigators used a case-study approach and gathered qualitative data during their interviews with residents in an Inuit community. Residents described their inability to travel regularly on the land or their inability to travel as devastating, depressing, sad, scary, and extremely stressful. Their inability to travel meant that residents spent more time at home with many family members in crowded conditions. This situation increased family stress. The investigators observed that the inability of the Inuit to engage in their traditional land-based cultural practices had an adverse effect on mental well-being and was associated with an increased use of addicting drugs and alcohol as well as an increased risk of suicide.

The third example underscores the relationship between climate changes and social and cultural breakdowns resulting in conflict and violence both at the personal and population levels. The fourth example illustrates the relationship between climate changes and the inability of a population to engage in traditional cultural practices resulting in poor mental well-being, increased use of harmful drugs, and increased probability of suicide. Since the evidence is that human-mediated climate changes are increasing globally, these associations between climate changes, social/cultural dislocations, and personal mental and physical well-being will only become stronger. How these relationships might alter the trajectory of human evolution is unknown, but these relationships will surely have an effect on human ageing patterns and longevity.

Perspectives

The evolution of the human lineage has taken place over millions of years and for almost all of that time, humans lived in an environmental setting very different than that of today. As human-specific culture evolved, the process of gene–culture co-evolution became the primary determinant of the evolutionary path. Animal domestication staring with the dog about 15,000 years ago and the emergence of agriculture approximately 10,000 years ago were key cultural events. Social groupings increased in size and early cities appeared about 7,000 years ago. The advent of agriculture made possible the expansion of the human population. Over the time course of human evolution, there has been a progressive increase in longevity with a significant increase in longevity occurring in the Upper Paleolithic Period. Lifespan at that time was probably long enough for social groupings to contain grandparents. By the pre-industrial early agricultural period, the ageing and mortality patterns were similar to those of modern humans. Approximately two-thirds of those who survived to age 15 years could expect to live to age 45 years. The average modal age of death, conditional upon reaching the age of 15 years, for these pre-industrial populations was 72 years. Hence, these pre-industrial societies had a significant number of older individuals, males, and post-reproductive females, who could provide grandparent care as well as other forms of intergenerational resource transfers. The causes of death, however, were much different in pre-industrialized societies compared to today's societies. In pre-industrial societies, illnesses, most likely infectious, accounted for about 70% of deaths, degenerative diseases made up about 9% of deaths, and accidents and violence almost 20% of deaths. By contrast, in Canada in 2009 (Statistics Canada, 2012), degenerative diseases accounted for about 62% of deaths and accidents only 4.3%.

Given that gene–culture co-evolution is the primary process shaping human evolution, in what directions will present-day culture drive human evolution? In particular, how will modern-day culture shape human ageing and longevity? The data of Gurven and Kaplan (2007) indicate that the major features of human ageing and mortality had evolved by at least early agricultural times. However, the causes of death between our earlier hunter-gatherer/forager-horticultural and present-day populations have changed considerable. Infections, trauma, and violence as major causes of death have been replaced by so-called degenerative diseases including cancer. What effects the four "troubling" cultural trends described in the previous paragraphs in addition to other cultural trends such as expanding urbanization and the increasing use of technology for social relationships and not just information transfers, will have on ageing and mortality patterns are difficult to predict.

CHAPTER 5

Pulling the Threads Together

As discussed in the **Introduction**, in this book, the ageing process has been treated somewhat unconventionally as a journey starting from a central point and traced by a series of concentric circles. The center point of the circles was a specific mutation in the gene encoding the proteins lamin A/C. Individuals who harbor this mutation express the premature ageing disorder, HGPS. These individuals make a truncated lamin protein known as progerin. Interestingly, it is now known that this protein is produced in small amounts by everyone; its physiologic function is unknown. As one explores the mechanisms underlying this disorder, a widespread interacting web of biological processes that connect with lamin biology soon becomes apparent. This web includes the process of alternative splicing, the regulation of telomere length and function, the mTOR pathway, molecular clocks, stem cell function, DNA damage and repair mechanisms, Wnt signaling, and the function of polycomb protein complexes. All of these biological domains interact with the biological functions of lamins and these interactions become evident as one delves into the mechanisms considered responsible for HGPS. The wider implication is that this web of biological processes gives insights into the ageing process in general and not just HGPS. These multiple biological processes comprise the innermost concentric circle. The second concentric circle comprises the individual's experiences. Two individuals exposed to the same objective experience will have different subjective experiences. Experiences can modify epigenetic marks and through these epigenetic mechanisms change gene expression patterns. Finally, the outermost circle will constitute the social and cultural context relevant to that individual. Culture has been a potent and unique force in human evolution, acting through the process of gene–culture co-evolution. The individual's cultural and social context will modulate that individual's ageing pattern.

In this chapter, the threads will be pulled together by reviewing the themes that were presented in the **Introduction.** The biological, genetic, and social and cultural data set discussed in this book are intended as a basis to develop the aforementioned themes and to establish a conceptual framework by which the ageing of an individual human can be understood. This framework includes the individual as a member of a species which has evolved a developmentally set ageing and longevity pattern. Mammalian species such as elephants, whales, humans, and other primates all have their characteristic lifespan and ageing trajectories. As a member of the human species, we cannot expect to live as long or age in the same manner as for example certain marine mammals or other species of mammals. However, within the evolved broad boundaries of our species-specific ageing and lifespan characteristics, there is extensive variation across individual humans with respect to these phenotypes. The data discussed in this chapter are mainly from previous sections, but additional data has been added as necessary.

Theme: The Biology of Ageing at the Species and Individual Levels

Developmental programs: Lifespan and ageing

Ageing and longevity at the level of the individual human is constrained by limits set at the level of the species. Mammalian species have characteristic ageing and lifespan patterns that are developmentally programmed. Evidence for this proposal has been presented by de Magalhaes *et al.* (2007). These investigators analyzed the relationships between developmental schedules, longevity, and ageing. Although their data set contained all vertebrate groups, this discussion will be mainly restricted to mammals. The database used by de Magalhaes and co-workers consisted of a compilation of the maximum longevity (*tmax*) for a large number of different vertebrate species. These investigators make the case that *tmax* is related to the rate at which a species physiologically ages, and hence, they have used this measure as a proxy for species ageing rate. Another variable that has been used as an estimate of ageing rate is the mortality rate doubling time (MRDT). The life history traits used in the analysis included; body mass, gestation time and age at sexual maturity. The variable *tsex* was defined as the time from conception to sexual maturity (age at sexual maturity plus gestation time). Maximum adult lifespan was defined as *tmax* minus the age at sexual maturity and postnatal growth rate (*K*) was calculated from published growth curves.

Across 1,456 species of mammals, birds, amphibians, and reptiles there was a positive linear log/log relationship between *tmax* and body mass. A similar relationship existed for 856 species of mammals only. Bats and primates were above the log/log mammalian regression line indicating that these animals had a greater *tmax* for a given body mass than mammals in general. These relationships demonstrated that larger animals lived longer than smaller animals and for the combined data set of 856 mammalian species, body mass explained 66% of the variation in *tmax*. MRDT, another measure of ageing, also showed a positive linear log/log relationship with body mass across a combined set of 26 species of mammals, birds, and reptiles. Other life history traits showed significant correlations. There was a positive linear log/log relationship between body mass and *tsex* indicating that larger animals take longer to reach maturity and postnatal growth rate (*K*) showed a negative linear log/log relationship with body mass across 204 mammalian species.

Maximum adult lifespan showed a positive log/log linear relationship with *tsex*, a measure of developmental timing, across 606 mammalian species. A similar relationship was also found when primates alone were considered. These relationships still held when body mass was removed as a confounder which indicated that independent of body size, developmental time is strongly associated with maximum adult lifespan. There were also strong negative correlations between *tsex* and *K* as well as between maximum adult lifespan and *K* across 204 species of mammals.

The longer lifespans of larger animals is thought to reflect their reduced susceptibility to predation by virtue of their size. Bats and primates live longer than equivalent-sized mammals because they occupy less hazardous ecological niches; bats can fly and primates have large brains and better cognitive capacities than most mammalian species.

The observations of de Margalhaes *et al.* (2007) reveal the relationships between developmental programs and ageing and longevity phenotypes. Time to sexual maturity, body size, and postnatal growth rate are the products of developmental programs. Further, larger mammals live longer and age slower than smaller animals. Larger animals also take longer to develop (reach maturity) and a longer developmental time is associated with a longer lifespan. A more rapid postnatal growth rate is associated with smaller body size, shorter time to maturity, and decreased longevity. The finding of relationships between developmentally set life history traits and measures of longevity and ageing, even though these relationships are indirect (a large body size does not directly cause a long lifespan), would support the proposal that at the species level, ageing and longevity are also developmentally set. In summary, the data of de Magalhaes and co-workers demonstrate that when life history traits are examined over many species of mammals, clear relationships are apparent between developmental programs that set these life history traits and measures of ageing and longevity. Developmental programs, ageing, and longevity are evolutionarily linked species characteristics.

Leigh (2004) looked at relationships between brain size and growth and life history traits in primates. Eight species were in the data base including chimpanzees and humans. For each species, Leigh plotted brain mass as a function of postnatal age. Some species such as baboons and Rhesus macaques showed little brain growth in the postnatal period, indicating that most of the growth occurred prenatally. By contrast, chimpanzees and humans showed rapid brain growth in the early postnatal period (first three years) and then a slower growth phase. In humans, maximum brain mass was achieved by about six to seven years of age. There was considerable diversity in patterns of prenatal and postnatal brain growth among species of primates. Leigh found across these eight species that there was a strong positive relationship between adult brain size and age at reproductive maturation, but there was no relationship between age of reproductive maturation and age at which brain growth ceased, i.e., duration of brain growth. These results indicate that species-specific rate of brain growth is the life history trait that is developmentally set. A similar conclusion was reached by Robson and Wood (2008) in a comparison of chimpanzee and human. The absolute rate of brain growth was much faster in the human than the chimpanzee, whereas their relative rates of growth were quite similar. Both the chimpanzee and the human achieved 90% of adult brain size by the same age.

The positive correlation between adult brain size and age of reproductive maturation across primate species was achieved through different rates of brain growth not through differences in duration of brain growth. A species-specific rate of brain growth appeared to be the developmentally set parameter. One other observation based upon Leigh's study is important to note. Although across primate species, adult brain size was established by developmentally set differences in brain growth, within a given species there was considerable individual variation in adult brain size. Each primate species has its own characteristic mature brain size, but within a species there is considerable individual variation in adult brain size around that characteristic value.

Somel *et al.* (2013a) have reported brain developmental gene expression studies. Seventy percent of genes expressed in the human, chimpanzee, and macaque brain undergo changes in mRNA abundance during postnatal development. For most of these expressed genes, the timing of expression differed among these species, but on average, timing differences scaled with species' differences in maturation rates. However, Somel *et al.* (2013a) noted that gene expression differences across species are quite complex and genes showing human-specific changes in expression trajectories are involved in various synaptic functions.

The data of de Magalhaes *et al.* (2007), Leigh (2007), and Robson and Wood (2008) indicate that each mammalian species has a characteristic ageing pattern, maximum lifespan, and mature brain size. Species-specific rates of ageing and maximum lifespan correlate with body mass, age at which sexual maturity is attained, and postnatal growth rate; these three life history traits are set by developmental programs. Species-specific adult brain size correlates with age at which reproductive maturity is reached, but this association is actually based on species-specific rates of postnatal brain growth during the maturation period, a developmental program. However, even though ageing, longevity and mature brain size are the end results of species-specific developmental programs, there will still be extensive variation in these variables across individuals within a given species. For example, Leigh's study showed extensive variation in adult brain size within a species and longevity across humans shows considerable variation as is commonly observed.

The extreme diversity of developmental programs across species was illustrated by data presented by Jones *et al.* (2014). These investigators presented age related mortality and fertility patterns for 48 species which included mammals, other vertebrates, invertebrates, and even trees and plants. For each species, mean standardized mortality and fertility were calculated by dividing age specific rates by the weighted mean for the whole adult population sampled. The age range used for these calculations was the adult lifespan defined as starting at the mean age of reproductive maturation and ending when only 5% of adults were still alive. Across the 48 species, there was considerable variation in the mortality and fertility patterns, but I will restrict comments to the 13 mammalian species included in the data set. All of these species showed a rise of mortality with advancing age but the steepness of the rise was quite variable and was most pronounced for humans. The steepest rise was observed in Japanese women in 2009. Similar but less steep mortality curves were seen in a Swedish human cohort born in 1881 and in the Ache of Paraguay, a hunter-gatherer population. The shallowest curves were seen in the tundra vole and the yellow-bellied marmot. The fertility trajectories for humans (Japanese women, Swedish cohort, and Ache) were bell shaped and concentrated at younger ages. Other primates had bell-shaped fertility curves but the curve was spread over more of the course of life. The vole and marmot had fertility curves that were relatively flat and similar to their mortality curves.

One measure of senescence can be derived from the shape of the mortality trajectory over the adult life span. A senescence parameter can be calculated as the ratio of mortality at the terminal age (only 5% of adults still alive) to the average level of adult mortality. Using this parameter as a measure of senescence or ageing, the data of Jones *et al.* (2014) showed no correlation between length of life and degree of senescence. This result is at odds with the observation of de Magalhaes *et al.* (2007) who reported that large mammals lived longer and aged slower than small mammals.

Nonetheless, the species-specific mortality and fertility trajectories reported by Jones *et al.* (2014) represent the phenotypic expression of underlying species-specific developmental programs. Interestingly, in their paper, Jones and co-workers showed standardized mortality trajectories for laboratory rats and laboratory mice. The trajectories for these two species were derived from different strains and populations of rats and mice and the figure in this paper displayed these individual strain and population trajectories. Individual strain and population trajectories showed considerable variation within each species.

Genetic underpinnings of developmental programs

The very beginnings of animal development are directed by the maternal genome (Tadros and Lipshitz, 2009). In mammals, fertilization leads to egg activation and alterations in maternal gene regulation and in the early zygote, maternal transcripts direct basic biosynthetic processes. In the mouse, the maternal-to-zygote transition occurs at the first or second cleavage cycle of the zygote. This transition, which spans a period rather than being a point in time, consists of the degradation of maternal mRNA and activation of the zygote genome with production of zygote transcripts (Tadros and Lipshitz, 2009). There is evidence that histone H1, which is involved in the regulation of chromatin compaction, is also involved in the timing for activation of the zygote genome (Siriaco and Tamkun, 2013). However, the developmental timing of zygote genome activation appears to be a plastic process as reviewed by Blaxter (2014). The maternal-to-zygote transition was compared in two nematodes: *Ascaris suum* and *Caenorhabditis elegans*. Activation of the zygote genome is viewed as a two-stage process. Initially, a few zygotic genes are activated and then these genes regulate the transcriptional activity of additional waves of zygotic genes as well as mediating the destruction of maternal RNA transcripts. In *A. suum*, zygotic gene transcription was detected in the fertilized zygote prior to fusion of the two pronuclei and was identified in single-cell-stage embryos. By contrast, in *C. elegans*, maternal transcripts were found through to the 28-cell gastrulation stage. Hence, these two nematodes with similar developmental processes differed considerably in the length of time during which maternal transcripts directed development. *A. suum* showed early and extensive zygote gene expression with rapid degradation of maternal transcripts following fertilization. This observation indicated that a "silent" period with respect to zygote genome activity during which maternal transcripts controlled development was not necessary. In the next paragraphs, the pattern of gene expression during vertebrate embryogenesis will be reviewed using data obtained with zebrafish and humans.

Mathavan *et al.* (2005) measured the expression of 16,416 zebrafish genes in embryos at various developmental stages: unfertilized egg, blastula, gastrula, segmentation, pharyngula, and hatching (Kimmel *et al.*, 1995). Three thousand, six hundred, and fifty-seven genes showed differential expression (compared to reference RNA levels) with single peaks during the course of development. Of these 3,657 genes, 622 showed a maximum level of transcript abundance at the unfertilized egg stage and these transcripts are maternal in origin. The remaining 3,035 genes are zygotic in origin. Transcripts from maternal genes had different rates of degradation. About 35% of the transcripts were present only at the unfertilized egg stage and were degraded prior to commencement of the blastula stage. Forty percent of maternal transcripts persisted into the blastula stage and 24% into the gastrula stage and beyond. Hence, after fertilization, the earliest developmental processes are directed by maternal transcripts with the zygote genome programming subsequent development from about the mid-blastula stage. During development, clusters of zygotic genes displayed maximum expression during different stages of development, i.e., different clusters of genes are involved in different developmental stages. Of the 3,035 zygotic genes, 609 showed peak expression during blastula, 1,006 during gastrula, 688 during segmentation, and 732 during pharyngula stages. The genes activated during different developmental stages regulated different biological functions. Maternal gene-derived transcripts coded for cyclins, CDKs, proteosome components, and ubiquitins — functions involving cell-cycle dynamics, cell proliferation, and cell differentiation. Some of these maternal transcripts persisted into the pharyngula stage. Somitogenesis and organogenesis occur during the segmentation and pharyngula stages (Kimmel *et al.*, 1995). These are morphological processes regulated by zygotic genes. During these stages, muscle specific proteins and various somitogenesis transcription factors were expressed. Transcript accumulation from ribosomal protein coding genes began at the blastula stage and the expression of these zygotic genes increased continuously during subsequent stages.

De Mendoza *et al.* (2013) looked at transcription factor evolution and its relationship to developmental programs. Complex multicellular organisms with embryonic development have very rich transcription factor expression patterns that evolved through innovations at two time points: one at the stem cell shared with unicellular ancestors and another at the transition to multicellularity. The investigators categorized transcription factors into phylogenetic clusters such as a paneukaryotic group and a metazoan group. In metazoans, particularly bilaterians, homeobox, zf-C2H2, and bHLH classes of transcription factors constituted almost 50% of the total number. These classes of transcription factors are involved in patterning and cell differentiation. As an example, the investigators looked at transcription factor activity during various stages of zebrafish development. During embryogenesis, the zebrafish showed two peaks in transcription factor activity: one during gastrulation and a smaller one in the pharyngula stage. The expression peak at gastrulation reflected transcription factors belonging to phylogenetic groups, opisthokonta, and metazoan while during the segmentation and pharyngula stages, transcription factors belonging to the phylogenetic paneukaryotic group showed the highest expression. These results indicate that embryogenesis in a complex metazoan such as the zebrafish was orchestrated by the differential temporal expression of a repertoire of transcription factors that are shared with other phylogenetic groups. This is another example of the economy of evolutionary processes. Transcription factors that are shared by phylogenetically distinct groups are used in developmental programs in species-specific patterns.

These observations show that in the zebrafish, development is orchestrated by the temporal expression of clusters of genes. The very earliest developmental processes were programmed by maternal gene-derived transcripts and involve primarily cellular proliferation and differentiation. At about the mid-blastula stage, zygotic genes became activated and the sequential temporal expression of clusters of these genes were responsible for subsequent developmental stages. One important question concerns the nature of the role played by the maternal genome in activating the zygote genome. Several candidate mechanisms have been proposed to account for zygotic genome activation (Tadros and Lipshitz, 2009).

In an analogous study, Yi *et al.* (2010) reported a genome-wide expression analysis during weeks 4–9 of human embryogenesis. The investigators chose this developmental stage to study since between weeks 4–9 post-fertilization, the embryo switches from mainly rapid cell proliferation to the development of organs. Total RNA was extracted from three embryos individually for each week between weeks 4 and 9 of development. The investigators detected 28,761 expressed transcripts in the embryos of which 5,358 displayed significant changes in expression levels during weeks 4–9 of development. These 5,358 transcripts were grouped into 42 clusters according to their expression patterns. Examples of different patterns included transcripts which showed increasing expression over the 4–9-week period or decreasing expression over the same period, and transcripts which showed an "arch" type pattern with expression highest at the mid-period of 5–7 weeks. The investigators also referenced transcriptome data in human oocytes and pre-implantation embryos. The human oocyte has been found to contain 5,331 significantly upregulated transcripts (Kocabas *et al.*, 2006). Of these maternal transcripts, the expressions of 793 were regulated during weeks 4–9 of embryogenesis. Of 1,461 transcripts highly expressed in three-day-old embryos, the expressions of 354 were regulated during weeks 4–9. Finally, 1,424 transcripts with high levels in the oocyte could not be detected during the 4–9-week period. The 793 oocyte transcripts and the 354, three-day-old embryo transcripts that were regulated during weeks 4–9 had their highest expression during weeks 4–5 and were down regulated by week nine. The highly expressed 5,331 oocyte transcripts could be classified into three groups: those that are absent by the four-week development stage, those that showed falling expression levels during weeks 4–9, and those that were expressed constitutively during weeks 4–9. These groupings reflected different degradation profiles for these maternal transcripts and were similar to profiles described in the mouse and zebrafish (Mathavan *et al.*, 2005; Tadros and Lipshitz, 2009). The first two groups of maternal (oocyte) transcripts are involved in membrane utilization, metabolism, and cell-cycle regulation. Transcripts in the third group were likely related to housekeeping genes.

The transition from the early cell proliferation phase (weeks 0–4) to organogenesis (weeks 4–9) was associated with a reduction in number of stem (progenitor) cells since these cells commenced differentiating into tissue- specific cell types. Stem cell specific genes were highly expressed during weeks 4–5 but down regulated by week nine. Genes associated with organ development were upregulated during weeks 4–9 with the highest levels of expression occurring during week nine. Development of the nervous system during weeks 4–9 was associated with sequential expression patterns for different gene clusters.

Yi *et al.* (2010) found that genes that participated in the same biological processes tended to have similar expression profiles during developmental weeks 4–9. This observation allowed Yi and co-workers to assign probable biological functions to genes given their expression profiles; i.e., a gene with an unknown function could be assigned a similar function to that of a gene of known function if both genes had similar expression profiles during the 4–9-week period. The observations of Yi *et al.* (2010) paint a picture of human embryogenesis that is critically dependent on the correct temporal expression of maternal and zygotic transcripts; maternal-derived transcripts direct cell metabolism and proliferation primarily before week four, but about 15% of maternal transcripts persist and regulate these biological functions beyond week four of development. The details concerning the coordinated interactions between maternal and zygotic genomes remain an active research field.

The developmental program is regulated by a sophisticated array of regulatory elements including enhancers. Nord *et al.* (2013) studied the role of enhancers in the embryogenesis of the mouse focusing on three tissues: forebrain, heart and liver, and seven developmental stages: three embryonic and four postnatal time points. The investigators used the chromatin modification H3K27ac (acetylation of lysine 27 on histone H3) as a marker of an active enhancer. Across the three tissues and seven time points, they identified 105,394 H3K27ac-enriched regions including 89,169 distal enhancers (more than 1 kb from a transcription start site) which were considered to be candidate developmental enhancers. The three tissues displayed different temporal profiles for distal enhancer activity and enhancers were also tissue specific. Forty-eight percent of enhancers were active in only one of the three tissues, 31% were active in two tissues, and 21% were active in all three tissues. The investigators were able to link candidate developmental enhancers with tissue specific biological processes. For example, enhancers found to be active in early forebrain development were enriched near genes involved in neural progenitor cell proliferation and in the process of axonogenesis. In addition, the investigators noted that the number of regulatory enhancers was far greater than the number of coding genes being regulated, underscoring the importance of regulatory DNA in determining developmental programs. Since coding genes are generally conserved, the precision of developmental programs is based on specific spatial–temporal gene expression patterns. These patterns are controlled by genomic regulatory elements and the critical role of such elements underscores the notion that regulatory DNA is a primary target for the action of evolutionary processes.

Ong and Corces (2012) in their review highlighted the cell-type specificity of enhancers in their role as regulators of developmental programs. Enhancers can be functionally categorized in terms of the specific pattern of associated chromatin marks, and there appears to be a complex interplay between chromatin marks, specific DNA sequences, and the methylation state of DNA. Chromatin marks such as H3K4me1 and H3K27ac are associated with active enhancer-mediated gene transcription, while the marks H3K27me3 and H3K4me3 are associated with repression of enhancer activity in relation to genes "poised" for future transcription. Enhancers appeared to function in a cell-type specific fashion which is determined by distinct patterns of chromatin marks. By contrast, promoters are much more generic in their function and their chromatin marks are largely invariant. It is widely accepted that nucleosome packing controls the binding of transcription factors to specific DNA sequences; open packing favors binding whereas tight nucleosome packing prevents access of transcription factors to binding sites. In summary, distinct patterns of chromatin marks intersecting with specific DNA sequences and the methylation status of associated CpG sites control the cell-type specificity and activity of enhancers, and by so doing, regulate developmental programs. The important role of epigenetic marks in determining the activity and specificity of enhancers would account for the plasticity of developmental programs, i.e., their modification by environmental cues.

DNA methylation is not only involved in determining enhancer activity and specificity, but also plays an additional role in the regulation of developmental programs by temporarily imprinting certain classes of genes. Rutledge *et al.* (2014) used a mouse model to study the fate of differential CpG methylation between alleles derived from the oocyte and sperm. The investigators identified genes that were methylated in the oocyte (none of these were significantly methylated in sperm) and genes methylated in the blastocyst. Most of the genes methylated in the blastocyst (prior to implantation) were also methylated in the oocyte. Rutledge and co-workers classified these genes as oocyte gametic differentially methylated (gDMG) genes. Two gene classes were over-represented in the oocyte gDMGs: genes associated with brain function and genes associated with testis function. The investigators summarized their observations as follows. The oocyte has fully methylated brain-specific and testis-specific genes which are passed on to the zygote. These two classes of genes were essentially unmethylated in the sperm. At the pre-implantation blastocyst stage, these two gene classes still showed persistence of differential methylation between the maternal and paternal alleles. Post-implantation, the paternal allele underwent *de novo* methylation and the differential methylation status between the maternal and paternal alleles was lost.

The extremely complex nature of developmental programs can be illustrated by just considering a single example. Van Heesbeen *et al.* (2013) reviewed the development and maintenance of dopaminergic neurons. Dopaminergic neurons in the substantia nigra are essential for the coordination of motor functions, whereas dopaminergic neurons in the mesocorticolimbic system are involved in the regulation of emotions and reward-seeking behaviors. Impaired functioning of this class of neurons is associated with a wide range of disorders: Parkinson's syndrome, addictions, and depression to name a few. As outlined by van Heesbeen *et al.* (2013), epigenetic mechanisms play a major role in the development of dopaminergic neuronal precursors and their differentiation into mature dopaminergic neurons. Epigenetic mechanisms are also involved in maintenance of this neuronal class as well as in the regulation of the enzyme tyrosine hydroxylase. This enzyme catalyzes the conversion of tyrosine to DOPA which in turn is converted to dopamine by the enzyme DOPA decarboxylase. The important point is that the developmental program for this one neuronal system is highly complex and subject to considerable variation due to the susceptibility of the involved epigenetic mechanisms to environmental influences. Given the complexity of development for this one system, it is difficult to comprehend the sum total complexity across all the interconnected developmental programs that result in the "creation" of an individual human.

In summary, developmental programs are orchestrated by a complex set of epigenetically controlled genomic regulatory elements. These programs are plastic in that they can be modified by environmental cues and for an individual organism, the developmental program sets the general ageing pattern and longevity for that organism. In the next section, some of the possible mechanistic links between development and ageing and longevity are discussed. Finally, two other aspects related to the genetic underpinnings of developmental programs are worthy of note.

Firsty, in a recent commentary, Zaret (2014) drew an interesting analogy between the processes responsible for genomic silencing during mitosis followed by its reactivation after mitosis with the mechanisms regulating genomic activity during developmental cell fate specification. During mitosis, chromosomes condense, RNA polymerases exit the chromatin, and transcription stops. Amazingly, within about an hour post-mitosis, transcriptional programs specific to that particular cell are faithfully reactivated. During mitosis, the bulk of chromatin acetylation marks are lost whereas repressive methyl histone marks are retained. Further, a subset of transcriptional factors and chromatin-binding proteins are also retained. These factors and proteins function as "bookmarks" that target genes coding for transcriptional regulators that are essential for cell fate maintenance. At the time the cell exits from mitosis, these bookmarking factors help to reset specific transcriptional programs. The retention of bookmarking transcriptional factors and repressive chromatin marks during mitosis form the core components of the mechanisms responsible for silencing transcriptional activity during mitosis and faithfully restarting transcriptional activity as the cell exits mitosis. Zaret (2014) compared the regulation of transcription activity across mitosis to the regulation of transcription during developmental cell-fate specification. During development, so-called pioneer transcription factors bind to silent genes in progenitor cells imparting the potential of gene activation upon the reception of specific signals. He compared these pioneer transcription factors to the bookmarking factors found during mitosis. Cell-fate specification during development is also regulated by repressive chromatin complexes which function in an analogous role to the repressive marks retained during mitosis. This similarity between regulation of transcriptional activity during mitosis and during developmental cell-fate specification is another example of the economy of evolution; — using similar mechanisms for multiple functions.

Secondly, there are data supporting the role of non-protein coding RNAs in developmental programs. One class of these RNAs is lncRNAs. Initially these non-translated RNA transcripts were considered transcriptional noise although these lncRNAs originated from widespread regions of the genome and were expressed at different times during differentiation and development (quoted in Mattick, 2013). In mammals, most lncRNAs are expressed in the brain and many are primate specific. Recently, Rinn and co-workers selected 18 lncRNA genes in the mouse genome that had no protein-coding capacity and did not physically overlap with known protein coding genes (Mattick, 2013). KO mutants of these genes were engineered and abnormal developmental phenotypes were expressed in five of these 18 mutant mice. Three of the mutant strains exhibited embryonic or early postnatal death, while two strains which survived showed developmental defects in multiple organs. Other mutant strains expressed abnormalities in brain formation. These experimental observations support the proposal that lncRNAs do not represent transcriptional noise but are important players in mammalian development, brain function, and physiology (Mattick, 2013).

We all share comparable developmental programs and therefore we all develop into recognizable humans with similar characteristics. However, as discussed in the next theme section, each of us has a unique genome and epigenome and developmentally regulated genes are a part of each individual's unique genetic background. Developmental programs will not be identical between individual humans. Just as our species has evolved a characteristic maximum lifespan and rate of ageing, each of us will have inherited our own maximum potential lifespan and rate of ageing, but within the evolutionary boundary constraints set by our species. None of us can escape ageing or death.

Developmental programs: Physical processes

Although developmental programs are clearly orchestrated by the temporal and spatial regulation of gene expression patterns, physical processes are equally important in the biology of morphogenesis (Keller, 2012). The results of studies a number of years ago led to the concepts of cell recognition and selective affinity; processes considered important in the mass movements of embryonic tissues as different structures were formed. The three primary embryonic tissues, ectoderm, mesoderm, and endoderm, were found to recognize one another as alike or different (cell recognition) and they were observed to have a positive or negative "affinity" for one another. Later, the role of embryonic cell motility patterns was considered and linked to the finding of an internal cytoskeleton and cell surface receptors with both adhesive and signaling functions (Keller, 2012). Specific cell motilities generated mechanical forces between cells and between cells and extracellular matrix that helped to shape the developing embryo at different stages. Physical processes are generic in that they act not only on living tissues but also on nonliving viscoelastic materials (Newman, 2012). These physical processes are combined with hierarchical gene expression programs to create the programs responsible for morphogenesis (Newman, 2012).

One of the links between physical processes and genetic programs is the regulation of target genes by the secretion of morphogens by certain cell populations (Morata, 2014). Examples of secreted morphogens in *Drosophila* include the proteins Hedgehog and Wingless (part of Wnt signaling pathway). It is believed that morphogens act at a distance and that morphogen concentration gradients are transduced nonlinearly into signal gradients (Kicheva *et al.*, 2012). (As an aside, there is now evidence that in *Drosophila* the morphogen Wingless can also act at short distances (Morata, 2014; Struhl, 2014)). The signaling effectors activated by the morphogen control downstream target genes that cross regulate each other to form a gene regulatory network (Kicheva *et al.*, 2012). Hence, the expression of a target gene is not determined solely by a signaling effector but also by input from members of the gene regulatory network (Kicheva *et al.*, 2012).

Morphogenesis involves the spatial and temporal regulation of stem cells which have the capacity to robustly proliferate and to differentiate. The fate of a stem cell is regulated by signaling that occurs within its niche or microenvironment and Furusawa and Kaneko (2012) have found that at the single-cell level, gene expression oscillations are necessary for cells to maintain their "stemness," i.e., the potential for proliferation and differentiation. In summary, morphogenesis results from a complex interplay between biophysical processes including morphogen concentration gradients that control the behaviors of cell aggregates and biological processes determined by specific temporal and spatial gene expression patterns. The economy of evolution is apparent from the observations that the same genetic toolkit is used to mediate similar morphogenetic processes in all animal phyla and the recurrent appearance of a limited set of morphological patterns in all animal body plans and organ forms (Newman, 2012).

Several mechanistic links are apparent between developmental programs and organismal ageing. The Wnt signaling pathway is involved in many aspects of morphological development and as discussed in **Chapter 1**, telomerase is involved in the regulation of Wnt signaling. Telomere biology, which includes the action of telomerase to maintain telomere length, as discussed in **Chapter 1**, is believed to be an important determinant of the ageing trajectory. Hence, Wnt signaling is involved in both development and ageing. In addition, as discussed in a subsequent part of this chapter, although cell senescence is considered to contribute to organismal ageing, recently cell senescence has been implicated in the process of developmental morphogenesis. Hence, mechanisms involved in development are also involved in the ageing process consistent with the notion that developmental programs and ageing patterns and longevity are tightly linked.

Several model organisms have been used to explore the link between developmental programs and ageing and lifespan. One organism is the zebrafish (Sasaki and Kishi, 2013). Zebrafish have a lifespan of about 3–5 years and show many phenotypic aspects of ageing similar to that of humans. Sasaki and Kishi (2013) have examined for "developmental" genes that are also involved in the ageing process. The worm, *Caenorhabditis elegans*, is another commonly used model organism for studying the ageing process. Old worms move slowly, become flaccid, and accumulate an age-related pigment in their intestines called lipofuscin. Histologically muscle degeneration has been documented in old worms. In an interesting set of experiments, Budovskaya *et al.* (2008) studied the link between development and ageing in this model.

In these experiments, gene expression patterns were examined in three strains of *C. elegans*: wild type, dauer larvae (developmentally arrested worms), and mutant strains with either an extended or shortened lifespan. In wild-type worms, 1,254 genes were identified that changed expression during normal ageing. The investigators found a significant correlation between expression changes observed in these genes during normal ageing and in dauer larvae. Genes that were upregulated in old age were upregulated in dauers and vice versa. Furthermore, genes that increased expression during ageing were increased in expression in long-living mutants and decreased in expression in short-lived mutants. These observations described a shared gene expression pattern between normal ageing, the dauer state, and genetically induced alterations in lifespan.

A common feature of these genes was that the promoter regions were enriched for a common DNA sequence that was recognized by GATA transcription factors. *C. elegans* has 14 GATA transcription factor genes and the investigators chose one of these factors, elt-3 GATA transcription factor, as a representative for further study. Elt-3 regulates the expression of a number of downstream genes that change expression with normal ageing. GATA factors form a hierarchical transcriptional network — elt-3 expression is activated by elt-1 and repressed by elt-5 and elt-6. Elt-3 is found in a number of the tissues of *C. elegans* and its expression declines with age. The age-related decrease in elt-3 expression was not the result of extrinsic factors since experimentally induced heat shock, oxidative stress, or irradiation-mediated DNA damage did not change expression of elt-3. The investigators found that two of the elt-3 repressors, elt-5 and elt-6 increased with age. In summary, the GATA transcriptional network is a key player in development and normal ageing in *C. elegans*. During embryonic development, elt-3 expression is activated by elt-1 and repressed by elt-5 and elt-6. With normal ageing, elt-5 and elt-6 expression increases leading to repression of elt-3 and changes in the expression of elt-3 downstream target genes. Changes in the activity of the GATA transcriptional network as the worm passes through the life cycle stages of development, adulthood, and ageing do not appear to be due to extrinsic factors but rather reflect some type of intrinsic developmental mechanism.

Although *C. elegans* is a much simpler organism than a human, the same general evolutionary principles would apply. In *C. elegans*, developmental programs and ageing share mechanisms and the ageing and longevity trajectory of the organism is broadly set during the developmental stage of the life cycle.

Intrauterine environmental influences on developmental programs

Perhaps the most dramatic example illustrating the effects of intrauterine environment on developmental processes was an experiment carried out by Storrs and Williams in 1968. The subject of this experiment was the nine-banded armadillo. The reproduction of this animal involves the regular production of monozygous quadruplets. A single zygote develops to the blastocyst stage of development, becomes implanted, and then four primordial buds are formed in two stages. The four buds develop independently to form four individual animals. The important point is that the four buds which are the precursors of four individual animals are genetically identical. Storrs and Williams included 16 sets of quadruplets in their study. The newborns were sacrificed at birth and subjected to 20 morphological and biochemical analyses which included body weights, organ weights (adrenals, brain, heart, kidney, and liver), and levels of various chemicals present in these organs. The investigators compared these variables between individuals within the quadruplet sets. For all 20 variables, there were always differences between members of the same quadruplet. However, differences within the same quadruplet varied in magnitude. Various explanations were proposed by the investigators and others (Culliton, 1968; Storrs and Williams, 1968), but the important finding of this experiment was that phenotypic differences between members of a quadruplet at birth could not be due to genetic differences present at the time of conception. Some type of gene–environment interaction took place within the intrauterine space that differentially modified the developmental processes among the four embryos which started out genetically identical. Variables such as body weight and organ weights that were measured in this experiment are traits which have a polygenic basis. For example, within different sets of quadruplets, body weights between individual members varied by as little as 2.5% and as much as 25%. Hence, gene–intrauterine environment interactions can lead to a significant effect on complex developmental processes that involve multiple genes.

It is becoming increasingly clear that exposure of the developing embryo and fetus to various intrauterine events can significantly alter the developmental process and induce lasting phenotypic changes. Some of these intrauterine events are the result of maternal behaviors: assisted reproduction techniques, alcohol and tobacco consumption, maternal exposure to chemical toxins, and maternal malnutrition. The effects of elevated levels of glucocorticoids during fetal (and early postnatal) development were recently reviewed by Monaghan (2014).The postulate is that these environmental insults epigenetically reprogram the expression of placental genes which in turn alters placental functions. Altered functioning of the placenta would modify the development of the embryo and fetus leading to phenotypic changes (Koukoura *et al.*, 2012). In **Chapter 2**, data were reviewed which showed that pregnant female laboratory rodents fed a high-fat diet gave birth to offspring with epigenetic-modified expression of certain hepatic genes. Suter *et al.* (2011) found that smoking during pregnancy altered placental DNA methylation in a CpG site-specific manner and that this change in DNA methylation pattern resulted in meaningful alterations in gene expression along important cell signaling and metabolic pathways. There is also very preliminary evidence that intrauterine events by altering fetal brain development can result in children with aggressive behaviors (Hall, 2013).

The relationship between intrauterine events and developmentally directed phenotypes has been categorized under the rubric of developmental plasticity; a given genotype (or in this case a given developmental program) can give rise to different phenotypes depending upon environmental (in this case intrauterine) conditions (Bateson *et al.*, 2004). For example, the developmental program can be significantly altered by the nutritional state of the mother; a poor maternal nutritional state during gestation can result in adverse long-term health consequences for the offspring (Bateson *et al.*, 2004).

The placenta is directly responsible for bringing maternal and fetal blood systems into contact, facilitating nutrient exchange, and determining maternal–fetal resource allocation. With respect to the genetics of placental functioning, most of the data have focused on the role of gene imprinting. In the human, a number of genes are imprinted and many of these imprinted genes are expressed in the placenta (Frost and Moore, 2010; Yamazawa *et al.*, 2008). Genomic imprinting refers to the silencing of one parental allele in the zygote resulting in monoalleleic expression of this gene in the offspring (Koukoura *et al.*, 2012). Genomic imprinting is accomplished through epigenetic mechanisms that involve non-coding RNAs, histone modifications, and differential methylation of DNA cytosines in imprinting control regions (see next paragraph). These epigenetic changes silence one allele of a gene in a parent-of-origin manner (Fowden *et al.*, 2011). A well-studied imprinted gene pair is *IGF2* and *H19*, both found on human chromosome 11. On the paternal chromosome, *IGF2* is expressed and H19 is silent while the reverse is true for the maternal chromosome. IGF2 is a growth-regulating protein while H19 codes for a long noncoding RNA. The RNA encoded by *H19* inhibits the translation of *IGF2* mRNA, and the expression balance of these two genes is involved in regulating embryo and fetal growth.

The epigenetic control of these two genes has recently been elucidated (Reik and Murrell, 2000). A region of chromosome 11 upstream of the gene *H19* (imprinting control region; ICR) is methylated on the paternal chromosome but not on the maternal chromosome. A protein factor CTCF binds to this DNA sequence but only when it is unmethylated. Hence, CTCF binds to the maternal chromosome but not the paternal chromosome. CTCF acts as an insulator and prevents access of enhancers downstream from *H19* to the gene *IGF2* which is upstream from *H19*. Hence, on the maternal chromosome, the enhancers interact only with the promoter of *H19* and cannot interact with the promoter of *IGF2*; *IGF2* remains silent. On the paternal chromosome, CTCF is not present and hence the enhancers have access to the promoter of *IGF2*; *IGF2* is expressed. The *H19* promoter is methylated and therefore the paternal *H19* gene is silenced.

Aberrant expression of placental *IGF2* and/or *H19* has been implicated in several growth disorders (Fowden *et al.*, 2011; Yamazawa *et al.*, 2008). Reduced methylation of the ICR on chromosome 11 which controls the imprinting of *IGF2* and H19 has been found in placentas derived from pregnancies with poor fetal growth and those complicated by preeclampsia (Koukoura *et al.*, 2012). In the Silver–Russell syndrome (SRS), which is characterized by severe pre- and postnatal growth, about one-third of affected individuals have hypomethylation of the ICR on chromosome 11 (Koukoura *et al.*, 2012). According to the aforementioned mechanism, this would lead to binding of CTCF on both maternal and paternal chromosomes with a reduction in *IGF2* expression and severe growth restriction. Hypermethylation of the ICR on chromosome 11 would result in biallelic expression of *IGF2* (with reduced expression of maternal *H19*). The increased transcription of *IGF2* would result in a pre- and postnatal overgrowth disorder. Current evidence indicates that the methylation status of the *H19* ICR in placental tissue is sensitive to environmental factors such as maternal nutrition, alcohol consumption, or treatment with glucocorticoids (Fowden *et al.*, 2011; Koukoura *et al.*, 2012).

In an interesting study, Yamazawa *et al.* (2008) studied 60 individuals with the growth restricted phenotype characteristic of SRS and measured the methylation status of the *H19* ICR on chromosome 11. These measurements were made in leukocytes of all 60 patients and in a few of the cases also in samples of paraffin-embedded placental tissue. In 20 individuals with SRS, leukocyte *H19* ICR showed hypomethylation, while in the other 40 patients, the methylation status of *H19* ICR was within the normal range. Placental samples from three patients with hypomethylated leukocyte *H19* ICRs showed decreased expression of *IGF2*, while in two patients with normally methylated *H19* ICRs, placental tissue showed normal *IGF2* expression. *H19* expression was normal in placental samples from all five cases. The observations of Yamazawa *et al.* indicate that about one-third of cases of SRS were associated with hypomethylation of the *H19* ICR in the fetus (leukocytes) and that in some of these cases there was also hypomethylation of the *H19* ICR in the placenta. The origin of these alterations in methylation status (epimutations) was not investigated but other evidence would suggest that at least for the placenta, these epimutations are probably environmentally induced.

In summary, developmental programs that regulate embryogenesis and fetal maturation can be significantly altered by the nature of the intrauterine environment. Evidence suggests that the intrauterine space can be modified by environmental cues, many of which are directly mediated by maternal behaviors during pregnancy. One of the underlying mechanisms linking environmental events and changes in the intrauterine environment involves epigenetic reprogramming of the expression of placental genes, many of which are imprinted. Changes in placental gene expression would alter placental function and thereby modify the developmental trajectory of the maturing fetus.

Developmental programs are not "restricted" to the life cycle period during which embryogenesis and organismal maturation take place. Several examples illustrate this point. Roensch *et al.* (2013) compared the sequence of events in limb regeneration and limb development in the salamander. During limb development, progenitor cells are specified in a proximal-to-distal sequence and these investigators observed the same pattern in the salamander during limb regeneration following amputation. Driskell *et al.* (2013) described the fibroblast lineages that were involved in skin development and in the repair of a full-thickness skin wound. The lineages were the same as were their differentiation sequences in both skin development and repair. Evolutionarily designed animal biology is very economical. Biomolecules and biological processes generally serve multiple functions. The findings of Driskell *et al.* (2013) and Roensch *et al.* (2013) indicate that developmental processes function in a repair situation, e.g., regeneration of a salamander limb or repair of a skin wound.

Another less-appreciated example involves the mammalian arterial tree. The vessels of the high pressure arterial system are subject to unusual hemodynamic forces at sites of high curvature and regions with bifurcations. At these regions, alterations in shear stress occur which through mechanosensor-transduction mechanisms lead to the production of focal intimal thickenings (Singer, 2013). Such focal intimal thickenings are particularly prevalent in the coronary arterial system and such thickenings have been observed in human fetuses and infants (Singer, 2013). These thickenings can be considered part of the arterial developmental process and are a consequence of the high pressure and complex anatomical arrangement of arteries that has evolved within the mammal (Singer, 2013).

These examples pose a simple question: why should "Mother Nature" create a new process to grow another limb, repair a skin wound, or protect a vascular surface from the effects of altered shear stress when developmental processes that can fulfill those functions already exist?

Biology of ageing

The biology of ageing was presented in **Chapter 1** as a journey with HGPS as a starting point. In reality, this premature ageing disorder can be considered as an extreme example of dysregulated developmental programming. The rate of ageing and lifespan of individuals with this disorder in lamin biology are drastically altered. The rate of ageing is greatly accelerated and longevity is greatly reduced. As the biology of HGPS was explored, it became apparent that abnormalities in lamin functions associated with this disorder were just part of a larger network of intersecting processes which together comprised the basis for the phenotypic manifestations of this developmental disorder. The production of progerin, the truncated lamin found in HGPS, was dependent upon the utilization of an alternative splice site. This change in alternative splicing appeared to be triggered by abnormalities in telomere structure and function. But other biological processes are also involved in HGPS and they include: the DDR, alterations in gene expression patterns and in the differentiation of progenitor cells, Wnt and mTOR signaling pathways, and the function of molecular clocks. The interactions between these various processes and lamin structure and function were described in detail in **Chapter 1**. Recently, Lopez-Otin *et al.* (2013) categorized what they believed to be the cellular and molecular "hallmarks" of the ageing process. These "hallmarks" include genomic instability, telomere attrition, epigenetic alterations, loss of proteostasis, deregulated nutrient sensing, mitochondrial dysfunction, cellular senescence, stem cell exhaustion, and altered cellular communication. These investigators defined the basis of ageing as the time-dependent accumulation of cellular damage with these "hallmarks" representing the molecular processes responsible for this damage. They also pointed out the interconnectedness between these various molecular processes. Two of these "hallmarks", loss of proteostasis and cellular senescence, will be further explored to bring out additional complexities.

Loss of proteostasis

Proteostasis involves mechanisms designed to stabilize correctly folded proteins, to restore the structure of misfolded proteins, and to degrade and eliminate proteins that cannot be so salvaged. These mechanisms prevent the accumulation of misfolded proteins which tend to aggregate and lead to cellular dysfunction and/or death. Proteostatic mechanisms become less efficient with ageing and the accumulation of protein aggregates is considered one of the "hallmarks" of the ageing process (Lopez-Otin *et al.*, 2013).

Coelho *et al.*, (2013) looked at ageing in a single cell organism, the yeast *Schizosaccharomyces pombe*. In unicellular organisms, ageing is defined as an increase in the time between consecutive divisions and an increase in the probability of cell death the more times the cell divides. This type of replicative ageing was described many years ago in the budding yeast, *Saccharomyces cerevisiae* which divides asymmetrically. Mother cells were capable of generating about 25 daughter cells and then cell growth slowed and death ensued (Moseley, 2013). Ageing and death were due to the asymmetric budding program which resulted in retention of ageing factors by the mother. Coelho *et al.* decided to test the importance of asymmetric division in cellular ageing by examining for replicative ageing in the symmetrically dividing fission yeast, *S. pombe*. Although the frequency of death in *S. pombe* was higher than that in *S. cerevisiae*, *S. pombe*, in contrast to *S. cerevisiae*, did not show evidence of ageing; i.e., the time between consecutive divisions (division time) did not increase over the one to two cell divisions occurring prior to cell death. In *S. pombe*, death appeared to be due to a catastrophic failure in some process rather than due to the more gradual decline of ageing. There was a correlation in *S. pombe* between the amount of protein aggregate found in a cell and the probability of that cell dying. Cells inheriting a large amount of protein aggregate, above a certain threshold, at birth had a high probability of death at the time of the next cell cycle.

In another set of experiments, Coelho and co-workers subjected exponentially growing cells to two types of environmental stresses: heat (40° C) and exposure to an oxidizing agent (H_2O_2). These stresses increased the rate of formation of protein aggregates and interestingly, cells which inherited large aggregates showed an increase in division time (ageing) for one to two cycles prior to death. In the case of heat stress, the increase in division time occurred only in the cell cycle preceding death, whereas for oxidative stress, the increase in division time occurred consecutively over three divisions before death. This difference in replicative ageing between the two stresses was probably due to the rate of aggregate formation. Aggregate accumulation appeared to occur more rapidly after heat stress than after oxidative stress so that the ageing period would be shorter after heat stress than after oxidative stress.

In summary, under favorable conditions, *S. pombe* showed no evidence of ageing. On occasion, a cell inherited a large protein aggregate and if the amount was above a certain threshold the cell died without any preceding increase in division time (i.e., no prior ageing). When exposed to environmental stresses, the rate of protein aggregate formation was significantly increased and cells that inherited a large amount of protein aggregate under these conditions displayed ageing prior to death. The results reported by Coelho *et al.* (2013) underscore two important observations. First, the segregation of protein aggregates between daughter cells at the time of division appeared to be a random (stochastic) process. Under favorable growth conditions, when a daughter cell randomly inherited a large protein aggregate above threshold size, that cell died without any prior evidence of ageing. Secondly, ageing in this unicellular organism (increase in division time prior to death) was induced when an environmental stress was added. Under this set of conditions, a daughter cell that inherited a large protein aggregate showed evidence of ageing prior to death. The addition of heat or oxidative stress could have induced ageing by either changing the nature or composition of the protein aggregate or its kinetics of formation and aggregation. However, the implication is that protein aggregates above the threshold size do not induce ageing unless an additional environmental stress is present. Whether this implication holds for cells within a multicellular organism is not known.

Cellular senescence

Our concepts concerning cellular senescence are changing. Early studies showed that human fibroblasts in culture could only divide a fine number of times before assuming a state of replicative senescence (Banito and Lowe, 2013; Lopez-Otin *et al.*, 2013). More recently, it has been demonstrated that various stimuli such as oncogenes, DNA damage, and oxidative stress could induce cellular senescence (Storer *et al.*, 2013). Senescent cells generally share a combination of signatures: characteristic morphologic features, senescence associated beta-galactosidase activity (SAB-gal), enhanced expression of cell-cycle regulatory proteins (p53, p21, p16/RB group), and the senescence-associated secretory phenotype (SASP) (Storer *et al.*, 2013). The SASP is associated with the secretion of various growth factors (e.g., fibroblast growth factors, FGFs), cytokines, chemokines, and extracellular-matrix-remodeling proteins by the arrested cells. These secreted proteins have multiple effects which include recruitment of immune cells (which facilitate removal of senescent cells), induction of angiogenesis, and in some cases, promotion of cell proliferation and epithelial-to-mesenchymal transformation (tissue repair and remodeling). Since the number of senescent cells increases with ageing, the notion has arisen that cellular senescence is an important contributor to organismal ageing. However, cellular senescence is not a generalized feature of all tissues. In aged mice, senescent cells were found in liver, skin, lung, and spleen but not in heart, skeletal muscle, and kidney (Lopez-Otin *et al.*, 2013). Cellular senescence has been largely viewed as a stress response program (Banito and Lowe, 2013) which functions primarily as an anti-cancer mechanism (Campisi, 2013) and more recently to facilitate tissue repair. However, a new study by Storer *et al.* (2013) described observations suggesting that cellular senescence plays a physiological role in developmental programs.

Storer *et al.* (2013) examined mouse and chick embryos of mid- to late stages of development and used a marker for SAB-gal activity to identify senescent cells. Distinct areas of positively staining cells were detected in both mouse and chick embryos. To verify that these areas contained senescent cells, additional testing was done on the developing mouse limb bud at embryonic day 11.5. The bud contains an apical ectodermal ridge (AER) which is a major signaling center directing limb outgrowth and patterning. The AER secretes fibroblast growth factors to an area of mesenchymal proliferation beneath the AER. This mesenchymal area reciprocally maintains AER function. The cells of the AER were essentially non-proliferative (measured by BrdU incorporation) and stained positive for SAB-gal 99% of cells) and p21 (about 70% of cells) but not p53. The protein p21 functions to inhibit the cell cycle and to prevent apoptosis. The adjacent zone of proliferating mesenchyme was negative for these senescent markers. Similar observations were made for the roof plate of the hindbrain, an area that directs neural crest activity and closing of the neural tube. This area also contained non-proliferating senescent cells that were positive for SAB-gal activity as well as expression of p21, but negative for expression of p53. Gene expression analysis of the AER showed that there was considerable overlap between genes highly expressed by the AER and genes highly expressed by senescent human fibroblasts used a reference population of senescent cells. Many of the genes common to the AER and senescent fibroblasts had developmental functions including p21 and p15.

Storer *et al.* (2013) addressed the role of senescent cells in embryogenesis by studying mouse embryos deficient in p21, a mediator of both senescent and AER cells, at embryonic day 11.5. AER in p21-deficient embryos showed a reduced number of SAB-gal-positive cells compared to wild type mouse embryos. The AER of p21-deficient embryos also showed reduced expression of FGF-4 and FGF-8, two important signaling proteins by which the AER directs cell proliferation in the underlying mesenchyme of the limb bud. In addition, the expression of developmental patterning genes in the mesenchyme of the limb bud was altered in p21-deficient embryos. The AER and underlying mesenchyme engage in reciprocal signaling which initiates the development of limb growth at embryonic day 9.5. Limb development proceeds through well-defined stages to reach maturity by embryonic day 17.5. The AER is a transient structure which disappears at the later stages of development. The SAB-gal staining of senescent cells in the AER followed the same distribution and temporal pattern as the AER. The clearance of AER SAB-gal-positive cells was accomplished by a process of redistribution to the interdigital mesenchyme followed by apoptosis and macrophage-mediated removal. Storer and co-workers also demonstrated that the underlying mesenchyme of the limb bud through phospho-ERK pathways maintained AER function. Inhibition of these pathways resulted in a marked decrease in the number of SAB-gal-positive cells in the AER.

Cellular senescence appears to be a key component of the developmental program that instructs growth and patterning through the secretion of various factors disruption of which significantly impairs normal morphological development. There are distinct differences between stress-induced cellular senescence in a mature organism and developmental senescence in an embryo (Banito and Lowe, 2013). The former is initiated by DNA damage and oncogenic signals and functions primarily as a tumor-suppressive process. The cues triggering developmental senescence are unknown and the primary function of developmental senescence is to serve as an additional layer of regulation in the developing embryo (Storer *et al.*, 2013). Stress-induced and developmental senescence share a number of characteristics including SAB-gal activity and a similar secretory phenotype. But there are differences between the two senescent states. For example, developmental stress is not triggered by p53 but instead is mediated by p21 in a p53-independent manner (Banito and Lowe, 2013). The speculation is that senescence in the adult evolved from a primordial tissue-remodeling program. Developmental senescence evolved to optimize embryogenesis and that its beneficial postnatal functions of tumor suppression and tissue repair emerged more recently during evolution (Banito and Lowe, 2013; Campisi, 2013).

In summary, the process of cellular senescence appears to play important roles in embryogenesis and in postnatal regulation of cell proliferation. This is another example of the "economy" of evolution. A process that evolved initially to regulate morphological development was later seconded (with minor modifications) to fulfill important postnatal functions.

These examples illustrate that whether loss of proteostasis or cellular senescence are considered "hallmarks" of ageing depends very much on the context in which these biological processes are examined. In *S. pombe*, protein aggregates only induce ageing if there is a superimposed environmental stress and in the mouse, (and presumably in other mammals) cellular senescence plays an important regulatory role during embryogenesis. These experimental observations are consistent with the theme that it is impossible to dissect the ageing process from the living state itself.

One of the themes of this book is that there are no biological processes that are specific to ageing. Even the very circumscribed ageing disorder HGPS involves a vast web of interconnecting biological processes. This interconnectedness of biological processes follows from evolutionary principles. Natural selection acts on phenotypes. A phenotype that conveys a fitness advantage to an individual organism will be favored and if that phenotype is based on a heritable genetic variation it can be passed onto future generations. With spread of this genetic variation throughout the population, a species evolves. However, species evolution does not occur in a vacuum. A favorable phenotype is one that conveys a fitness advantage within the environmental pressures that an individual organism faces. Evolution is thus a dynamic process; environmental pressures change and so the phenotypes that are favorable in one environmental context may not be favorable in a subsequent environmental context.

Darwinian evolution is slow however, and evolution, as already noted, is a master tinkerer. The key features of body designs that have proven to be successful are retained and minor changes are introduced at the "margins." This notion is consistent with the finding that there are a limited number of protein-coding genes shared by vertebrates and that in general they are well conserved. Tinkering is accomplished through variations in regulatory DNA elements rather than through major changes in the sequences of protein-coding genes. Evolution in general proceeds in small steps, not giant leaps. Since natural selection acts on phenotypes not genotypes, evolution strives to optimize the functioning, i.e., fitness of the whole organism, within its environmental context. Fitness is a feature of the whole organism and not of its individual parts. If we examine only a single biological process, we may conclude that it is not optimally designed, but that is not the appropriate perspective. The design of that single biological process can only be understood within the context of the whole organism; it cannot be truly studied in isolation. Since evolution must tune organismal fitness within the constraints of a changing environment, the design of individual biological processes will vary with time. In humans, the speed of Darwinian evolution has been accelerated by the co-existence of cultural evolution and the emergence of the process of gene–culture co-evolution.

Evolution is also economical. Individual biological processes are designed to fulfill multiple functions and not just a single function. Given all of these considerations, it is not surprising that cellular "damage" occurs as we age. I use the word 'damage' in quotes since we are generally unaware of the multiple functions that biological processes fulfill. For example, cellular senescence has been implicated as contributing to the damage of ageing. Yet now, we know that cellular senescence is a very important regulator of embryonic development. A biological process cannot be "equally good" at everything it does. Evolution involves trade-offs; cellular senescence is an important developmental regulator and a protective mechanism against unbridled cell proliferation under certain circumstances. The accumulation of senescent cells is the trade-off for these other functions. Telomeres function to protect the ends of chromosomes from being perceived as double-strand DNA breaks requiring repair. Telomeres also prevent the loss of important coding sequences during DNA replication. Telomeres shorten with repeated cell divisions and when a critical length is reached the cell will cease to divide. However, replicative cellular senescence is the trade-off for these other protective functions. All of this discussion is intended to support the notion that there are no biological processes that are specific to ageing. Ageing at the cellular level is simply a consequence of the multipurpose nature of biological processes and the various functional trade-offs that are required. In addition, at the cellular level, biological events can be stochastic, adding yet another dimension to the ageing paradigm. As discussed in a previous paragraph, the segregation of protein aggregates between daughter cells in the fission yeast *S. pombe* is stochastic. The daughter cell which by random receives a large protein aggregate at the time of birth has a very high probability of dying in a catastrophic manner. As discussed in the next section, gene transcription at the single-cell level is a stochastic process.

As we learn more about biological processes, their myriad of interconnections and multiple functions, ageing will be viewed as the "cost of living." Within this context, we will have to rethink our concept of diseases and this topic is discussed in **Chapter 6**.

In summary, developmental programs are initiated at the time that a zygote is conceived. At the very beginning, these programs are directed by maternal transcripts, products of the maternal genome, but early on, the zygote genome and epigenome take control. These programs are specific for each species and by the end of the developmental process, the zygote has developed into an individual with the morphology and behaviors characteristic of that species. Species-specific development sets the parameters for that species — maximum potential lifespan and ageing pattern. However, within the parameters set by these species-specific evolved programs, there is room for considerable individual variation. There are a number of sources of variation between individuals. The maternal transcripts responsible for the earliest stages of development will differ between individuals; they are a product of the maternal genome. The genome and epigenome inherited by the zygote will contain many variants with respect to DNA sequences and chromatin marks that are unique for that individual zygote. Events occurring within the intrauterine environment during the developmental process will have significant effects on these processes. Many of these events are mediated by maternal behaviors. Maternal care and events occurring during critical time windows within the immediate postnatal period will also modify the developmental and maturation process. All of these individual-specific variations will determine the potential lifespan and ageing trajectory for that individual within the broad limits set by that individual's species-specific programs. Humans cannot live as long as certain marine mammals, mice cannot live as long as humans, and even among primates there is considerable variation in lifespan. Within this construct, ageing really starts at the time of conception; the programs that are responsible for embryogenesis and fetal development determine that individual's potential lifespan and the broad outlines of that individual's ageing trajectory. However, ageing during this early part of the individual's life cycle is completely overshadowed by the rapid growth characteristic of the developmental and maturation stages. As discussed in the next sections, potential lifespan and ageing patterns are further modified by environmental cues (experiences) that occur during childhood and adult life and by that individual's social and cultural context. The response of an individual to environmental cues and an individual's interaction with his or her social and cultural context will be determined and shaped by that individual's unique genome and epigenome.

At the level of the species, the notion that developmental programming sets maximum lifespan and ageing trajectory has been challenged. Curtsinger *et al.* (1992) referred to this postulate as the limited lifespan paradigm, i.e., individuals are assumed to be born with a maximum lifespan that is genetically fixed, and rejected this postulate on the basis of experiments performed in the fruit fly (*Drosophila melanogaster*). Vaupel (2010) reiterated this position stating that species-specific fixed maximum lifespans had been experimentally refuted. However, the data of Jones *et al.* (2014) and reports by Baudisch and Vaupel (2012) and Vaupel *et al.* (2004) underscored the diversity of mortality and fertility patterns that different species had evolved over countless years. These diverse species-specific mortality and fertility patterns reflect differences in underlying development programs and as the data of de Magalhaes *et al.* (2007) showed, species-specific developmental programs are linked to that species maximum lifespan and ageing.

Curtsinger *et al.*, in their 1992 study, used four highly inbred lines of fruit flies which were then crossed within and between lines to produce 10 genotypes. Only males were studied and they were collected within 12 h of emergence. Mean lifespans for samples of the 10 genotypes were measured in three experimental blocks. There was considerable variation in mean lifespan between populations of flies with different genotypes and between different experimental samples of flies with the same genotype. Estimates of age-specific daily probabilities of death were measured for samples of flies with each of the 10 genotypes. The pattern was one of a gradual increase in death probability with no increase occurring at advanced ages. A more detailed study with one of the genotypes showed that the mortality increased until about day 30 and then remained relatively constant. In a more recent study, Khazaeli and Curtsinger (2010) used populations of fruit flies that originated from a cross between a long-lived variant and a control inbred population. They measured survival rates by determining the numbers of deaths in each cage and fecundity rates as the number of eggs deposited per female. Mean lifespans were greater in males than females and showed great variation between populations of flies. Groups in the top quartile had lifespans about 20 days longer than groups in the bottom quartile. Age-specific mortality increased exponentially in earl life but reached a plateau by day 20–30 similar to the pattern reported by Curtsinger *et al.* (1992). Age-specific fecundity rates showed considerable variation with three patterns predominating: a single early peak a few days after emergence, a single fecundity peak later in adult life, and a third pattern consisting of two peaks, one early and one in midlife. There was no correlation between the early peak fecundity rate and lifespan, but there were positive correlations between lifespan and midlife peak fecundity rate and total fecundity rate. In general, flies in the highest quartile of lifespan showed both early and midlife fecundity peaks whereas flies in the lowest quartile for lifespan showed only the early fecundity peak.

These experimental results do not refute the proposal that species have developmentally programmed maximum lifespans. The species *Drosophila melanogaster* is not immortal. Curtsinger *et al.* (1992) reported flies surviving more than 100 days after emergence, hence what the actual maximum lifespan is for this fly is not known. In the paper by Jones and co-workers, 95% of a cohort of fruit flies would be dead by 52 days. What is clear from the experiments of Curtsinger *et al.* (1992) and Khazaeli and Curtsinger (2010) is the extensive variation in lifespan and fecundity patterns that is present between groups of fruit flies. In addition, lifespan and peak fecundity timing and rate are to some extent independent of each other. The source of this extensive within species variation in lifespan and fecundity rate is both genetic and environmental in origin although details concerning mechanisms are unknown. The basis of this variation fits under the rubric of some type of gene–environment interaction. These results, however, do illustrate quite dramatically the considerable variation in both lifespan and fecundity pattern that can be observed within a single species.

Theme: Genetic Uniqueness of Each Individual Underlies that Individual's Unique Ageing Pattern

Evolution is a master tinkerer (see **Introduction**). In the case of vertebrates, general morphologic patterns are established by a basket of orthologous core coding genes. Morphologic diversity is created through the regulation of the spatial–temporal expression of these genes. Most probably a similar construct holds for behaviors. General behavioral patterns would be established by a set of orthologous core coding genes but behavioral diversity is the result of both the regulation of the spatial-temporal expression of these core genes as well as the addition of learned behaviors through environmentally induced epigenetic mechanisms. The conservative "tweaking what is successful"nature of evolution means that the genomes of different mammalian species will have many shared functional features. Even with today's advanced DNA sequencing techniques, it is still very problematical connecting genotypes with phenotypes. For example, in **Chapter 4**, the question of what genomic features underlie the phenotypic differences between humans and our closest relative the chimpanzee was considered. Although the outward differences between a human and chimpanzee in terms of anatomy and behaviors are obvious, the genomic underpinnings are not so discernible. There are documented genomic differences between humans and chimpanzees, but the functional implications of many of these differences are unknown. In the comparison between a chimpanzee and a human, defining genotype–phenotype relationships is not straightforward.

One goal of ageing research is to understand the genetic underpinnings of the ageing process. It is evident just through observation that there is extensive variation in the ageing pattern across individual humans. Although linking genotypes with phenotypes is difficult, we do know that identified genomic variants can have significant functional consequences. It has been well documented that somatic mosaicism for chromosomal structural variants is linked to ageing and cancer (Jacobs *et al.*, 2013; Kaiser, 2013; Laurie *et al.*, 2013; O'Huallachain *et al.*, 2012) and neurological disorders (Poduri *et al.*, 2013). Germline mosaicism with or without associated somatic mosaicism has also been implicated in a number of different "diseases" (Zlotogora, 1998). GWASs have identified genomic polymorphisms (SNPs) that are statistically associated with a variety of common "diseases" such as autoimmune disorders, type 2 diabetes mellitus, and coronary artery disease (Church, 2013; Manolio, 2010). Within this context,

I believe it is reasonable to use genetic variants as a proxy for variation in the ageing process; i.e., if two individuals harbor different genetic variants then they will most likely manifest differences in their respective ageing patterns. I realize that this approach is dependent upon certain assumptions. The principal assumption is that many (but not all) genetic variants will have functional consequences. This assumption is necessary since for most identified genetic variants phenotypic correlates have not been defined and currently, it is impossible to predict if a particular genetic variant is or is not associated with a functional effect. A second assumption is that the sum total of genetic differences between individuals is responsible for phenotypic differences including ageing and longevity characteristics. Within the context of these assumptions, what is currently known about genetic differences between individual humans? In the next paragraphs, I will first review genetic variations in DNA sequences and chromatin states that occur between individuals. I will then consider genetic variations that occur within an individual — somatic mosaicism and cell-to-cell genetic variation.

Genomic and chromatin variants between individuals

One important byproduct of the Human Genome Project has been the documentation that distinct genomic differences exist between individual humans (Macosko and McCarroll, 2013). For example, the genome of the average individual of European ancestry carries 97 loss-of-function DNA variants with 18 present in a homozygous state (MacArthur *et al.*, 2012). Human genomes contain many DNA structural variants; between any two individuals, there are upward of 2,500 duplications, inversions, and insertions and upward of 3,000,000 SNPs (O'Huallachain *et al.*, 2012). These DNA sequence variants make each human genome unique (Singer, 2013) and as a result, identifying so-called "disease-causing variants has become a needle-in-a-haystack problem — how to distinguish variants that are truly causally linked to a disorder from the enormous number of variants (of unknown significance) that are found in every genome (Macosko and McCarroll, 2013).

There is also extensive variation in chromatin states between individuals and the evidence is that patterns of chromatin modifications and transcription factor binding are heritable (Kasowski *et al.*, 2013). Kasowski *et al.* (2013) studied the chromatin state of lymphoblastoid cell lines derived from 19 individuals of various ancestries. Five histone marks (H3K27ac,

H3K4me1, H3K4me3, H3K36me3, and H3K27me3) and the SA1 subunit of cohesin and CTCF (a transcription repressor) were measured. SA1 and CTCF, DNA-binding proteins, were used as marks of gene bodies. Across individuals, genomic regions that were enriched for these histone marks, CTCF, SA1, and gene expression were identified. About 25–30% of regions which were enriched for the active chromatin marks H3K27ac, H3K4me1, or H3K4me3 or the repressive mark H3K27me3 showed variability across individuals, whereas only 5–15% of regions enriched for CTCF or SA1 or enriched for expressed genes (RNA expression) showed variability. The active marks H3K27ac and H3K4me3 showed higher variability if the region was an enhancer than a promoter.

The genome of each individual (cell line) was segmented into 15 chromatin states based on the combinatorial pattern of chromatin marks and their functional correlates. Examples of chromatin states included active promoters, strong transcription, weak transcription, active enhancers with H3K4me3, poised promoters, poised enhancers, and CTCF-enriched regions. Enhancer states, particularly poised enhancers, displayed the most variability across individuals. The investigators also examined conversions between chromatin states and found that the most common switches were between active states (such as enhancers, promoters, or transcribed regions) and weakly active or repressed states. A substantial fraction of the enhancers that showed variability across individuals did not cause detectable differences in gene expression perhaps due to enhancer redundancy or compensatory regulatory effects.

Regions that were variable across individuals were also enriched in SNPs compared to non-variable regions. The number of SNPs increased with the variability. Variable regions enriched with active chromatin marks contained on average more SNPs than variable regions enriched with CTFC or SA1. The investigators also found that variable regions enriched for active chromatin marks were also enriched for motif disrupting (transcription factor binding) SNPs. Variable regions enriched for H3K27ac showed the greatest enrichment for motif-disrupting SNPs. There were significant associations between each motif disruption score and H3K27ac signal differences for each variable region. Two trios (parents and daughter) of genomes were examined for evidence of inheritance. SNPs that were homozygous in both parents were chosen. There was a strong correlation between allele-specific signals between daughters and parents especially for SA1, CTCF, and enhancer and promoter chromatin marks. This observation suggested that specific alleles, patterns of chromatin modifications, and transcription factor binding are heritable.

Finally, variable regions and allele-specific SNPs are enriched for DNase I sensitive QTLs (enhancers and promoters), expression QTLs (genomic loci that regulate gene expression), and GWAS SNPs, indicating that chromatin variability has widespread functional implications.

In summary, Kasowski *et al.* (2013) showed that between individuals, there is considerable variability in chromatin marks most marked for active chromatin marks and the repressive mark H3K27me3. Active chromatin marks showed the most variability at enhancers compared to promoters. Variable regions were in general enriched with SNPs and motif-disrupting (transcription-binding) SNPs were particularly enriched in variable regions of active chromatin marks, particularly H3K27ac. There was evidence of parent to offspring inheritance of allele-specific SNPs linked to active chromatin (enhancer, promoter) marks. These observations indicate that heritable genomic structural variants by changing transcription factor-binding motifs modulate variation of histone chromatin marks. These observations are consistent with the results of McVicker *et al.* (2013) discussed in **Chapter 2; (epigenetics and ageing)**.

Hence, between two individuals, there are significant differences in DNA structural variants and patterns of chromatin histone marks. However, it is impossible to translate individual differences in DNA structural variants and chromatin histone marks into individual phenotypic differences. In fact, an even more general question is, how much of the human genome is functional (Ponting *et al.*, 2011)? Results of the ENCODE project showed that although only 1–2% of the human nuclear DNA was devoted to protein coding, more than 80% of the noncoding DNA was transcribed, indicating that most of genomic DNA was biochemically active (Singer, 2013; Ward and Kellis, 2012). However, the functional significance of most of these transcripts is unknown. Another measure of DNA function based upon evolutionary considerations has concluded that only about 6.7–10% of the human genome is functional (Meader *et al.*, 2010). This measure is based upon the hypothesis that since most nucleotide changes (mutations) in functional DNA sequences will be detrimental; such changes will be removed by purifying (negative) selection. This process results in the creation of evolutionarily conserved sequences. This evolutionarily based technique has generally involved comparing the human genome with genomes from other mammalian species. The proportion of the human genome that is considered conserved, i.e., evolutionarily constrained, based on this comparison is considered a measure of the proportion of the genome that is biologically functional (Meader *et al.*, 2010). The discrepancy between these two measures of how much of the genome is functional (ENCODE versus evolutionary constraint) has not been resolved (Ward and Kellis, 2012) and remains a very controversial issue (Bhattacharjee, 2014). One possibility is that although most of the DNA is transcribed, much of this biochemical activity may be of no significant consequence. Another possibility is that part of the purifying selection is lineage specific and applies to only some species in the comparison. This possibility will generally result in an underestimate of how much of the human genome is functional.

Clearly, then, there are extensive genomic differences across individuals, but what about the genomic landscape within a single individual?

Genetic variants within an individual: Somatic mosaicism

One view has held that the genome was the same in all somatic cells of an individual and that cells of various tissues differed by virtue of gene expression patterns that were specific for each tissue (Macosko and McCarroll, 2012). This view has now been challenged and as discussed in **Chapter 3**, within a specific tissue each cell probably has its own "personal" genome, i.e., somatic cell mosaicism.

Evidence for somatic mosaicism has come from recent technological advances that have made possible single cell DNA sequencing (Owens, 2012). Owens summarized a number of studies in which single-cell DNA sequencing was performed. For example, single cells taken from a breast cancer cell line that was maintained in the laboratory showed significant differences in DNA structural variants between these single cells. Individual human neurons have been shown to contain between 80 and 300 unique genomic insertions (Owens, 2012). Poduri *et al.* (2013) discussed the association of somatic mosaicism with neurodevelopmental and neurodegenerative disorders.

Zong *et al.* (2012) performed single-cell genome sequencing in a SW480 cancer cell line. These investigators measured both copy number variants (CNV) due to insertions, deletions, or duplications and SNPs. In one set of experiments, individual cells were taken from the SW480 cancer cell line and CNVs were identified in each cell genome. Although the majority of CNVs were similar between the single cells, there were distinct cell-to-cell variations in CNVs — evidence for somatic mosaicism. In a second set of experiments, the investigators took a single cell from a heterogeneous population of the SW480 cancer cell line and cultured this cell for 20 generations. The vast majority of this single cell clonal expansion was used to extract DNA for bulk sequencing which gave the investigators the original founder cell's genome. A single cell from this clone was chosen and cultured for four additional generations to obtain kindred cells. Three of these kindred cells were individually sequenced. These three cells each contained 35 SNPs not present in the genome of the original founder cell. These 35 SNPs were acquired during the 20 cell divisions of clonal expansion. Hence, the clone derived from the original founder cell displayed somatic mosaicism. The youngest clonal cells contained 35 SNPs that were not present in the original ancestral cell. Assuming a detection efficiency of 72% for SNPs, this result translated into a mutation rate of about 2.5 SNPs per generation.

Several groups of investigators have explored the relationship between somatic mosaicism, ageing, and cancer (Macosko and McCarroll, 2012). Jacobs *et al.* (2012) extracted DNA from blood or buccal samples from 57,853 individuals: 31,717 cancer cases and 26,136 cancer-free controls. They broadly defined mosaic chromosomal abnormalities as the presence of both normal karyotypes as well as karyotypes with large structural DNA variants in distinct and detectable subpopulations of cells. The large structural variants identified included CNVs with gain or loss of genomic material and copy-neutral loss of heterozygosity (LOH). The analysis, which was based on the assessment of copy number changes and allelic imbalance, was set to detect clonal mosaic variants greater than 2 Mb in size. Across individuals in whom large structural variants were detected, the proportion of cells containing copy-neutral LOH or copy number losses varied between 7% and 95% of the total cell population, and for copy number gains the proportion of cells containing this variant varied between 22% and 88% of the total cell population. In cancer-free individuals, the incidence of mosaic chromosomal abnormalities was 0.74%. In individuals with cancer, the incidence differed according to the type of cancer. For non-hematological cancers, the incidence was 0.96%, whereas for hematological cancers, the incidence was 16.52%. Mosaic abnormalities also increased with age. For cancer-free individuals, those under 50 had an incidence of 0.23% which increased to 1.91% for those between 75–79 years. For individuals with cancer, the incidence at these two ages were slightly higher than for the cancer-free population. Detectable mosaic abnormalities increased the risk of that individual developing cancer. For non-hematological cancers, the odds ratio (OR) was 1.27 whereas the OR for mosaic individuals developing lymphocytic leukemia was 35.4. The observations of Jacobs and co-workers have important implications. Somatic mosaicism for large structural DNA variants was present in about 0.74% of individuals and in these individuals, the proportion of cells with these variants varied between 7–95% of the total cell population tested. The incidence of somatic mosaicism increased significantly with age. In addition, individuals with detectable somatic mosaicism were at a much higher risk for developing hematologic malignancies such as leukemia compared to individuals with no detectable somatic mosaicism.

In a companion paper, Laurie *et al.* (2012) looked at the prevalence of clonal mosaicism using DNA extracted primarily from peripheral blood in subjects genotyped for the GWAS consortium. The methodology used by Laurie and co-workers could detect a minimum estimated proportion of abnormal cells amounting to 5–10% of the total population and a minimum estimated proportion of normal cells of 20–30% in the total population. Three types of large genomic variations were measured: chromosomal deletions, chromosomal duplications and copy-neutral mixtures of biparental and acquired uniparental disomy (copy-neutral LOH). The variants detected were large — median lengths 34.1 Mb for duplications, 3.8 Mb for deletions, and 34.1 Mb for copy-neutral LOH. The observed prevalence of clonal mosaic anomalies increased with the age of the subjects. In those younger than 50 years of age, the percent occurrence was less than 0.5%. In subjects who were 65–69 years old and greater than 80 years of age, the frequencies were 1.77% and 2.69%, respectively. This age effect was specific for mosaic anomalies and the analysis indicated that most of these mosaic anomalies appeared later in life. The origins of these mosaic chromosomal variants were unknown, but the late-life age prevalence suggested that the origins may be different for chromosomal variants that occurred in subjects less than 50 years of age compared to those that occurred later in life. Interestingly, the clonal mosaic variants described in this study were similar to many of the variants that have been described in hematological malignancies. In addition, in individuals with detectable clonal mosaicism, the risk of hematological cancer was estimated to be 10-fold higher than for individuals without mosaic anomalies. As discussed in the previous paragraph, the data of Jacobs *et al.* (2012) indicated that individuals with mosaic anomalies had a 35-fold higher risk of a hematologic malignancy.

Laurie and co-workers speculated as to possible genomic mechanisms responsible for the late-life increase in the occurrence of mosaic anomalies. One possibility is that these variants resulted from replicative errors that increased cumulatively with age plus an age-related decline in genomic maintenance mechanisms. Another possibility is that some of the variants could have resulted in a loss of function of genes that regulate cellular proliferation such as genes encoding for tumor suppressor proteins. Loss of such regulatory genes could lead to the proliferation (clonal expansion) of cells containing that responsible chromosomal variant.

These two reports described the presence of extensive genomic mosaicism in blood cells and document that the occurrence of such somatic mosaicism significantly increased in later life. In addition, individuals harboring blood-cell genomic mosaicism are at a much higher risk for developing a hematological malignancy. Somatic mosaicism appears to be a widespread phenomenon as described in the next two reports and will probably be considered in the future as part of the human condition (Macosko and McCarroll, 2012).

Abyzov *et al.* (2012) identified genomic CNVs in the fibroblasts of human skin. The approach they used was indirect. Skin fibroblasts were collected from seven members of two families (seven fibroblast parental cultures) and 20 human-induced pluripotent stem cell (iPSC) lines were "produced" from these fibroblasts. Each iPSC line was clonally derived from just one or at most a few fibroblast cells. They compared the genome of each iPSC line with the genome of its fibroblast cell population of origin and were thus able to identify CNVs that were present in the iPSCs and not in the fibroblasts. On average, each iPSC line had two CNVs not apparent in the fibroblasts from which the iPSC was derived. These line-specific CNVs constituted only a small fraction of the total number of CNVs in an iPSC line. The investigators then questioned the origin of these line specific CNVs. Had they occurred *de novo* in the iPSCs as a byproduct of the reprogramming process or were they actually present in the fibroblasts but at very low allele frequencies? Twenty line-specific CNVs were tested for cell of origin and 10 were detected in the parental fibroblast. These somatic CNVs were present in the respective parental fibroblast cell populations varying from close to 0–14.6%. In summary, about 50% of CNVs manifested in iPSCs could be traced back to the original skin fibroblast population. These low-frequency CNVs were present in as many as 14.6% of the original fibroblast population, indicating extensive somatic mosaicism for CNVs in the genome of human skin fibroblasts.

Human neurons are another tissue in which somatic mosaicism has been demonstrated. McConnell *et al.* (2013) in their study used neurons from two different sources: neurons derived from human hiPSCs and human post-mortem frontal cortex neurons. These neurons were subjected to single-cell genome analysis. Of forty single neurons derived from hiPSCs, 13 had unique genomes. The investigators identified seven whole chromosome gains, four whole chromosome losses, and twelve sub-chromosomal CNVs in these 13 hiPSC-derived neurons. Each CNV was identified in merely one neuron. The investigators also performed single-cell genome sequencing on 110 post-mortem frontal cortex neurons from three different individuals. The majority of the neurons had either no CNVs (59%) or one or two CNVs (25%). The vast majority of CNVs were sub-chromosomal and ranged in size from 2.9 to 75 Mb. The CNV load was much skewed in distribution across the 110 cells with seven cells accounting for nearly half of the CNVs. The origin of this somatic mosaicism in the nervous system is unknown. These somatic DNA structural variants could occur during neuronal lineage development or as a result of DNA damage with imperfect repair.

The studies of Abyzov *et al.* (2012) and McConnell *et al.* (2013) add to the list of tissues (skin, nervous tissue) in which somatic mosaicism has been demonstrated.

In summary, there has been identified significant variation between individual human genomes in terms of DNA sequence variants (e.g., SNPs, CNVs, insertions, and other chromosomal and sub-chromosomal rearrangements) and chromatin histone marks. In addition, within a given individual, many if not most tissues also show genomic mosaicism; i.e., there is genomic variation across cells within a single tissue. The extent of somatic mosaicism, at least in peripheral blood cells, increases significantly with advancing age, i.e., after 50 years of age (Jacobs *et al.*, 2013; Laurie *et al.*, 2013).

Variation in transcriptional and translational processes in single cells

The studies reviewed in the previous sections describe structural genomic variation between individuals and within the same tissue of a given individual. There is also a "deeper" layer of genomic variation that occurs at the single cell layer — variation in transcriptional and translational processes between cells. This variation in genomic processes between cells will give rise to different cellular phenotypes. An example will illustrate this type of variation (Zenobi, 2013). Consider a hypothetical situation in which all cells in a clonal population are synchronized and grown under identical conditions. After some time, different cellular phenotypes will emerge within this clonal population of cells as a result of stochastic biological processes. This type of observation is not new but with the development of various single-cell analytical technologies, the basis of this cell-to-cell variation is becoming defined.

Zenobi (2013) reviewed available analytic technologies for studying single-cell metabolomes. The metabolome is defined as the array of small molecular weight (MW < 2 kD) metabolites found in a given cell, i.e., compounds that are the building blocks of proteins and polysaccharides, membrane lipids, neurotransmitters, energy substrates, and intermediate products of catabolic and anabolic reactions. The metabolome reflects the molecular cogs of the cell's metabolic machinery and as such can be considered the cell's functional phenotype. Metabolomic studies have examined for example, the catabolism of glycosphingolipids by individual neuronal cells. Vastly different levels of metabolites were found between cells indicating that individual neuronal cells have different functional phenotypes (Zenobi, 2013).

The underlying basis for such phenotypic differences between cells appears to be rooted in the stochasticity of genomic processes. Stochastic in a biological context refers to noise or irregularities in the rates of biochemical reactions and is more pronounced in reactions that involve a small number or reactants (Zenobi, 2013). The amount of protein produced by a gene will be a function of its expression and the expression of a given gene will vary from cell to cell. This variation or "noise" occurs because of the stochastic nature of the transcriptional mechanism. Stochastisticy can be classified as intrinsic or extrinsic (Elowitz *et al.*, 2002). Extrinsic "noise" arises from random fluctuations in the concentrations and locations of molecules that regulate the transcriptional process of a given gene, while intrinsic "noise" arises from random fluctuations in the molecular events (and their order of occurrence) that comprise the actual components of that gene's transcriptional process.

Elowitz *et al.* (2002) created strains of *Escherichia coli* containing two alleles of the reporter gene *GFP* which encodes the jellyfish green fluorescent protein. Each allele encoded a color variant of the GFP. Each allele was controlled by identical promoters. Cells that produced the same amount of protein from each allele appeared yellow, whereas cells expressing more of one fluorescent protein than the other appeared red or green. Extrinsic and intrinsic noise could be distinguished as follows. Variation in mean fluorescent intensities between cells of the same color was a measure of extrinsic noise; different cells were producing different amounts of protein. Variation in color between cells was a measure of intrinsic noise; different cells were producing different amounts of one or the other protein. Cells containing unrepressed promoters showed low noise levels, both intrinsic and extrinsic. These cells have high transcriptional activity so the investigators reduced the transcription rate by repressing the promoters. With a reduction in transcription activity, both intrinsic and extrinsic noise increased significantly. These observations of Elowitz *et al.* (2002) indicated that at low transcriptional rates, stochastic processes can give rise to variation (noise) in protein production by a particular gene between cells. This noise can be the result of stochasticity (fluctuations) of the gene's specific transcriptional machinery or the result of stochasticity in the processes regulating the gene's transcriptional activity.

Yu *et al.* (2006) explored the stochasticity of gene expression in single *E. coli* cells modified such that the investigators could track the production of single-protein molecules. The investigators engineered a chimeric gene that encoded a fusion protein consisting of a color variant of the GFP fused to a transmembrane protein. Since the fusion protein was tethered to the cell membrane, its diffusion was slowed and it was more easily detected by its fluorescent signal. The chimeric gene was incorporated into the *lac* operon and its promoter (*lac* promoter) was strongly repressed by the activity of the *lac* repressor. Yu *et al.* measured fluorescent fusion protein production using time-lapse photography and with this technique, individual fusion protein molecules could be identified. The investigators found that protein molecules were produced in bursts and that for a given cell, these gene expression bursts occurred at random times within the cell cycle. Each expression burst was based upon a single mRNA molecule which was translated into a variable number of protein molecules. The average number of bursts per cell cycle was 1.2 with a range from 0 to 4 and on average, 4.2 protein molecules were produced per burst. The observations of Yu *et al.* (2006) complement those of Elowitz *et al.* (2002). Hence, the transcriptional and translational processes showed variation between cells. Each cell showed variation in the number of gene expression bursts, the timing of these bursts, and the number of protein molecules produced per burst. These results confirmed the stochastic nature of gene expression at low levels of transcriptional activity.

Sanchez and Golding (2013) have reviewed recent single cell gene expression studies. Consistent with the observations of Elowitz *et al.* (2002) and Yu *et al.* (2006), Sanchez and Golding underscored that gene expression within a single cell was inherently a stochastic process. Initiation of gene transcription required the arrival by diffusion of multiple regulatory factors to specific DNA sequence binding sites as well as the subsequent occurrence of a number of biochemical steps. These biochemical reactions were essentially single-molecule events which amplified the stochastic nature of the reactions. Two kinetic models of mRNA transcription have been described. In one model, mRNA transcription from a given gene occurred as random uncorrelated events with a probability that is constant over time, so-called Poisson kinetics. A second kinetic pattern was one in which a gene displayed bursts of transcriptional activity (as described by Yu *et al.*, 2006) in which multiple mRNAs were transcribed within a short time interval followed by a period of promoter inactivity. In yeast, the kinetics of gene expression were strongly influenced by the DNA sequence of the promoter and both Poissonian and burst-type kinetics were observed with different promoters. For example, promoter activity for the gene *MDN1* was Poissonian while promoter activity for the gene *PDR5* was "bursty." When promoter DNA architecture was deliberately altered in yeast, there resulted significant changes in gene expression "noise" as measured by burst size. Hence, at least in yeast, stochastic transcription kinetics for a given gene were strongly influenced by promoter architecture. For animal cells, including cultured mammalian cells, burst-type kinetics appeared to be the rule. Sanchez and Golding noted that burst-type kinetics were very widespread in eukaryotic cells and perhaps transcription bursts were unavoidable. But this appeared not to be the case, since there are a number of examples of Poissonian kinetics that have been found in mammalian and *Drosophila* cells. The suggestion has been raised that transcriptional bursting may be an evolutionarily advantageous trait. However, more experimental data will be needed in order to understand what constraints are in place to limit or buffer stochasticity in the process of gene transcription.

Another stochastic aspect of the transcriptional process has been described by Deng *et al.* (2014), i.e., random monoallelic gene expression. These investigators studied transcriptome profiles from single mouse embryonic cells (from oocyte to blastocyst pre-implantation stages) as well as single adult mouse fibroblast and liver cells. The zygote and the early two-cell stage contained only maternal transcripts, but by the four-cell stage, maternal transcripts had been cleared and the zygotic genome activated. During the pre-implantation developmental stages four-cell to late blastocyst, 54% of genes on average showed monoallelic expression. The more highly expressed genes were studied between the four-cell and late blastocyst stages. For mRNAs, 12–24% of cells displayed monoallelic expression split almost equally between paternal and maternal alleles. Interestingly, when all the cells from the same embryonic stage (four-cell to late blastocyst) were pooled, the pooled mRNA at each stage of development showed only biallelic expression. Cell-specific random monoallelic expression was masked. Experimental results indicated that expression from the two alleles of the same gene occurred independently of one another. For a specific gene, in some cells that gene was expressed from both alleles and in other cells, that gene was expressed from only one allele. Monoalleic expression was also examined in adult mouse liver cells and fibroblasts. For both single liver cells and fibroblasts, about 24% of cells showed monoallelic expression with paternal and maternal alleles being equally represented.

The transcriptional picture at the single-cell level can be described as follows. Gene transcription is a stochastic process that occurs in random bursts. For a specific gene, transcription bursts from the maternal and paternal chromosome will occur independently and stochastically. At any given time, both or only one of the alleles will be expressed. When gene expression is studied as an average over populations of millions of cells, the random pattern of monoallelic or biallelic gene expression at the single cell level is masked. Or phrased in a different way: in cells with an identical genome, fluctuations of regulator molecules and stochastic gene expression can cause significant deviation of individual cells from the population average (Junker and van Oudenaarden, 2014).

In a complementary study, Gendrel *et al.* (2014) found that about 19% of autosomal genes in the mouse genome were expressed monoallelically. In about 5% of these genes, monoallelic expression was determined by an associated DNA sequence variant and 2.5% of the genes were found to be randomly monoallelic expression. Most of these monoallelically expressed genes were involved in cell-adhesion functions and organ development. The monoallelic expression state was found to be very stable over multiple cell divisions and was preserved during the process of cell differentiation. Epigenetic mechanisms appeared to be involved in maintaining this stability of the monogenetic expression state across cell divisions and during cell differentiation. Gendrel *et al.* (2014) also explored the pattern of monoallelic gene expression in the development of the inner ear, eye, and kidney in the mouse and the role of such genes in developmental disorders involving these organs.

In summary, there is extensive genomic and epigenomic variation between individuals and within any one individual, there are considerable genomic differences between cells within the same tissue. In addition, at the single-cell level, the gene transcriptional process is stochastic both in terms of activity and whether one or both alleles are expressed and this stochasticity results in considerable transcriptional variation between cells. All of this variation accounts for the phenotypic differences between any two individuals, although it is not possible to equate specific genomic and epigenomic variants with specific phenotypic differences. Phenotypic variation across individuals would include variation in developmental programs (embryogenesis and fetal maturation) which set individual-specific maximum potential lifespan and ageing patterns. As reviewed in the next section, genomic and epigenomic differences across individuals also determine individual-specific responses to environmental cues or experiences and these experiences mediate additional phenotypic variation (including ageing trajectory and lifespan) across individuals.

Genomic and epigenomic variants mediate phenotypic differences by altering gene expression patterns and this is accomplished by modifying the regulatory mechanisms which control expression. Regulation of gene expression is accomplished by a highly complex and multilevel system that is still poorly understood. Some of the component regulatory processes in this system have been discussed in previous sections of this book and are briefly summarized here. This list of processes is by no means exhaustive: (1) at the level of the cell, chromatin–nuclear lamina interactions and the nuclear spatial compartmentalization of "active" and "inactive" chromatin can modify gene expression; (2) alternative splicing of pre-mRNA represents a process by which a single gene can potentially encode multiple proteins; (3) factors such as microRNAs and RNA-binding proteins can regulate translation and allow for "independence" between mRNA transcription and the translation of its corresponding protein; (4) at the level of the nucleosome, modification of histone tails, histone–DNA contacts, and the nature of DNA sequences can alter different aspects of RNAP II activity kinetics; (5) establishment of epigenetic marks (methylation, acetylation) on histone tails as well as DNA methylation on cytosines at CpG sites can enhance or repress gene expression; and (6) DNA promoter and enhancer sequences, to which specific protein transcription factors bind, regulate gene expression, and hence, variations in these sequences will alter expression. DNA loops function in this process to juxtapose specific enhancer and promoter regions.

These and other regulatory mechanisms are coordinated such that "biologically" correct temporal and spatial patterns of gene expression take place. It is important to note that experiments in which regulatory mechanisms have been studied generally present results averaged over millions of cells. At the single-cell level, gene transcription and mono- or biallelic expression are inherently stochastic processes. What contribution this cell-to-cell random variation in gene transcription makes to phenotypic variation is presently unknown.

Acquired environmental cue-induced genetic variations; experiences

In **Chapter 2,** changes in behavioral and molecular phenotypes induced by environmental cues (experiences) were described. Environmental cues can be considered in terms of their component sensory inputs e.g., visual, auditory, and touch. It is through such sensory inputs that an individual experiences the environmental cue. These cues result in the establishment of chromatin and DNA (epigenetic) marks which change gene expression patterns. Although the sequence of molecular events has not been well defined, some of these events include activation of region-specific central neurons and upregulation of IEGs which mediate the epigenetic reprogramming of regulatory DNA sequences in target genes. In addition, there is evidence that an individual's "epigenetic response" to an environmental cue is, at least partially, genetically determined. Relatively transient environmental cues can induce changes in behavioral phenotypes that are quite long lasting, and many of these phenotypic changes will influence the ageing pattern of that individual. Two individuals exposed to the same objective environmental cue will have very different subjective experiences; hence, it is quite plausible that the cumulative "epigenetic" effects of an individual's exposure to environmental cues (life experiences) will help shape that individual's ageing pattern.

A recent study by Dias and Ressler (2014) revealed several features of experience-induced phenotypic changes. These features included the specificity of the sensory component of the experience, the underlying epigenetic mechanism, and the evidence that the experience was heritable. The subjects in this study were two-month-old male mice who were conditioned to the odor of acetophenone — the F0 generation. Several control groups were used including mice not exposed to any odors and mice exposed to a different the odur, that of propanol. A second study group was the adult male offspring of the male F0 mice — the F1 generation. Mice possess the M71 odorant receptor in their olfactory epithelium which specifically detects the acetophenone odor. Fear conditioning was done by training the F0 male parent generation to associate acetophenone with foot shocks. The investigators found that odor-naïve male F1 offspring of acetophenone-conditioned F0 males also showed a fear-potentiated startle response to acetophenone even though these F1 males had not been previously exposed to this odor. The F1 odor-naïve male offspring did not show a fear response to the odor of propanol but only to the odor their fathers had been conditioned to. The F1 male offspring of acetophenone-conditioned F0 males showed enlarged M71 specific glomeruli in their olfactory bulbs compared to control group F1 offspring. These enlarged glomeruli contained an increased number of M71 olfactory sensory neurons. Cross-fostering experiments showed that the acetophenone fear-potentiated startle response in the F1 offspring was due to a biological type of inheritance and not a social mode of inheritance. When F1 males from acetophenone -conditioned F0 fathers were mated, the F2 odor-naïve male offspring also showed a fear-potentiated startle response to acetophenone, indicating that this behavioral phenotype persisted until at least the F2 generation. The F2 male offspring also showed larger M71 glomeruli in their olfactory bulbs than control mice.

The *Olfr151* gene encodes the M71 receptor which is expressed by olfactory sensory neurons in the main olfactory epithelium. Sperm from F0 males showed reduced cytosine methylation of the *Olfr151* gene particularly at one locus at the 3′ end. Hypomethylation of the *Olfr151* gene was also found in the sperm of F1 male offspring of F0 acetophenone-conditioned males. However, the olfactory epithelium of F1 and F2 male mice from F0 acetophenone-conditioned males did not show hypomethylation of the *Olfr151* gene.

These observations are consistent with transgenerational germline inheritance of epigenetic marks, in this case the methylation status of the *Olfr151* gene in sperm. This epigenetic gene reprogramming was induced by a specific fear-conditioned sensory experience, the odor of acetophenone. However, the investigators were unable to explain how olfactory stimulation was linked to epigenetic changes in the sperm of the mouse. The inheritance of this epigenetically altered gene was manifested as an increased number of specific M71 odorant receptors in the olfactory epithelium of the offspring and this neuroanatomical change was presumably responsible for the inherited phenotype, fear-potentiated startle response to a specific odor.

In summary, these data of Dias and Ressler show how sensory pathway signaling can mediate a behavioral phenotypic change through epigenetic gene reprogramming. In addition, both the behavioral phenotype and the changed epigenetic marks were inherited between generations. This is an important observation and certainly needs to be corroborated. This mechanism would allow an individual of one generation to pass the epigenetic effects of an environmental cue to members of descendant generations. For example, could fears of various types be inherited through this type of mechanism?

This study raises other questions (Hughes, 2014). The impact of environmental exposures during pregnancy has tended to focus attention on the mother as the primary agent mediating changes in developmental programs. However, the observations of Dias and Ressler point to the father as an equally important agent in mediating modifications in developmental programs through epigenetic changes in sperm (Hughes, 2014). Developing sperm cells undergo several rounds of epigenetic reprogramming so to pass down an epigenetic mark through multiple generations would require this mark to survive multiple rounds of such reprogramming. However, imprinted genes appear to escape periods of reprogramming so passage of a sperm-associated epigenetic mark across many generations is certainly a possibility. The observations of Dias and Ressler also raise the possibility that Lamarkian-type inheritance of acquired phenotypic traits is biologically possible.

The transmission of experiences through the maternal germline was explored by Zaidan *et al.* (2013). The experimental design of their study was as follows. Female rats were divided into two groups. One group was exposed to a chronic stress and the other group served as a control. Both groups were mated 14 days later. On postnatal day 90, the offspring of both groups were subjected to behavioral testing. Changes in stress are mediated by the HPA axis. CRF is secreted by the hypothalamus and acts on the pituitary to secrete ACTH protein which regulates adrenal gland production of corticosteroids. The effects of CRF are mediated by the receptors CRF-1 and CRF-2. Zaidan *et al.* measured CRF-1 mRNA levels in oocytes and various brain regions as part of their study design.

Prior to mating, the oocytes of the pre-stressed female group had mean *CRF1* mRNA expression levels about 19-fold that of the control female group. The pre-stressed group also showed greater *CRF1* mRNA expression in the frontal cortex of the brain than the control group. Newborn offspring of pre-stressed females also had greater *CRF1* mRNA expression in the brain than offspring of control females. Behavioral testing in the offspring at adulthood showed gender-specific changes in anxiety and fear learning tests. In addition, offspring displayed gender-specific changes in *CRF1* mRNA expression that was experience modulated. Offspring given a low-stress experience showed no changes in brain *CRF1* mRNA expression regardless of whether their mothers had been pre-stressed or were "control" mothers. Offspring given a high-stress experience showed gender-specific and brain-region specific changes in *CRF1* mRNA expression. Female but not male offspring from pre-stressed mothers showed increased mean frontal cortex mRNA expression compared to offspring of control mothers, while both male and female offspring from pre-stressed mothers showed decreased mean *CRF1* mRNA expression in the amygdala compared to control mothers. This study is consistent with other observations that a parental (in this case maternal) adverse experience can influence the manner in which offspring cope with adverse life events in their own generation. Here, maternal stress prior to pregnancy modified the response of offspring during adulthood to anxiogenic stimuli and situations. The mechanism appeared to be transgenerational transmission through the maternal germline.

Each human at conception acquires a unique set of genomic variants and chromatin histone marks. These individual genomic and epigenomic variants will determine the trajectory of that individual's development and maturation. As embryogenesis proceeds, tissue-specific genomic mosaicism arises and further modifies ongoing developmental processes. Since DNA transcription is a stochastic process, another layer of genetic variation is added to the developmental mix. During the intrauterine period, environmental events have the potential to epigenetically reprogram placental genes and change their expression pattern. Changes in placental gene expression can alter placental function and thereby modify embryogenesis and fetal development. By the time of birth, the unique developmental trajectory of that neonate has set in place the broad limits of its maximum potential lifespan and the broad features of its ageing trajectory. After birth, the life experiences of that individual will superimpose additional unique changes in his/her pattern of chromatin and DNA epigenetic marks. These environmentally induced epigenetic changes will result in reprogramming gene expression patterns. All of the genomic and epigenomic variants unique to that individual will determine his/her unique phenotypic features. Those unique phenotypic features will include that individual's maximum potential lifespan and the pattern of that individual's ageing process. In addition, there is now a body of evidence suggesting that the epigenetic changes induced by life experiences can be passed on to descendant generations.

Theme: Ageing and an Individual's Social and Cultural Environment

Genes and culture represent two streams of inheritance and their interaction has been defined as gene–culture co-evolution. Cultural evolution is much more rapid than gene-based Darwinian evolution and as such has been an important driver of the recent evolution of humans. In this section, the process of gene–culture co-evolution will be reviewed and its relevance to human ageing and longevity explored. Gene–culture co-evolution is considered by many the dominant form of evolution in humans (Laland, 2008).

Gene–culture co-evolution

In the last 100,000 years, humans have experienced many pivotal events. There was extensive radiation of humans from East Africa into all regions of the globe. This geographic spread exposed humans to a variety of extreme climates. Agricultural practices were introduced and animals were domesticated. With growth of the global population, cities were built and urbanization occurred. These are just a few of the changes that have taken place and these changes have been driven by cultural evolution and gene–culture co-evolution. One specific type of gene–culture co-evolution has been termed 'niche construction.' Niche construction describes processes by which organisms through their actions and choices make long-term changes to their environments, and these changes modify the selection pressures on themselves and their descendants as well as other organisms within that environment (Kendal *et al.*, 2011; Wollstonecroft, 2011). The defining characteristic of niche construction is modification of the relationship between an organism and its environment (Laland *et al.*, 2008). This characteristic is critical since it underlies the link between culture and genes. An animal is not a passive inhabitant of its environment; through niche-constructing practices, an animal actively interacts with and modifies its environment. An animal's environment is not simply the physical space it occupies; that environment also includes the cultural practices of its inhabitants and the effects of those practices. By modifying the environment through niche-constructing activities, animals thereby change the selection pressures within that environment. This modification of selection pressures can establish an environmental context in which certain alleles become adaptive and undergo positive selection, whereas prior to a culturally mediated change in selection pressures, these same alleles were not so favored.

Animals build a variety of shelters such as nests, burrows, holes, and webs in order to control or at least modify environmental parameters such as temperature and humidity. However, humans have developed a capacity for very complex niche construction. We have clothes to protect us from rain and cold. Our houses are well insulated and the microenvironment within our houses is regulated by heating and air-conditioning systems. Our capacity for sophisticated niche construction has allowed us to occupy regions of the globe that would not normally be habitable. A testament to our skills at niche construction is that we can travel to and walk on the moon, which has no atmosphere, and we can live and work for extended periods in an orbiting space module. Agricultural practices fall under the rubric of niche construction. Farming and the pasturing of livestock represent modifications of our environment in this case not for shelter but for ensuring a reliable food supply. The definition of niche construction is quite broad and includes changes in the local environmental due to information, knowledge, and behavioral practices passed down from previous generations as well as modifications in the material and energy resources of our local environment made by our ancestors (Kendal *et al.*, 2011).

The human capacity for sophisticated niche construction leads to modification of environmental selection pressures. Farming and other agricultural practices remove the uncertainty of food supplies and housing and appropriate protective clothing allow us to live in otherwise uninhabitable locations. These agricultural and shelter-building cultural practices modify the environmental selection pressures we face and in this way modify the course of evolution. The transmission of information and knowledge between generations can be considered as modifications of our local environment because these transmissions allow us to deal more effectively with our local environmental pressures based on the learning experiences of our ancestors. Niche construction is perhaps the most important example of gene–culture co-evolution with respect to the trajectory of human evolution. Niche construction is adaptive because it is informed and directed. For example, the occurrence of food shortages would spur the development of new agricultural practices and innovations to alleviate those shortages. In this way, niche construction removes the selective pressures that would have been created by a food shortage. As succinctly stated by Laland (2008): "the gene–culture leash tugs in both ways." Culture may be shaped by genes but the architecture of the human genome has been profoundly shaped by culture.

Farming and livestock domestication are examples of cultural niche-constructing practices. The clearing of forests by the Kwa-speaking yam cultivators of West Africa exposed this population to a new pathogen, the malaria protozoan. This exposure modified natural selection pressures in favor of an increase in specific alleles such as the sickle-cell S allele. In the heterozygous state, the S allele confers protection against malaria (Laland, 2008). In this case, a human modification of the environment through an agricultural practice triggered selection favoring certain genes. This example is similar to the role of dairy farming, another niche-constructing practice, in promoting the positive selection of the alleles for lactase persistence. These examples illustrate the link between niche-constructing practices and positive selection for specific genomic variants. These genomic variants were adaptive only within the environmental context created by the introduced cultural practice.

The principles of niche-constructing theory have been defined in a number of papers, e.g., by Kendal *et al.*, 2011; Laland, 2008; and Laland *et al.*, 2008. Briefly, these principles can be summarized as follows. Genes and cultural practices represent two types of inheritance systems; genes are passed from one generation to the next and ancestral niche-constructing practices pass "modified" environments from generation to generation. Each generation inherits genes from previous generations and a local environment that had been modified by the niche-constructing practices of that generation's ancestors. These two streams of inheritance can be termed 'genetic inheritance' and 'ecological inheritance.' These two streams are not separate and interact in a reciprocal fashion. Natural selection acts on heritable phenotypic traits leading to the transmission of fitness-promoting traits within that environmental context to the next generation. Ecological inheritance presents to the current generation, a local environment that has been modified by previous generations. This modified environment will have different selection pressures than the original environment and hence the action of natural selection will be altered in this new environment. Evolution of the species is a function of these two inheritance streams which are linked through a reciprocal relationship with natural selection. Within a given local environment, natural selection "chooses" which genes are passed onto the next generation, while ecological inheritance by determining the nature of that local environment will influence the choices that natural selection makes. The previously discussed example of lactase persistence clearly illustrates the reciprocal relationship between genetic and ecological inheritances. Prior to the introduction of the niche-constructing practice of dairy farming, alleles that encoded for lactase beyond the weaning period had no fitness advantage. The development of dairy farming, which represented a change in the local environment, altered selection pressures within that modified local environment such that alleles associated with lactase persistence into adulthood now carried a fitness advantage. These alleles would be subject to positive selection and would spread throughout the population during subsequent generations (genetic inheritance). The new local environment of dairy farming would also be passed down to subsequent generations (ecological inheritance) and further modified by these generations.

The two classic studies that are generally quoted as illustrating the role of cultural practices in the course of human evolution are the spread of lactase persistence after the introduction of dairy farming and the spread of the sickle-cell allele after introduction of yam cultivation in Africa. Both of these studies have already been discussed. In **Chapter 4**, several "troubling" cultural trends were noted. These trends can be re-examined within the framework of niche construction theory with a view to defining their possible impact on human evolution.

One of the "negative" trends involves the enormous wastage of food as it proceeds along the supply chain. The losses amounted to about 40% of food produced and most of the losses were incurred at the level of the consumer. Wollstonecroft (2011) looked at the role of food processing in human evolution using a niche construction framework. Food processing is a cultural practice that involves food extraction, storage, and preservation. Our ancestors passed down to subsequent generations the skills and knowledge of food processing they had acquired (ecological inheritance). Descendant generations expanded and added to this knowledge and skill set (cultural evolution). The knowledge and skills of food processing allowed the local human population to use a greater variety of food resources with a better bioavailability of energy and nutrients. In essence, this cultural practice modified the local environment since humans could now make use of a wider spectrum of food sources within that local environmental space, i.e., food-processing niche construction gave our ancestors greater dietary choices (Wollstonecroft, 2011). According to the niche construction model, modification of the local environment would alter selection pressures in that environment, thereby changing the action of natural selection and the evolution of humans within that modified environment.

Has the cultural practice of food processing led to genetic changes as would be expected according to the niche construction model? The data of Perry *et al.* (2007) would suggest the answer is yes. Food processing increased dietary choices and as a result starch became an increasingly prominent component of the diet among agricultural societies. Perry and co-workers examined the evolution of the salivary amylase gene (*AMY1*) in humans and non-human primates. First, they found that diploid *AMY1* gene copy number showed a positive correlation with salivary amylase protein level. They then measured *AMY1* copy numbers in three populations with a high-starch intake and four populations with a low-starch diet. Mean diploid AMY1 copy number was greater in the group with high-starch intake compared to low-starch-intake populations. About 70% of individuals from the combined high-starch sample had at least six *AMY1* copies, whereas for the combined low-starch group the proportion was 37%. Additional analysis by Perry *et al.* (2007) supported the proposal that the greater *AMY1* copy number in the high-starch populations was due to positive selection. The investigators suggested several mechanisms whereby greater salivary amylase protein expression would confer a fitness advantage to individuals ingesting a high-starch diet. In comparison, chimpanzees and bonobos had only two diploid copies of *AMY1* consistent with their low starch intake. In summary, the observations of Perry and co-workers strongly suggested that the pattern of human specific *AMY1* copy number variation was driven by dietary related positive selection. The acquisition of niche-constructing food-processing practices by early humans gave them access to starchy foods and led to the positive selection of increased salivary amylase gene copy number in those populations with a high-starch intake.

Can the observations of Wollstonecroft and Perry *et al.*, with respect to food processing, be applied to the current practice of food wastage? The following discussion is speculative but I believe reasonable. The wastage of food by consumers is no doubt linked to a number of other cultural practices and behaviors. In the developed world, the purpose of food processing has changed from one of enhancing and preserving the nutritional value of food to one driven by the marketability of food. The emphasis is now on palatability and the chemistry of food processing is directed at our neural reward pathways. The wastage of food by consumers devalues the significance of food as an essential metabolic resource and this practice implies that food is unlimited in supply. The wastage of food by consumers means that more food must be produced than what is really required. Unnecessary food production will consume more of the resources used in production plus consume more of the resources used in disposal of the extra waste. Within the framework of niche construction, the practice of food wastage modifies the environment since the environment is not simply physical space, but also includes the cultural practices of its inhabitants and the effects of those practices. The practice of food wastage is allied to the practice of food processing which as noted has been changed in modern societies to fulfill different goals. Resources used in the production and disposal of extra quantities of food will be lost and not available for other purposes. Food behaviors and choices have been altered by a food-processing system that now stresses the taste of food rather than its nutritional value. If this practice of food wastage persists, it will be transmitted to descendant generations, i.e., ecological inheritance. All of this is to say that the cultural practice of food wastage has modified our environment and the selection pressures within that environment. Different selection pressures will alter the action of natural selection and thereby human evolution. In **Chapter 4**, a report by Byars *et al.* (2010) looked at the action of natural selection in the population enrolled in the Framingham Heart Study. Their results indicated that women were being selected for increasing weight and decreasing height. Perhaps the trajectories of these trait changes reflect the effects of niche-constructing practices such as food processing.

In summary, cultural practices have completely changed our relationship with food. Other animals eat according to metabolic need. Not humans. Our food behaviors have many determinants other than metabolic need and perhaps one of the strongest is reward-seeking behaviors. In general, within developed countries, our supply of food is reliable and abundant such that our current cultural practice is to actually waste about 40% of food produced. For other animals, food availability and food type are important environmental selection pressures that drive evolutionary change. Our cultural practices around food production and processing have so modified our ecological niche, that food type and food availability are no longer environmental selection pressures. In our culturally modified environment, a different set of food-related selection pressures will help determine the course of human evolution.

As discussed by Wells (2012), cultural practices are the major driver responsible for the high prevalence of obesity in our industrialized countries. Wells examined the role of cultural practices within an evolutionary context. Storing energy in the form of fat is a characteristic of all vertebrates. Adipose tissue is not simply an inert fuel dump but is an active metabolic tissue secreting a number of cytokines and growth factors as well as being responsive to a variety of signaling molecules. Wells developed the construct that adipose tissue evolved as a metabolic form of risk management which enabled humans to respond to changes in food availability. Wells posed the question: at what level of biology is the disease of obesity located? Does obesity reside within an individual, a characteristic of their genome and physiology, and hence amenable to a possible pharmaceutical treatment, or is it external and the consequence of environmental factors, specifically cultural practices? The empirical evidence is strong that behavioral treatment and prevention programs at a population level have not been successful in curbing the obesity "epidemic." The prevalence of obesity has been characterized by an upward trend in many industrialized countries and even more so in countries undergoing rapid economic development. In addition, obesity is rare in traditional pre-industrialized societies and many non-human species, such as captive primates and dogs, become obese when exposed to the human industrialized cultural niche. The view that obesity at the individual level is the consequence of too much food and too little activity, implicating personal responsibility, is over-simplistic.

Wells emphasized the potent role of cultural practices in shaping individual food behaviors. The food industry has manipulated food processing to create energy dense products that are marketed for our current frenetic lifestyle. By manipulating the salt, fat, and sugar content of food products, these food products can induce addictive-type food behaviors. Food behaviors have become divorced from metabolic and physiologic requirements. Within a given population, there can exist economically disadvantaged groups, suffering from chronic under-nutrition juxtaposed to groups, rich in material resources, suffering from obesity. Wells concluded by noting that obesity can be considered a disease external to the body due to an inappropriate food supply and a marketing system fostering non-physiologic eating behaviors. Within this cultural niche, individuals will vary in their susceptibility to these cultural practices based on their unique mix of genes and experiences.

A second negative trend noted in **Chapter 4** is the intensification of social inequalities that exists in many developed countries. Social inequalities including wealth inequalities have long been implicated in having an adverse effect on population health status. Shennan (2011) has examined the consequences of property and wealth inequality through the lens of the niche construction model. In societies that practiced transfers of material resources such as land and/or animals between generations, i.e., heritable private property, the conditions under which these intergenerational transfers took place changed the relative success of different reproductive strategies. The data indicated that in these societies, it was important for individuals to give priority to maintaining a "high investment quality" to these intergenerational transfers even at the expense of immediate reproductive success. Often, high compared to low investment strategies did not yield a beneficial reproductive outcome, measured as the relative number of offspring per generation, until the fifth or sixth generation. Within the context of agricultural intensification, the niche-constructing practice of passing land and/or animals to offspring, i.e., the inheritance of private property changed the nature of reproductive competition. The data suggest that accumulation of private property was not an end in itself. The conditions under which private property were transferred to descendant generations was instrumental in selecting which reproductive strategies would lead to future success. In this example, ecological inheritance drives Darwinian evolution, i.e., reproductive fitness.

In a complementary study, Pettay *et al.* (2007) examined the selection pressures on female life history traits as a function of socio-economic class. The study population was pre-industrial Finnish women born between 1702 and 1863. The women were divided into three wealth classes: rich, middle-class, and poor, based on the husband's occupation. In terms of lifetime survival, the ranking was rich > middle-class > poor. For all three wealth groups, fecundity (number of offspring born to a woman during her lifespan) was the life history trait under the strongest selection. The more offspring women gave birth to, the higher was their fitness. The direction of natural selection favored later age at last reproduction for rich and middle-class females and earlier age at first reproduction for females in the poor class; i.e., rich and middle-class females increased their fitness by having children later in life, while poor females increased their fitness by having children earlier in life. Offspring survival to adulthood was under strong natural selection for all three wealth classes. The interactions between age at first and last reproduction and offspring survival to adulthood affected female fecundity differently among the three wealth classes, whereas fecundity had similar effects on fitness across the three wealth classes. In summary, the data showed that selection led to substantial differences in fitness gains for several key life history traits depending upon the resources (wealth) of the female class. However, directional selection pressures can be modified by other constraints. For example, natural selection favored an earlier age for first reproduction in poor females, but the actual mean age at which poor females had their first offspring was greater than that for rich and middle-class females. A sampling of mean life traits for the three groups of females (rich, r; middle-class, m-c; and poor, p) included: fecundity (number of children born to a woman during her lifespan) r = 7.38, m-c = 6.33, and p = 4.60; lifespan (years) r = 62.4, m-c = 60.64, and p = 57.6; percent offspring survival to adulthood r = 60%, m-c = 58%, and p = 49%; and number of grandchildren r = 15.8, m-c = 11.2, and p = 6.08.

The data of Shennan (2011) and Pettay *et al.* (2007) underscore that the action of natural selection on individual reproductive fitness is modified by the amount of resources available to individuals. Individuals that have access to greater material assets have a higher probability of reproductive success. For example, returning to the study by Pettay *et al.* (2007), lifetime reproductive success (number of children who survived to age 15) correlated with female wealth and social status; mean values were rich 4.27, middle-class 3.5, and poor 2.01. Societies where there are heritable private property and assets, can maximize individual reproductive success by controlling the conditions under which intergenerational transfers of resources take place, i.e., by regulating the passage of resources between generations. Hence, wealth inequalities within a society reflect a situation in which some individuals by virtue of possessing greater material resources have a reproductive fitness advantage. These inequalities represent an example where local environments differ between individuals and the individual is the agent for these differences (niche construction). Those with greater material assets are able to modify their environments, hence changing selection pressures and the action of natural selection within their altered environmental niche. As noted by Shennan, the reproductive fitness advantage gained by such environmental modification may not become apparent for six or more generations. In societies with extensive wealth inequalities, the action of natural selection will be very different for those with and those without resources. On a global basis, wealth inequality is staggering and as noted in a recent Oxfam Briefing Paper (2014), currently the wealth of the 1% richest people amounts to 110 trillion dollars or almost half of the world's wealth.

The selective effects of niche construction cultural practices that result in social and wealth inequalities can act within small populations as revealed by the Urban HEART @ Toronto Study (2014) discussed in **Chapter 4**. Variations in niche construction practices across different neighborhoods within the Greater Toronto area were associated with variations in longevity and ageing patterns across those same neighborhoods.

The two other "troubling" cultural trends referred to in **Chapter 4** revolve around the consequences of climate change. Climate change will cause uneven environmental modifications across different regions. These regional climatic modifications will lead to alterations in selective pressures within those regions and ultimately influence the course of human evolution. As discussed in **Chapter 4,** there is a clear association between climate changes and levels of personal and societal violence and strife. In the case of climate change, niche-constructing practices are being developed to counteract the effects of climate change. One such development being considered is the instillation of certain chemicals into the atmosphere to overcome the warming effects caused by the accumulation of greenhouse gases (climate geoengineering). What cumulative effects all of these processes, i.e., climate change and counteracting niche-geoengineering practices, might have on human evolution is impossible to predict.

Cultural evolution is much more rapid than Darwinian evolution and the evidence is that gene–culture co-evolution is also a rapid process. The introduction of dairy farming resulted in the spread of alleles for lactase persistence within populations engaged in this activity and this spread probably occurred over a 5- to 10,000-year time frame. Hunemeier *et al.* (2012) looked at the pace of cultural-driven phenotypic evolution in six populations from the Brazilian Amazon and Central Plateau region. Among these populations, two genetic and linguistic sister groups, Xavante and Kayapo, diverged about 1,500 to 2,000 years ago from a common ancestor. These two groups differ culturally in terms of marriage practices, clan and political organization as well as other social traits.

The phenotypic features measured in this study were eight head and facial structural variables, while genetic measurements were based upon mitochondrial DNA. Out of the six populations, the Xavante was the most differentiated in terms of the measured phenotypic features. This differentiation could not be accounted for by differences in climate or geography between the Xavante and the other populations. The investigators created a genetic tree linking all six groups and on this tree the Kayapo and the Xavante occupied sister branches diverging from a common ancestor. The investigators used the phenotypic data plus the genetic tree to calculate the pace of phenotypic change for the various branches of the tree. For the Xavante branch (which originated at the node common to the Xavante and Kayapo), the pace of morphological change was almost four times greater than the average pace for the other branches. This observation indicated that the phenotypic diversification of the Xavante population from the common ancestor with its genetic and linguistic sister population the Kayapo, was accelerated by some mechanism in comparison with the pattern observed in the remaining branches of the tree. The other mechanism appeared to be cultural evolution due to sexual selection. Some of the Xavante chiefs have been documented to have had notable reproductive success. In one village, 25% of inhabitants were offspring of one particular Xavante chief. The investigators concluded that sexual selection favoring wealthy men with high socio-economic status, rather than men with perhaps more desirable phenotypes, was the mechanism driving rapid microevolution in the Xavante society. This conclusion is consistent with the observations of Shennan (2011) and Pettay *et al.* (2007) as to the importance of material resources as a driver of reproductive success and human evolution.

In standard evolutionary theory, the individual organism is largely a passive actor in the process. Environmental pressures direct natural selection to "choose" phenotypic traits that are adaptive for a particular environmental context. If those traits are heritable they will be transmitted to descendant generations. Cultural practices and in particular niche construction add an additional dimension to the evolutionary process. Most animals modify their local environment by building shelters such as nests, burrows, or other structures and some animals display limited tool use. The environmental modifications generated by these activities are very limited compared to what humans are capable of doing. Human culture is much more sophisticated than that of other animals and humans are capable of making major modifications to their local environment. These environmental modifications change the selection pressures within the local environment altering the action of natural selection with respect to its "choice" of what constitutes an adaptive phenotype. As a result of cultural practices, humans became an active player in the course of their own evolution as well as influencing the evolution of other species living within these modified environments. Since cultural practices can be passed down to descendant generations, it is in actuality an inheritance system paralleling that of the genome.

Cultural practices by virtue of modifying environments can lead to genomic changes by promoting selective sweeps. Examples of such sweeps include the spread of alleles associated with lactase persistence, an increase in the amylase gene copy number, an increase in frequency of the sickle-cell hemoglobin allele in certain populations, and the positive selection of genes encoding products promoting an increase in brain size. Gene–culture co-evolution is probably now the dominant mode of human evolution given the sophistication of our cultural practices and the degree to which these practices can modify our local environments. Within this context, it is important to have a broad view of what constitutes our environment. Culture has expanded that definition. The environment is not simply the physical space that we inhabit. It also includes the structures we have built, the cars we drive, the practices we engage in such as agriculture, commerce, and health care, and the technology we have developed for communication and "entertainment." Our environment also includes intergenerational transfers of information and material resources. As discussed in the previous section, all of these cultural aspects of our environment are heritable, and by modifying the selection pressures within our environment, it drives the trajectory of human evolution including ageing patterns and lifespan. Gene–culture co-evolution being relatively fast has allowed humans to adapt to rapid external changes in their environment (including climate changes). Since we are now an active participant in our own evolution, it becomes extremely difficult to predict what the course of that evolution will be. Gene–culture co-evolution is a process unique to the human species.

As a final note, human niche construction practices are now having a global impact which is being manifested in many aspects of the earth's environment. The one aspect that is perhaps most uppermost is that of climate change, but there are a number of intersecting problems that have been well described in the literature, e.g., by Ehrlich and Ehrlich, 2013a,b and Gross, 2013a. As noted by Ehrlich and Ehrlich (2013a), the survival of the human species is not a preordained evolutionary program. Clearly, global trends in climate disruptions, reduction in animal and plant biodiversity, land degradation, and overpopulation to name a few are being intensified by human cultural practices. These trends will have uneven effects across different geographical regions and thereby have variable influences on the evolution of human populations in different areas. From the perspective of ageing and longevity, global environmental trends will alter both of these human phenotypic traits on both species and individual levels but not in a uniform manner across the globe.

Framework for the Ageing Process

The objective of this book is to develop a framework for the ageing process with the focus on the individual human. This framework is a pragmatic one rooted in experimental observations starting with observations related to HGPS reviewed in **Chapter 1**. I have purposely avoided a theoretical or mathematical model. Although natural selection acts on the individual, the species is the unit of evolution so our framework must begin at the species level.

Each species has a characteristic maximum lifespan and rate of ageing, and these two life traits are linked to the developmental program of that species. For mammals, there is a positive correlation between body mass and lifespan and a positive correlation between the length of development and lifespan (independent of body mass). Larger mammalian species generally live longer and take longer to develop than smaller mammals. As a general observation, longer-lived animals show a slower rate of ageing. Independent of body size, a longer development time is associated with a greater maximum adult lifespan. Species-specific developmental programs are driven by species-specific temporal and spatial gene expression patterns in conjunction with generic physical processes that direct the movements of cell aggregates into certain shapes at different developmental stages. Developmental programs are based upon highly conserved processes that utilize a shared genetic toolkit. As a result of these species-specific developmental programs, giraffes give birth to giraffes and humans to humans, etc. With respect to longevity and rate of ageing, these species-specific developmental programs set these parameters for each species. A giraffe has a certain maximum lifespan and rate of ageing as does a chimpanzee and as does a human.

Within a given species, including the human lineage, there is extensive across individual variation in the developmental process and hence, each individual is born with a unique morphologic and behavioral phenotype. The developmental process is the key to understanding ageing and longevity. The zygote inherits a species-specific developmental program, but the building blocks for that program are unique to that zygote. These building blocks are the genomic and epigenomic elements inherited from parental gametes and these elements include a set of DNA structural variants and chromatin marks specific to that zygote. In *Caenorhabditis elegans*, for example, there are numerous variants in genomic regulatory loci that affect the expression of genes during development (Waszak and Deplancke, 2013). The developmental program directs the "assembly" of an individual human with his/her own unique morphological and behavioral phenotypes, although of course, these phenotypes will conform to the general characteristics of the species. The plasticity of the developmental program (at the individual level) is evidenced by the observation that this program can be modified by intrauterine events as well as environmental cues acting in the early postnatal developmental period. Some intrauterine events are stochastic in nature but many are mediated by maternal behaviors during gestation. The individual's unique developmental program sets that individual's maximum potential lifespan and probable rate of ageing; the only constraints being the boundaries set for these life traits by the species.

In **Chapter 2** and in an earlier section of **Chapter 5**, a number of examples of developmental plasticity were discussed. Environmental cues acting in critical time windows either during gestation or in the early postnatal period can modulate developmental programs and through this modulation create new phenotypic traits that are both long lasting and expressed in later adult life. The uniqueness of an individual's developmental program will be a function of that individual's specific genome and epigenome as well as being a function of the modulating effects of specific environmental cues (experiences) during gestation and the early postnatal developmental stage.

This proposal has two implications: first, ageing starts at the time of conception when developmental programs are activated, but is initially "overshadowed" by the growth associated with both pre- and postnatal maturation; and secondly, lifespan (which is much easier to measure than rate of ageing) has heritability. The heritability of human longevity has been examined in a number of studies. Herskind *et al.* (1996) and McGue *et al.* (1993) used as their study population like-sex twin members of the Danish Twin Register born between 1870 and 1880 in the McGue study and between 1870 and 1900 in the Herskind study. Both study populations contained a mixture of monozygotic and dizygotic twins. McGue *et al.* (1993) calculated a lifespan heritability of 22%, while Herskind *et al.* (1996)

calculated lifespan heritabilities of 26% for males and 23% for females. Mitchell *et al*. (2001) used genealogical information derived from old order Amish (Pennsylvania, USA) born prior to 1890. They calculated a lifespan heritability of 25%. As reviewed by Cournil and Kirkwood (2001), a study on lifespan heritability was done using a large database of inhabitants of Iceland. The methodology used in this study did not allow a quantitative estimate of lifespan heritability, but the strong familial component to longevity noted in this study was of a modest size. Hence, most estimates of lifespan heritability, although highly significant, are in the order of 25% (Cournil and Kirkwood, 2001). This "amount" of heritability to human lifespan is consistent with the proposal described in the previous paragraphs. The potential lifespan of the zygote is set by its developmental program and the genetic information underlying that developmental program is inherited from its mother and father. However, since the genome and epigenome of the zygote will differ from that of either parent so will its developmental program and thereby its potential maximum lifespan. Hence, heritability contributes only a limited amount to the variation in lifespan across individuals.

This modest contribution of heritability to lifespan is consistent with the results of studies done on populations with a high prevalence of centenarians (Buettner, 2008; discussed in **Chapter 2**). These populations have a diverse genetic background and the major determinants of longevity were behavioral and cultural practices.

Within the context of this model, HGPS discussed in **Chapter 1** can be considered a developmental variant and as such, proof of principle that lifespan and rate of ageing are set by an individual's developmental program. If a zygote inherits a C > T nucleotide substitution at position 1824 in exon 11 of the gene (*LMNA*) encoding lamins A/C, its developmental program will be changed resulting in the premature ageing phenotype characteristic of HGPS. The morphologic phenotype in HPGS reflects the altered biology of lamins, due to the overproduction of a truncated lamin A known as progerin, plus associated altered interactions with other biological processes as discussed in **Chapter 1**. The phenotypic changes resulting from the inheritance of this particular structural variant in the *LMNA* gene involve mesenchymal cell lineages including skeletal system, dermis, adipose tissue, and vascular smooth muscle (Zhang *et al*., 2011). The brain is spared in HGPS. Scaffidi and Misteli (2008) demonstrated that progerin interferes with normal mesenchymal stem cell differentiation which is proposed to be the molecular basis of the HGPS phenotype. These individuals have a much shortened lifespan and an accelerated rate of ageing compared to individuals without this *LMNA* structural variant. Even though the HGPS mutation significantly alters the developmental program, all individuals with this DNA structural variant will still have individual-to-individual variation in their genomes and epigenomes. Hence, among this group of individuals there should still be variation in the developmental process. Such indeed is the case. Ogihara *et al.* (1986) described an individual with the HGPS phenotype who was alive at the age of 45. Their review of the literature indicated that although the median age of death was 13.4 years, the age range at death was quite significant varying from 7 to 27.5 years. In a more recent paper (referred to in **Chapter 1**). Olive *et al.* (2010) presented the post-mortem cardiovascular changes in two individuals with HGPS. Both individuals had confirmed mutations of the *LMNA* gene. One individual, a female, died at the age of 9.9 and the other individual, a male, died at the age of 14. Hence, within our framework, the "HGPS" structural variant in the *LMNA* gene alters the developmental program and thereby resets the maximum potential lifespan and rate of ageing; the maximum lifespan is significantly reduced and the rate of ageing significantly increased. However, as per this model, there is still considerable variation in these two life traits among individuals with this particular mutation.

Interestingly, as reviewed by Dauer and Worman (2009), other mutations in the *LMNA* gene can alter the developmental program to produce different phenotypes than that characteristic of HGPS. These phenotypes generally involve various combinations of abnormalities of adipose tissue, striated muscle, and peripheral nerve. For example, Emery–Dreifuss syndrome is characterized by a scapulo-humeral-peroneal muscular dystrophy distribution. There may also be an associated dilated cardiomyopathy. Other phenotypes based on *LMNA* mutations include Dunnigan-type lipodystrophy with associated insulin resistance and Charcot–Marie–Tooth type 2 peripheral neuropathy. Hence, different DNA structural variants within the *LMNA* gene can result in tissue-specific phenotypes as a result of differential changes in individual developmental programs.

Developmental programs establish tissue-specific transcriptomes, proteomes, and metabolomes, and in so doing set the maximum potential lifespan and ageing pattern for that individual. A corollary of this proposal is that there are no specific longevity and/or ageing genes and pathways. Ageing and maximum lifespan are a reflection of all the biological processes within that individual. Some of these biological processes have been classified as repair processes but as noted in **Chapter 5**, repair processes are simply developmental processes recruited for specific functions.

After birth, a person's interactions with his/her environment can lead to epigenetic-mediated alterations in gene expression patterns. Some environmental cues or experiences will be more biologically important than others and also certain stages of the life cycle are more sensitive to the effects of experiences. For example, the nature of maternal care in the early postnatal period (a developmental stage), as demonstrated in rodents, can significantly modify the stress response in the offspring and this phenotypic change is long lasting, persisting into adulthood. Social isolation in early childhood can have an adverse effect on brain functional development and thereby alter behavioral phenotypes. Since each individual has a unique genome and epigenome, the effects of the same objective experience will be different between any two individuals; i.e., the same objective experience will induce different subjective experiences across individuals. In **Chapter 2**, the effects of a variety of environmental cues were discussed in some detail. Some of the examples discussed included the effects of social environment and voluntary exercise on adult neurogenesis, social experiences and social isolation on the stress axis, the nature of food intake and its association with metabolic reprogramming, and the consequences of hedonic food behaviors. Epigenetic reprogramming of gene expression patterns was explored in **Chapter 2** as the most likely mechanism underlying the effects of these environmental cues and the links between these environmental cues and ageing were considered in **Chapter 2**. There is even evidence that sensory experiences can be transmitted between generations as described for the experience of olfactory fear conditioning in mice which was passed across generations from males in the F0 generation to male offspring in the F2 generation (discussed in a previous section of **Chapter 5**). Hence, the experiences an individual has can modify his/her developmentally set lifespan and rate of ageing most likely through epigenetic reprogramming of gene expression patterns.

The link between developmental programs and longevity has also been described for natural animal populations other than humans. Lee *et al.* (2013) studied an elephant population in Kenya and measured the relationship between early postnatal experiences and offspring growth and longevity. Elephants are a long-lived species with of lifespan of about 70 years. Growth occurs over a period of more than 50 years for males and more than 30 years for females. Lee and co-workers observed an interaction between parity, gender, and the occurrence of drought (an environmental experience) on offspring growth and longevity. Parity was used as a proxy for maternal experience in child rearing since mothers of first born offspring will be less experienced in caring for offspring than mothers who have already gone through child rearing. Firstborn sons who experienced drought, an adverse environmental cue, during the lactation period in early life were small for their age and had a reduced growth curve compared to male offspring not exposed to drought. For daughters, neither drought nor birth order had an effect on growth. The experience of drought during lactation decreased survival for both genders and firstborn males particularly had a high risk of death during the first year. In summary, being born during a drought period as a firstborn male to an inexperienced mother had adverse consequences for longevity, adult size, and reproductive potential. These factors — gender, the nature of maternal care during early life, and exposure to an adverse environmental cue (drought) during early life — all conspired to modify development resulting in negative effects on the longevity and growth of male offspring.

Humans live within specific social and cultural environments. The modern human lineage is about 200,000 years old (Hopkin, 2005) and over the past 15,000 to 20,000 years, there has been a rapid change in our cultural practices. Unlike other species, humans are an active participant in their own evolution. Humans have acquired knowledge and skills through the cultural inheritance system. Knowledge, information, and various skills are passed between generations with each generation adding to and modifying this basket of knowledge and skills. As discussed in a previous section in **Chapter 5**, we have the capacity to significantly modify our local environments and now even our global environment. By modifying our local environments we also modify the selection forces acting within those environments and by this means we redirect natural selection and the course of (human) evolution. Gene–culture co-evolution is considered the main driver of human evolution and is responsible for recent selective sweeps for a number of candidate genes and genomic loci. Gene–culture co-evolution by shaping human evolution will influence lifespan and ageing through several mechanisms. First, gene–culture co-evolution will modify the developmental program of our species and in this way modify our maximum potential lifespan and rate of ageing. But gene–culture co-evolution will also act at the level of the individual. Cultural practices involving intergenerational transfers of material resources can lead to wealth inequalities within a society and promote different reproductive strategies between those having and those not having material resources. Human-mediated climate changes have been linked to increases in violence on both the individual and societal level. Gene–culture co-evolution is unique to the human species and has allowed our species to rapidly adapt to a variety of fluctuations in environmental pressures. Further, gene–culture co-evolution represents a loop-like arrangement. Through niche-constructing practices, humans modify their local environment and the course of their evolution. Modification of the environment by a cultural practice leads to a reciprocal cultural practice to "deal" with the now-modified environment, etc. Clearly, gene–culture co-evolution represents a very dynamic interaction between two interacting inheritance systems.

As already emphasized, the key determinant of an organism's longevity and ageing trajectory is its developmental programs. Experimentally, the rate of ageing is generally measured as age-specific mortality rates or age-related changes in selected biological parameters; examples of such parameters include telomere length, DNA methylation patterns, and metabolomic profiles. The important point is that ageing is generally conceptualized in terms of quantifiable biological processes. However, developmental programs also set behavioral profiles at the species and individual levels. Behavioral profiles include social behaviors, cognitive capacities, emotional behaviors, stress responses, and personality to name a few; this bundle of behaviors can be defined as the behavioral phenotype. This behavioral phenotype is set at a species level and perhaps the best-studied example is the domestication behavioral phenotype discussed in **Chapter 3**. Dogs as a species have a developmentally based domestication behavioral phenotype. However, within the boundaries of this species-set phenotype, there is considerable variation in behaviors both across breeds and within a given breed. Likewise, humans have a developmentally based suite of behaviors that are specific for our species and which differ from the behaviors of other mammals including non-human primates. (One could argue that since behaviors are evolutionarily based, the differences between human behaviors and those of other mammals are more apparent than real.) However, as with other species, the human suite of behaviors shows considerable variation across individuals. This intra-species variation results from the plasticity of underlying developmental programs.

The plasticity of developmental programs that determine human behavioral profiles was studied by Sachser *et al.* (2013). These investigators chose one particular environmental cue, social experiences, and examined its modifying effects on the behavioral phenotype as a function of the life cycle stage at which the exposure occurred. Sachser and co-workers looked at three developmental stages: prenatal period, early postnatal period, and adolescence. During gestation, the maternal social environment affects fetal behavioral profiles through neuroendocrine mechanisms. Adverse or stressful maternal social experiences activate the maternal HPA axis with the increased production of stress hormones. It is these stress hormones that are implicated in modifying fetal brain development and thereby altering the developing behavioral phenotype. During the early postnatal period, the mother shapes offspring behaviors through the nature of care that the mother provides. In this critical time window, the infant–mother relationship is based on attachment and the security provided by the mother. One hypothesis is that the behavioral profiles shaped by the mother during gestation and in early life are adaptive because these particular behaviors help the offspring survive under conditions that match those experienced by the mother during these periods. For example, if the mother was exposed to stressful social experiences during pregnancy or in the early postnatal period, the offspring's behavioral phenotype will be modified to include behaviors that are adaptive for stressful social encounters that the offspring will presumably experience in its own lifetime. Simply put, the mother modifies the fetal behavioral phenotype under the presumption that the offspring will experience the same social environment in the future that the mother experienced during pregnancy or in the early postnatal period (Sachser *et al.*, 2013). The adolescent period is another critical time window in which the child's behavioral phenotype can be modified by social experiences. Animal studies have shown that social experiences related to the density of animals in the colony at the time of adolescence can determine the level of aggressive behavior manifested by males.

Studies such as those reviewed by Sachser *et al.* (2013) underscore the critical role that developmental programs play in shaping the behavioral phenotype that a human expresses in adult life. The model proposed by Sachser *et al.* (2013) is that the nature of maternal care during gestation and the early postnatal period is designed to induce a behavioral phenotype in the offspring adaptive for an environment similar to the one experienced by the mother during these developmental stages of the offspring. If the offspring as an adult does not experience the same social environment as their mother experienced during pregnancy or the early postnatal period, then the behaviors of that adult offspring will probably be maladaptive to their social environment. Such maladaptive behaviors will exert an influence on the ageing trajectory and longevity of that individual.

In summary, the maximum potential lifespan, most probable rate of ageing and the behavioral phenotype for an individual human is set by their unique developmental program. The program is determined by the genetic building blocks inherited from the parents of that individual. Intrauterine environmental cues can modify the program and in so doing shape lifespan, ageing pattern, and behavioral profiles. After birth, life experiences have the potential to epigenetically reprogram gene expression patterns. Whether gene expression patterns are actually changed, and if so, the nature of the change will depend upon the specific features of the life experience, the response of the individual, and at what stage in the individual's life cycle he/she is exposed to the experience. All of these processes and influences converge to determine each individual's unique ageing pattern, behaviors, and lifespan. Since developmental programs appear to be the key to ageing and longevity, future research should examine developmental programs with this notion in mind. In **Chapter 6**, the basis of the disease construct is examined within the context of the framework of ageing developed in this **Chapter 5**.

Two additional points need to be considered. The profound influence of culture and cultural inheritance in human evolution carries the implication that the results of laboratory experiments involving inbred animals may not directly apply to humans. Such experimental setups cannot mimic the human gene–culture co-evolution process. Secondly, stochastic variations in processes such as gene transcription that occur in the single cell are masked when one studies averaged gene transcriptional activity over millions of cells. Likewise, in many cases phenotypic variation between individuals has been masked in studies that looked at phenotypes averaged over large numbers of individuals. Statistical methods can certainly be illuminating but often these methods hide the uniqueness of individuals. As examples, the Physiome Project (Hunter and Borg, 2003) and Phenomics (Houle *et al.*, 2010) are studies designed to define the relationships between genotypes and whole-organism phenotypes. These are clearly important initiatives, but these types of studies by virtue of their design hide the variations present at the level of the individual organism. From a human ageing perspective, understanding and appreciating the uniqueness of the individual is paramount.

Understanding the nature of ageing as well as the observation that organisms have a finite lifespan are currently intense areas of study. Perhaps the only organism that appears to have escaped ageing and death is the hydra (Boehm *et al.*, 2012). The reason for the biological immortality of hydra, when reproducing asexually, is based on the indefinite self-renewal capacity of its stem cells. There are many challenges inherent in the study of ageing but two in particular need to be emphasized. The mechanistic links between development and ageing and longevity are still poorly understood. Developmental programs are plastic and can be significantly modified by environmental cues probably through epigenetic mechanisms. Defining how development and ageing are linked is critical but clearly this represents a significant challenge. Investigating these links can be approached through experimental studies with model organisms. However, understanding human ageing presents additional challenges. The cultural inheritance system is unique to the human species and has been a profound force in human evolution. Cultural practices have shaped our development, ageing, and lifespan. However, it is impossible to simulate the cultural inheritance system in any experimental study using model organisms. This limitation of experimental studies means that translating results from these studies to human ageing and longevity will be very difficult.

One of the themes developed in this book is that the biology of ageing and the biology of the living state are one and the same. Within this construct, there are no specific "ageing" or "longevity" genes. For example, individuals with HGPS have an accelerated ageing trajectory and a shortened lifespan. This phenotype is the result of a mutation in the *LMNA* gene. Does this mean that the *LMNA* gene should be considered a longevity gene since a mutation in this gene is associated with premature ageing? The answer is clearly no.

Recently, newer technologies have been used to study human ageing and Deelen *et al.* (2013) and Valdes *et al.* (2013) have reviewed these studies. Genomic and transcriptomic studies have revealed age-related changes, but often, these changes are tissue-specific. As reviewed in **Chapter 2**, DNA methylation patterns show age-related changes and changes in numerous metabolites also correlate with age (Valdes *et al.*, 2013). The heritability of human lifespan is about 25–30%, but GWASs have not provided convincing evidence for genomic loci that "control" human ageing and longevity (Deelen *et al.*, 2013). These results are not surprising. Biological processes change with age, but these are not biological processes specific to the ageing process. Age-related changes in DNA methylation patterns are widespread across the genome and not limited to a few selected regions. These observations confirm that there are no specific ageing and longevity genes or ageing-specific biological processes. The key is to understand how developmental programs determine not only the ontogeny but also the ageing pattern and longevity of an organism. In the human, there is strong evidence that embryogenesis and fetal maturation are plastic processes that can determine phenotypes that may not be expressed until later in adult life. Numerous examples of this phenomenon have been noted in this book. Describing the mechanistic links between ontogeny and programmed phenotypes that are not expressed until later life is a key research issue.

CHAPTER 6

The Disease Construct and Ageing

The concept of disease is a very old one in human culture. Presumably, this construct was "invented" for explanatory purposes, but it is clear that the disease concept has changed over time as our knowledge of human biology has expanded. In our modern society, the disease concept has been used for other than explanatory purposes. The disease construct has help shape societal policies including what types of biological research are funded and it has fostered a large pharmaceutical industry. In some cases, humans have begun to treat diseases as a military foe. We have declared war on cancer as well as on other disease entities. In this chapter, I will explore the disease construct and how this construct relates to the framework for ageing developed in this book.

The word 'disease' is difficult to define. Webster's New World Dictionary describes a disease as any departure from health or any destructive process in an organ or organism with a specific cause and characteristic symptoms. However, disease is clearly a concept that people are familiar with and seem to understand even though the definition is imprecise. Perhaps it is better to define the disease concept operationally and Duffin (2005) has analyzed the components that make up the disease construct.

She described a number of components comprising the disease construct. One component is the symptoms the individual is suffering from. The constellation of symptoms reported by the individual sufferer is known as the illness component. The individual is also part of the disease concept and if it is a contagious disease, the demographics of the affected population become important. The doctor is another component since the doctor records signs or objective indicators of the illness and as well makes a diagnosis and prescribes treatments. Other components include the diagnosis which is the name or label attached to the illness, the outcome or prognosis, and the cause and the treatment. From the individual's perspective, a disease is considered bad and the individual expects the disease to be discontinuous, i.e., go away. Another component that I have added is one of 'sameness' (Singer, 2013). The disease concept carries the implicit assumption that individuals with the same disease label have the same underlying molecular phenotype and that the results of a specific treatment can be generalized to all individuals having that disease label. This particular aspect of the disease concept was commented on by Kola and Bell (2011). They noted that human diseases are still diagnosed as if they were homogeneous entities and pointed out the negative consequences for clinical studies and medical practice of not recognizing disease heterogeneity.

History of the Disease Construct

Although by today's standards historical disease concepts appear very simplistic, it is important to consider previous disease constructs within the context of biological knowledge at that time. I suspect that in 50 to 100 years from now, our present-day understanding of animal biology will most likely be considered quite primitive.

During the Hippocratic Era, there was no distinction made between diseases in general and specific disease entities (Veith, 1969). A disease was a state that affected the body in general although it might manifest in one spot in the body. Hippocrates was one of the first individuals to remove "gods" from the concept of disease causation. Epilepsy was considered a divine or sacred disease, but Hippocrates dismissed the concept that epilepsy was a "sacred disease" (Veith, 1969). Several new frameworks were developed to account for disease causation and these frameworks focused on either disharmony among four fundamental substances (humors) or disharmony among the minute solid particles that maintained health. Within these frameworks, treatments were crafted to correct disharmony among these particles or substances and restore harmony. In the Middle Ages, there was a return to a disease concept in which causation was due to the action of evil forces, in this case witches. The idea that disease could be transmitted from one individual to another, the concept of contagion, was introduced in the 16th century. Fracastoro wrote about disease contagion at that time, but this concept was not accepted until the existence of microorganisms was substantiated (quoted in Veith, 1969). In the 17th and 18th centuries, disease states were rooted in symptoms. An individual could not have a disease unless that individual had symptoms (Duffin, 2005). In the medical community, great emphasis was placed on the characterization of patient's symptoms including their duration, sequence, pattern, and combination. One could not be sick without feeling sick. The concept of a subclinical or asymptomatic disease was unknown. The invention of the stethoscope by René Laennec in the 19th century allowed physicians to examine the "inside" of a living patient and ushered in the era of anatomic medicine. Today, we have begun to conceptualize diseases in terms of subcellular structures and specific biological processes. Diseases such as HGPS, which reflect impairments in the functioning of the nuclear membrane of the cell, are being referred to as nuclear envelopathies (Dauer and Worman, 2009), and neurodegenerative diseases are now thought to be the result of protein misfolding and aggregation — protein misfolding diseases.

As biological knowledge has expanded, disease taxonomies have become more sophisticated (Singer, 2013). In 1911, the forerunner of the present-day International Statistical Classification of Diseases (ICD) was created. In the ICD, diseases were categorized on the basis of various combinations of symptoms, histological features, laboratory tests, and imaging studies. Revising the ICD tends to be a ponderous process and in 2011, the National Research Council in the U.S. proposed the development of a more fluid taxonomy that would be responsive to the rapidly expanding field of molecular biology which encompassed genomics, proteomics, transcriptomics, and metabolomics — the "omics" (Singer, 2013). The goal is to have this new taxonomy dovetail with the concept of personalized medicine, a driving force in medicine since the sequencing of the human genome and the advent of the era of "omics." Over the past several hundred years, our disease concept has greatly changed. We have progressed from defining a disease entity on the basis of a constellation of symptoms to defining a disease entity on the basis of functional impairments in underlying subcellular structures and biological processes.

One might argue that examining how humans have considered the disease construct over the ages is of historical interest only. I would argue otherwise. The disease concept is a deeply ingrained human belief system. This belief system has been shaped by cultural evolution — the passage and modification of knowledge about diseases and medical practices regarding treatments and cures from generation to generation. It is helpful to consider the disease concept from two perspectives: that of the general population and that of the scientific community. Although humans no longer hold that "gods" or "evil forces" cause disease, most people would consider a disease as separate from themselves as a person. If I label someone as suffering from Alzheimer's disease, that person generally sees themselves differently once they have been so labeled. It is almost as if some alien process has hijacked that person's normal physiology and instilled a diseased state. As noted by Duffin (2005), humans generally consider a disease to be of limited duration and expect that the disease is at least treatable if not curable. The widespread reporting in the media of various "medical breakthroughs" adds to the public's expectations regarding the effectiveness of medical treatments and cures.

With respect to the scientific community, diseases are now framed in terms of impairments in molecular processes, and these impaired or altered biological processes are treated as being "different" than the unimpaired processes. Disease treatments are conceptualized in terms of manipulations that restore biological processes to their unimpaired state.

Hence, the general population and the scientific community have disease concepts that although differing, share a number of common features. Both consider a disease to be apart from the individual. To the public, a disease is some type of external distortion of their normal functioning. To the scientist, a disease is the result of complex interactions between impaired molecular processes which differ in a fundamental way from the unimpaired processes. For the public, an appropriate pharmaceutical or an appropriate intervention will correct the disease, while for the scientist, if the molecular processes can be successfully restored to their unimpaired state, a cure can occur. Another shared feature of the diseases concept between the general population and the scientific community is what I previously referred to as 'sameness.' For the general population, all individuals labeled as suffering from Alzheimer's disease or type 2 diabetes mellitus are put in the same box. The same is true for the medical community. In the basic design of drug trials, the disease group is considered homogeneous, even though this group would contain clusters of individuals with multiple different disease subtypes (Kola and Bell, 2011).

In a recent review article, Hopkins (2013) summarized current concepts concerning the molecular biology of the disease atherosclerosis. This very thorough review is over 200 pages long. Numerous biological processes are discussed and a number of these are similar to processes proposed by Lopez-Otin *et al.* (2013; see **Chapter 5**) as representing the hallmarks of ageing. Some of the processes common to both atherosclerosis and ageing include protein misfolding, reactive oxygen species-mediated damage, insulin signaling pathway, and inflammation. The point I am trying to make is that the number of putative pathways and processes that are considered to be involved in the genesis of atherosclerosis, many of which have been classified as signatures of ageing, is so vast and extensive that the atherosclerotic phenotype should really be considered a variant of ageing rather than a "disease" (Singer, 2013).

Although molecular biological data are accumulating exponentially, driven by the "omics," a few qualifiers need to be appreciated. As we learn more about biological processes at the single-cell and tissue level, more and more questions are raised. As our technologies advance, even more complexity is revealed. A few examples illustrate this point. The Human Genome Project raised as many questions as it answered (Singer, 2013). One surprise was the small number of protein-coding genes and the realization that most DNA appeared to be regulatory in function. Within an individual, there is extensive somatic genomic mosaicism which has been associated with neurodegenerative and neurodevelopmental disorders (Poduri *et al.*, 2013), ageing, and an increased risk for certain malignancies. At the single-cell level, gene transcription and translation have been found to be stochastic in nature. The relationships between stochastic genomic processes at the single-cell level and variations in phenotypes at the whole-organism level are unknown. All of these examples illustrate that molecular biological studies continue to uncover deeper and deeper levels of complexity which are difficult to integrate into our current disease construct. Finally, most laboratory experiments designed to examine disease causation use genetically inbred rodents (hence genetically similar) in an experimental setting that cannot simulate the social and cultural environment occupied by humans. The profound effects of cultural practices on the evolution of the human species had been discussed in **Chapter 5**. The use of genetically homogeneous laboratory animals in an unnatural environmental setting makes translating experimental observations to human biology problematical.

Personalized Medicine

The idea of personalized medicine is not a new one, but it was only with sequencing of the human genome that the concept of personalized medicine was transformed from an idea to a practice. In a position paper, the U.S. Food and Drug Administration (2013) outlined the regulatory steps it believes important in advancing this practice. In this position paper, the definition of personalized medicine is given; in fact, five definitions are presented as proposed by different medical organizations. Here are four definitions presented by four organizations:

- The use of new methods of molecular analysis to better manage a patient's disease or predisposition to disease. (*Personalized Medicine Coalition*)
- Health care that is informed by each person's unique clinical, genetic, and environmental information. (*American Medical Association*)
- A form of medicine that uses information about a person's genes, proteins, and environment to prevent, diagnose and treat disease. (*National Cancer Institute*)
- The tailoring of medical treatment to the individual characteristics of each patient. (*President's Council of Advisors on Science and Technology*)

These definitions underscore the specificity of each individual and the objective of directing medical care to the characteristics of that particular patient. An implicit assumption underlying these definitions is that diseases and disease risks are not uniform across all individuals. A disease is not a single entity but is made up of a collection of variations on that disease. After all, if a disease was a single entity and the same in each person, there would be no need for personalized medicine. If this train of thought is carried to its logical conclusion then a given disease should be considered as being different in each individual; i.e., each individual has their own personal variation of that disease. This conclusion is tantamount to saying that each individual's own disease variation is in reality a feature of their own ageing trajectory.

One of the disciplines that has embraced a "personalized" approach is the field of oncology. Genomic data have been used in several ways.

Specific tumors have unique genomic landscapes and one goal is to use this "uniqueness" as a target for the design of chemotherapeutic agents, i.e., a more personalized approach to cancer therapy. In addition, another goal is to use GWASs to identify genetic biomarkers that will predict the risks of specific cancers in an individual, i.e., a personalized approach to cancer prevention and diagnosis. Saunders *et al.* (2014) examined for genetic markers predictive of prostate cancer in a large population. Prostate cancer has a substantial heritable component, but to date identified risk-associated genomic variants carry only a limited predictive value. Ewing *et al.* (2012) described a germline mutation in the gene *HOXB13* (encodes for the transcription factor homeobox 13) in four families with a pedigree of hereditary prostate cancer. The mutation consisted of a substitution of adenine for guanine in codon 84 (rs 138213197).

Saunders *et al.* (2014) sequenced the *HOXB* locus in 20,440 individuals with prostate cancer and 21,469 controls. Five SNPs were identified with risk association for prostate cancer. Four of these variants, situated within or closely upstream of the *HOXB13* gene, had a frequency of 2.6–4.3 % in the control set and conferred a modest risk of prostate cancer; odds ratio 1.17 to 1.74. The variant rs 138213197 described by Ewing and co-workers was also identified. This variant had a much lower frequency in the control group (0.4%), but yet predicted a much higher risk of prostate cancer; odds ratio 3.88. Haplotype analyses indicated that the low-frequency variant was probably responsible for the "cancer risk" of this cluster of variants; the four variants with only a modest cancer risk are co-inherited with the low-frequency (rare) variant that carries a very high cancer risk. In this case, common SNPs associated with only a modest cancer risk are tagged to a rare variant that is associated with a high cancer risk.

The observations of Saunders and co-workers showed that the weak disease prediction of relatively common DNA sequence variants was because these variants were in reality just a tag for a nearby rare variant that is the actual carrier of the high-risk association. This study is apparently the first experimental evidence for what has been termed a synthetic association; a common SNP (or SNPs) acts as a tag, linked to a rare variant of high-risk prediction with the result that the tag variant considerably underestimates disease risk. What is the implication of these observations if they are generalized to other polygenic diseases? The disease construct assumes that individuals with the same label share a common molecular phenotype. These observations are not consistent with that construct. If the genetic basis of "polygenic" diseases is based on rare variants with large effect sizes, then all individuals with the same label will not have the same molecular phenotype. The underlying molecular phenotype will vary between groups of individuals depending on the number of rare variants involved. The fewer the number of rare variants involved, the more individual-specific the disease becomes; in the limit, the disease will become a variant of an individual's ageing pattern.

Despite the wealth of genomic data that has been amassed since the Human Genome Project was completed in 2000, there has been little progress in applying this data to advance the practice of personalized medicine. This issue was discussed by Katsios and Roukos in 2010. They point out that despite the spectacular advances in genome science in the first post-genome decade, the gap between basic research and clinical application has been widened. As Katsios and Roukos note, the more we learn, the bigger the problem becomes of understanding the mysteries of biodiversity, ageing, and complex diseases such as cancer. A major barrier to successfully using the expanding knowledge of molecular biology is our adherence to the disease concept. The disease construct locks our thinking into an unrealistic framework. As we gather more "omics" data, the complexity of the living state becomes more apparent; each new bit of information seems to generate more and more questions. Yet, we continue to try to "stuff" this expanding "omics" data into an outdated disease taxonomy. To help the fit, we have subdivided disease categories into subtypes. Thus far, we are not consistent in our approach to diseases. When we design clinical trials to test the effectiveness of a pharmaceutical agent, we treat the experimental arm of the trial as being composed of a group of statistically comparable individuals all suffering from the same disease.

Disease or Ageing?

One can learn a considerable amount about human ageing through personal observations. With ageing, hair color changes, becoming grey and then white; skin loses its elasticity and becomes wrinkled and sags particularly in the region of the face and neck; and mobility and balance become impaired, with older humans complaining of an increasing intensity of aches and pains, generally in areas related to the joints. These types of general changes occur in all individuals as they age, although there is extensive variation in the rate of these changes between any two individuals. Humans have no difficulty accepting these changes in external appearance, in certain behaviors (gait, mobility), and in certain sensations ("joint pain") as manifestations of ageing. However, humans do have difficulty accepting that similar changes due to ageing also occur to internal organs and to the myriad of biological processes that exist within the body. Instead, humans refer to ageing-related changes that occur to internal organs and biological processes as "diseases" with all the explicit meanings and connotations that the word 'disease' carries. Within this context and within the framework of the ageing process developed in this book, I want to re-examine three "diseases": atherosclerosis, a chronic degenerative disease; HGPS, a monogenetic disease; and autism spectrum disorder (ASD), a neurodevelopmental behavioral disorder.

Atherosclerosis

Whether atherosclerosis is a disease or a variant of the ageing process was discussed in a previous publication (Singer, 2013). I believe the evidence presented in that review supported the notion that atherosclerosis is actually a manifestation of ageing. In this section, I want to present additional observations consistent with that conclusion but based on links between atherosclerosis and developmental programs.

There are a number of observations that underscore the importance of developmental programs in determining arterial vessel morphology both at the time of birth and at a later age.

Velican and Velican (1979a) examined arterial vessel development in fetuses through to adults. Compared to other organ arteries, the coronary arteries during fetal development showed a rapid increase in intimal thickness with a progressive rise in the intima to media ratio. In addition, fetal coronary arteries showed large pads or cushions at branch points which coalesced after birth, giving rise to diffuse intimal thickening. In other arteries such as the basilar, renal, hepatic, and anterior cerebral, Velican and Velican found branch point pads only in postnatal life.

Neufeld *et al.* (1962) examined coronary artery development in fetuses, aged 27 weeks or older, through to adults. In three fetuses more than 34 weeks old, histological changes in the coronary arteries were noted. The internal elastic membrane showed localized areas of splitting and the intimal layer showed proliferation and fibrous thickening. In early infancy and childhood, these histological changes become more prominent.

Velican and Velican (1976) examined coronary artery histology in 50 fetuses (4–9-months-old), 50 neonates, and 100 infants and children up to 10 years of age. Even in four-month-old fetuses, the investigators noted histological changes particularly where the anterior descending branch emerged from the left main coronary artery. At this emergence zone, there was splitting of the internal elastic membrane and accumulation of medial smooth muscle cells, giving rise to an area of intimal thickening or cushion. With fetal development, these coronary artery cushions (or pads) became more numerous and were found at the bifurcation point and emergence region in several extramyocardial branches of both left and right coronary arteries. Full-term fetuses displayed much more extensive intimal pads in the anterior descending branch of the left coronary artery compared to the extent of intimal pads in the basilar and renal arteries. In addition, the anterior descending branch of the left coronary artery showed fragmentation of the internal elastic membrane, whereas this membrane was intact in the basilar and renal arteries. In summary, intimal thickenings which appeared as pads or cushions located at emergence zones of the coronary arteries appeared in the following sequence: anterior descending branch of the left coronary in four-month-old fetuses, left and right coronary trunks in five- and six-month-old fetuses, and in extramyocardial branches of both left and right coronary arteries during late fetal life.

Velican and Velican (1979b) examined coronary artery histology in neonates, infants, and children. All neonates and children up to 15 years of age examined had intimal branch pads (cushions) in the coronary circulation. Neonates did not show other types of histological changes. However, some children aged 1–5 years did show additional changes including microthrombi and diffuse intimal fibrosis. Some children aged 6–10 years showed fibrous plaques and focal intimal necrosis, while some children aged 11–15 years showed additional histological changes including fatty streaks and gelatinous plaques. Atheroma and complicated lesions were not found in the coronary arteries of children up to 15 years of age.

Weninger *et al.* (1999) studied the histology of the parasellar internal carotid artery (a frequent site of atherosclerosis in adults) in infants between three weeks and nine months of age. Essentially, all of the vascular samples showed hyperplastic intimal cushions although the location of these cushions within the path of the vessel varied. On average, cushions covered about 30% of the lumen circumference and the extent of lumen occlusion by these cushions varied from 6–18%. Although hemodynamic forces are generally incriminated as a causative factor in the development of intimal cushions, the investigators found a correlation between the developmental shape of the parasellar internal carotid artery and the frequency of occurrence of hyperplastic intimal cushions. Straight-running vessels had a much lower frequency of occurrence of cushions (about 40–55% less frequent in occurrence) than strongly curving vessels.

All of these studies indicate that during development of the arterial tree, focal hyperplastic intimal lesions are present in selected arterial vessels by mid-fetal maturation. The selected beds include the coronary circulation, particularly the anterior descending branch of the left coronary artery and the parasellar segment of the internal carotid artery. Intimal cushions have been noted in other fetal arteries, but their size was much less extensive. The relationship of these focal intimal thickenings in fetal arteries to the development of atherosclerosis in adults has not been established. However, atherosclerosis does begin as a focal process and atherosclerotic lesions do have a high predilection for the same arterial sites that intimal cushions are found in the fetus: the coronary circulation (particularly the anterior descending branch of the left coronary artery) and the parasellar section of the internal carotid artery. The occurrence of focal intimal cushions at arterial sites where there is an alteration in laminar blood flow such as regions of strong vessel curvature, vessel emergence zones, and areas of flow dividers (Velican and Velican, 1976; Weninger *et al.*, 1999) suggests that these cushions reflect vascular remodeling probably initiated by the vascular mechanosensory transduction mechanism in response to a reduction in mean positive shear stress (Singer, 2013).

Fetal and neonatal arterial intimal cushions, at sites of altered laminar blood flow, are a product of vascular remodeling, and vascular remodeling is the "second" developmental stage in the formation of the vertebrate cardiovascular system (Jones *et al.*, 2006). The cardiovascular system is the first organ system to develop in vertebrate embryos (Lucitti *et al.*, 2007) and the process of cardiovascular development occurs in two steps: vasculogenesis and then vascular remodeling (Jones *et al.*, 2006). During vasculogenesis, cell specification and differentiation occurs; endothelial cell precursors form lumenized tubes that coalesce to form an initial network for blood flow. During remodeling, this network is converted into a hierarchical vascular tree composed of arteries, veins, and capillaries. Other cell types are then added to the tree, including pericytes and vascular smooth muscle cells. The normal development of the cardiovascular system depends on a large number of genes. In addition, the phase of vascular remodeling requires the actual flow of blood which through generation of physical forces activates the genes and signal pathways necessary for the remodeling process (Lucitti *et al.*, 2007).

The heart and coronary circulation are formed by epicardial-derived cells which are the progenitor cell for endothelial cells, fibroblasts, and smooth muscle cells (Tomanek, 2005). Endothelial cells migrate into the developing myocardium and form vascular tubes. Later, smooth muscle cells are added as remodeling occurs and arteries and veins are differentiated. The protein factor VEGF (vascular endothelial growth factor) plays a major role in the process of vasculogenesis but multiple other growth factors are also involved. The two coronary artery stems are formed from a capillary plexus which in turns forms a ring at the base of the aorta. Penetrating vascular tubes enter the aortic wall at the sites of the right and left coronary cusps (Tomanek, 2005). Once coronary blood flow is established, vascular remodeling occurs. Smooth muscle cells are recruited to the two developing arterial channels at the aorta and the process of artery formation progresses toward the apex of the heart (Tomanek, 2005).

Lucitti *et al.* (2007) examined the role of blood flow in determining the process of vascular remodeling using a mouse embryo model. In the mouse embryo, a beating heart can be detected by the three-somite stage. The onset of a functional circulation is usually defined by the movement of erythroblasts, not just plasma. The investigators used a mutant mouse strain characterized by impaired cardiac contractions. This mouse showed a disruption in plasma and erythroblast circulation and impaired remodeling. Additional experiments demonstrated that plasma circulation alone was not sufficient to induce vascular remodeling. The increase in plasma viscosity imparted by erythroblasts was essential to trigger the remodeling process. The vascular tissue of the embryo expressed the components of the mechanosensory transduction signaling system. The investigators concluded that erythroblast circulation was a requirement for vascular remodeling to occur. The circulation of erythroblasts triggered vascular remodeling through changes in physical forces, primarily shear stress.

In conclusion, focal arterial intimal cushions, at sites of altered laminar blood flow, are a manifestation of the vascular remodeling which occurs during the development of the cardiovascular system in humans. These cushions have been found in almost all human fetuses examined. Over the first decade of postnatal life, the histological complexity of these cushions progresses in many of these children. In essence, the seeds of future atherosclerotic lesions are sown during embryogenesis; focal arteriosclerotic lesions are a product of developmental programs.

As an aside, similar arterial focal intimal cushions are observed in a variety of vertebrates such as fish, mammals, and birds, but rare in reptiles (Singer, 2013).

Developmental programs are plastic and subject to modification by environmental cues occurring during the gestational period. For example, the mother's diet during pregnancy can significantly modify development of the fetal cardiovascular system. Palinski *et al.* (2007) looked at developmental programming in both humans and an animal model. In the human study, 156 children aged 1–13 years were examined post-mortem for evidence of atherosclerosis. The children had normal cholesterol levels. When the extent of atherosclerosis in the children was plotted against age of the children, the slope of the relationship was much steeper in children of hypercholesterolemic mothers compared to normocholesterolemic mothers. In the animal study, normocholesterolemic female rabbits during pregnancy were fed one of three diets: normal chow or one of two other diets designed to induce modest and severe hypercholesterolemia respectively. At birth, the offspring were normocholesterolemic, but the extent of atherosclerosis varied considerable depending upon the mother's diet. Compared to controls, offspring of mothers with modest hypercholesterolemia had a 94% increase in cumulative size of aortic lesions, and offspring of mothers with severe hypercholesterolemia had a 253% increase in cumulative size of aortic lesions. The effect of maternal diet on extent of atherosclerosis in offspring was still evident in offspring aged 12 months. The observations reported by Palinski *et al.* (2007) showed that the composition of the maternal diet during gestation can modify the development of the fetal cardiovascular system.

Two other observations support the notion that atherosclerosis is a vascular remodeling variant of ageing.

As noted in **Chapter 4**, atherosclerosis (manifested as vascular calcification) has been found in ancient humans: a Tyrolean Iceman 5,300 years old and Egyptian mummies 1,650 to 4,000 years old. These ancient humans differ genetically from modern-day humans as well as being very different in terms of cultural practices. Does the atherosclerosis identified in these ancient individuals constitute the same "disease" we suffer from? I suggest that the vascular calcifications in these individuals are more likely a sign of ageing, not disease.

The second observation relates to the premature ageing syndrome HGPS. As discussed in **Chapter 1**, individuals with HGPS have a shortened lifespan as well as early onset atherosclerosis (Olive *et al.*, 2010).

Neither of the two patients reviewed by Olive *et al.* had traditional risk factors for atherosclerosis; neither was diabetic nor hypertensive, but both had slight elevation of triglycerides. HGPS is a developmental disorder resulting from the biological ramifications of a mutation in the *LMNA* gene. The altered developmental programs in this disorder set a shorter lifespan and an accelerated ageing trajectory in these individuals. Atherosclerosis appears at a very early age in these individuals since it is linked to the altered developmental program. Since these individuals have an accelerated ageing course and since atherosclerosis is an ageing variant, its appearance at an early age in individuals with HGPS is expected.

In summary, the evidence is quite strong that atherosclerosis represents a variant of ageing. Focal arterial remodeling begins during fetal development in regions of the arterial tree that are subject to disturbances in laminar blood flow. Changes in shear stress in these regions, acting through the mechanosensory transduction mechanism, induce a remodeling process that leads to focal arterial hyperplastic intimal cushions. The vascular developmental program is further modified by environmental cues occurring during the gestational period. The most prominent cues revolve around maternal behaviors such as dietary intake, alcohol consumption, and tobacco usage. As well documented (Palinski *et al.*, 2007; Velican and Velican, 1979b), the extent and complexity of atherosclerotic lesions continues to progress after birth in many children and young adults. Cultural practices after birth will also direct the course of arterial remodeling, e.g., food choices by the individual, level of chronic stress, and degree of physical activity. Within this context, the review article by Hopkins (2013) on the molecular biology of atherosclerosis is in reality simply a snap shot of all the known biological processes implicated in arterial remodeling. The review does not frame the process of vascular remodeling within the life cycle stages of the individual and hence does not include the profound influence of developmental programs or the significant role that cultural practices play.

HPGS

Should HPGS be classified as a disease, albeit a rare one (Gordon *et al.*, 2014), or a variant of ageing? The answer to this question has to be considered within the context of our current understanding of genetics (Gordon *et al.*, 2014).

Genetic variation is one of the essential ingredients of the evolutionary recipe. Each copy of the human genome is unique and differs in sequence from any other copy in the population by about one nucleotide in every 1,250 nucleotides (Reich *et al.*, 2002). This variation in DNA sequence is the basis for variation in individual phenotypes. Natural selection acts on individual phenotypes and a trait will be favored if it confers a fitness advantage to that individual within the local ecological niche. Such a trait will be passed onto descendant generations if it is based on a heritable genomic variant. Mutations (DNA structural variants) do not occur randomly throughout the genome; certain regions appear to be particularly susceptible to the occurrence of mutational events. For example, most meiotic crossover events are clustered into short regions of the genome known as recombination hot spots (Myers *et al.*, 2008). Such hot spots are a general feature of the genome and have a major influence on human phenotypic variation (Reich *et al.*, 2002). Myers *et al.* (2008) identified specific DNA sequence motifs that are enriched at hot spots, but the molecular basis linking specific DNA sequences and recombination hot spots has not been defined. Within the evolutionary recipe, genomic variants linked to adaptive phenotypic traits will be subject to positive selection and will increase in frequency throughout subsequent generations. However, most mutations are probably neutral or deleterious and the deleterious ones are subject to elimination by purifying selection.

HGPS is most commonly based upon a specific *de novo* mutation in the *LMNA* gene, and the phenotypic expression (premature ageing) of this DNA structural variant is the result of a quantitative rather than a qualitative change. The mutation facilitates the use of a cryptic alternative splice site leading to the production of a truncated lamin A protein known as progerin. However, progerin is also made, albeit in much smaller quantities, by normal cells. The biological function of progerin is unknown, but clearly too much of this protein is detrimental. Is HGPS a disease? HGPS is the phenotypic expression of a deleterious mutation in a protein-coding gene. This mutation is not subject to purifying selection since it generally arises *de novo*. Mutagenesis is an intrinsic characteristic of the living state. It gives rise to genetic variation which is the very basis of evolution. Since we accept the reality of mutagenesis, why have we chosen this particular deleterious mutation and labeled its phenotypic expression as a disease? How do we then decide which mutations are simply part of the "cost of living" and which mutations are considered disease-causing? Is this judgment based on the nature of the phenotypic consequences of the mutation? If so, how do we decide which phenotypes are within the continuum of human variation and which phenotypes are outside "the box" and represent diseases? I would contend that this judgment is arbitrary and not based on defined scientific principles. Within this context, the phenotypic expression of the *LMNA* gene mutation is not a disease, but rather a variant within the ageing continuum albeit an extreme variant.

Autism spectrum disorder (ASD)

ASD represents a highly complex neurodevelopmental "disease" with a wide range of phenotypes and a large patient base (Stessman *et al.*, 2014). One of the best known individuals with ASD is Temple Grandin who has had a very successful academic career (Gross, 2013b). Grandin has discussed her particular type of consciousness which integrates input from all her sensory inputs, but does not lead to integration through symbolic language or connection with emotions (Gross, 2013b). She has described her consciousness as "thinking in pictures" (Gross, 2013b). Due to the extreme phenotypic heterogeneity of ASD, Stessman *et al.* (2014) suggested a genotype-first approach to defining the subtypes of this disorder; i.e., varieties of ASD would first be classified according to the underlying genetic mutational landscape and then phenotypic correlates would be developed. The genomic basis of ASD is very complex as reviewed by Stessman and co-workers, and I propose that the genomic and phenotypic features of ASD fit best with this set of disorders being considered variants of developmental ageing and not as a disease construct. One of the implications of the framework for ageing developed in **Chapter 5**, is that ageing actually starts at the time of conception with the activation of developmental programs. Stessman *et al.* made the following observations about ASD. The number of genes associated with these disorders is vast, with more than 500 being implicated. Even a specialist dealing primarily with ASD would unlikely see the same genetic cause twice in 25 years of practice. Stessman and co-workers concluded that as the number of patients diagnosed with ASD increases, the variety of subtypes of ASD will also increase as will the number of implicated DNA structural variants. ASD is becoming a more and more heterogeneous set of disorders and this extreme heterogeneity of both phenotypes and (underlying) genotypes is more consistent with this set of behaviors being variants of developmental ageing rather than diverse manifestations of a "disease."

As already stated, the objective of this book is to develop a framework for ageing relevant at the level of the individual human. This framework is constructed from multiple unique components: each human has a unique genome and epigenome, each human has a unique developmental program within broad-based species-set boundaries and this developmental program is plastic. Individuals have unique subjective experiences which shape their ageing pattern and longevity, and these experiences occur within the context of a cultural inheritance system which itself will also modify each individual's ageing pattern and longevity. All of this "uniqueness" is incompatible with the disease construct which is based upon "sameness" — all individuals with the same disease label are put in the same box. In addition, the disease construct is dichotomous, e.g., an individual has diabetes or does not, an individual is hypertensive or is not.

This "sameness" feature of a disease is part of the basis for randomized clinical trial studies. Individuals with a disease label are placed into two groups: a control group and an experimental group, which is to receive a study drug. The two groups are statistically matched and considered comparable although in reality, since each human is unique, the construct of a statistical human is purely an artifact. An adverse disease outcome is chosen and the effect of the study drug on the rate of occurrence of this outcome is measured and compared to the rate of occurrence of the same adverse outcome in the untreated control group over the same time period. The effectiveness of the drug is judged on the basis of arbitrary statistical parameters. One such parameter that is used is the number of individuals that need to be treated to prevent one adverse disease outcome. In the case of cardiovascular disease, Abramson *et al.* (2013) reviewed the literature with regard to the use of statin-type cholesterol-lowering drugs in preventing adverse cardiovascular outcomes such as heart attack, stroke, or all-cause mortality. In low-risk individuals, statin treatment did not result in any reduction in overall mortality between treated and control groups. With respect to specific adverse cardiovascular outcomes (heart attack or stroke), 140 low-risk individuals must be treated with a statin drug for five years to prevent one major coronary artery event or stroke, without any reduction in all-cause mortality. Although this result is considered statistically significant, I would submit that this result is clearly not biologically significant. The problem with these types of studies is that they are based on the false assumption that all individuals carrying the same disease label are biologically equivalent.

The design of randomized clinical trials raises the issue of external validity: can the results of a particular study be generalized to other environmental contexts and other populations or are the results limited to the specific population and experimental conditions used in that study (Wurbel, 2000)? If standardization of experimental conditions and the study group were fully effective, then variation within the study group would be reduced to zero and reproducibility of effects between experiments would increase to 100%. However, each experiment would be equivalent to a single case study and the results would not be generalizable (Wurbel, 2000). Rothwell (2005) reviewed evidence that randomized clinical trials were subject to poor external validity. The baseline characteristics of control and treatment groups are statistically matched which tends to mask the variation between individuals in these groups. The experimental setting of the trial is standardized such that this setting may not be comparable to the "real-life" clinical settings in which medical practice occurs. These and other criticisms suggest that the utility of clinical trials may be limited by the lack of generalizability of the results.

In summary, diseases are not biological realities, but are creations of human culture that have been passed down and modified across multiple generations.

Disease or Ageing: Is It Simply Semantics?

The ageing process and the disease construct are fundamentally different concepts. Within the context of current medical practices, giving an individual a disease label such as Alzheimer's disease or hypertension puts them into the same box as other individuals with that disease label. The disease label strips an individual of part of their uniqueness, part of their personal identity. A disease is sometimes conceptualized as if it somehow existed outside the individual and it is treated as if it were a separate entity; e.g., a disease such as cancer is considered a military foe and we wage war on cancer.

Another downside of disease labels is that this type of categorization can rigidify thinking about specific "disease" entities. A case in point is schizophrenia which has been considered a chronic deteriorating psychiatric disorder characterized by rapid cognitive disintegration, so-called dementia praecox. There is a growing body of evidence that individuals with schizophrenia express an accelerated ageing phenotype

(Anthes, 2014) and that schizophrenia should not simply be viewed as a circumscribed psychiatric disorder. Kirkpatrick *et al.* (2008) summarized the ageing-type traits found in individuals with schizophrenia: reduced lifespan, rapid age-related decline in specific cognitive functions, higher prevalence of metabolic abnormalities, and an abnormal neurodevelopmental trajectory. Tang *et al.* (2009) examined gene expression profiles in post-mortem samples of human prefrontal cortex derived from normal subjects and individuals with schizophrenia, both groups ranging in age from 19 to 81 years. Although a number of genes displayed age-related differential expression in both normal subjects and in individuals with schizophrenia, there was very little overlap between these two gene sets, about 2.5%. In contrast, about 34% of genes differentially expressed in young versus old control subjects were also differentially expressed in individuals with early-stage schizophrenia (diagnosed within preceding four years) compared to age- and sex-matched controls. In other words, about one-third of genes which displayed age-related differential expression in normal subjects also showed differential expression in individuals with early-stage schizophrenia compared to matched controls. The implication is that normal ageing and early-stage schizophrenia share a common genetic landscape. The observations reported by Kirkpatrick *et al.* and Tang *et al.* suggest that schizophrenia should no longer be considered strictly a "psychiatric disease," but should instead be viewed as a variant of the ageing process.

The disease construct has had a major influence on the structure of both medical practices and medical education. One aspect of taxonomy has involved classifying diseases according to organ systems: cardiovascular diseases, endocrine diseases, neurological diseases, etc. This type of disease taxonomy fostered the development of specialist physicians: cardiologists, endocrinologists, neurologists, etc. Medical school curricula have embraced a systems approach to educating future physicians: cardiovascular block, endocrine block, neurological block, etc. Each block is generally composed of a review of the normal physiology and anatomy of the particular organ system plus a review of the diseases "specific" to that organ system. Unfortunately, the specialist approach to medical practice and the systems approach to medical education are at odds with a number of realities — at odds with the reality of an individual human as an integrated whole, at odds with the concept of integrated phenotypes, at odds with the multifunctionality of biological processes, and at odds with the reality of hierarchical gene regulatory networks.

The distinction between ageing variants and diseases is clearly not simply a matter of semantics. In addition, dispensing with the disease construct is not equivalent to therapeutic nihilism. People experiencing quality-of-life issues with their own ageing pattern will still need to be treated with technical interventions or pharmaceutical agents as "medically" indicated (Underwood, 2014). But public expectations will be more realistic if the emphasis is on managing ageing rather than aiming for disease cures. However, the objective of this book is not to attempt to undo ingrained cultural practices, but rather to put forward the evidence that our current understanding of molecular biology is not consistent with our cultural construct of diseases.

Epilogue

The complexity of ageing makes it impossible for a book on this topic to be complete. For example, I have not discussed the role of the gastrointestinal microbial flora in human health and ageing. In a sense, each individual and their associated individual-specific microbial populations can be considered a unique micro-ecosystem. Ageing is not based on specific biological processes. Ageing and the living state are one and the same. In this book, I have relied heavily upon genomics. However, despite all the advances since the completion of the Human Genome Project, we still do not understand the genomic basis of most phenotypic traits. This challenge will become greater as we begin to look at whole organism integrated phenotypes (phenomics) and not just selected phenotypic traits. The field of genomics is exponentially expanding, but each new "discovery" uncovers deeper layers of complexity and raises many additional questions. Even though most human DNA is biochemically active, we do not know how much of this biochemical activity is functionally significant. We do know that there is extensive genomic variation between individuals and extensive genomic variation within a single individual. We also know that at the level of the single cell, fundamental genomic processes such as gene transcription and translation are stochastic. We have not yet been able to relate these genomic variations between individuals and within a given individual to variations in integrated phenotypes.

What distinguishes humans from other species is their system of cultural inheritance. Although culture is an evolutionarily based activity, human cultural practices are far more sophisticated and complex than those of any other species. Humans can now inhabit almost any area of the Earth and also have the capacity to live on other locations of our solar system such as the moon. Through niche construction practices, humans can and do make major changes in their local environments, and as a result, humans have become active participants and drivers of their own evolution. This sequence can be viewed as a loop-type arrangement: cultural practices > evolution > cultural practices. As a corollary, humans have unwittingly begun to drive the evolution of other species as a result of human-mediated global climate change. In fact, understanding how species cope with rapidly changing environmental conditions has become an important research topic in biology (Nemeth *et al.*, 2013).

Within the framework developed in this book, human ageing begins at the time of conception. Human-specific developmental programs, the expressions of which will vary considerably across individuals, set the most probable ageing trajectory and potential lifespan for each individual. Developmental programs are plastic and can be modified by intrauterine and early postnatal environmental cues acting during critical time windows. After birth, the potential lifespan and ageing pattern of an individual are further modified by life experiences (environmental cues) that can be considered under the rubric of sensory experiences since experiences can only "enter" a person through sensory receptors and specific neural pathways. The cultural context of an individual is pivotal in determining the actual effects of these experiences.

We no longer need to invoke a disease construct; diseases are simply manifestations of the ageing process. These manifestations of the ageing process will show extensive variation in expression across individuals. However, the disease concept is a deeply ingrained human belief system which has shaped our medical practices and spawned a large pharmaceutical industry. The disease concept also represents an important part of our cultural inheritance. This concept has been passed down from generation to generation with modification by descendant generations in parallel with the expansion of biological knowledge. Humans are now at a juncture at which advances in molecular biology have raised public expectations for disease "cures." I believe that these expectations are unrealistic given that diseases are in reality variants of ageing and not separate discrete distortions of our underlying "normal" physiology. Members of the biomedical scientific community must be leaders in disseminating information about this new reality. We have acquired an extensive molecular database for living organisms, but how far have we actually come? It is interesting to compare disease concepts down through the ages. For example, if we consider a disease such as HGPS, is there a huge conceptual difference between considering the causation to be the acquisition of an evil spirit or force as compared to the acquisition of a "bad" gene? As noted by the medical historian Veith (1969): "There will always be a search for an answer to all our ills — perhaps in the hope that one day one school of thought will find the one true concept that has power over all disease."

This comment by Veith is consistent with recent assessments of public attitudes toward science-based topics. Although in modern societies information is widely available and easily disseminated, public understanding of scientific knowledge is still poor. A survey conducted by GfK Public Affairs and Corporate Communications (GfK, 2014) assessed public acceptance of scientific facts using 1,012 adult respondents in the U.S. Eighty-two percent of adults surveyed agreed that smoking caused cancer. Fifty-four percent of respondents agreed that the universe was so complex that there must be a supreme being guiding its creation. Interestingly, only 31% of surveyed adults agreed that life on Earth, including human beings, evolved through a process of natural selection. A smaller percentage (27%) of respondents accepted that the Earth was 4.5 billion years old. Hence, evolution and the known geologically based age of the Earth are scientific facts accepted by only a minority of those surveyed, while creationism is accepted by more than half of the respondents.

Even though the U.S. is one of the most technologically advanced countries, there exists a clear disconnect between the widespread availability of information about established scientific facts and public acceptance of those facts.

The disease concept will probably persist because it fulfills a cultural need by allowing humans to put aside the reality of ageing and embrace the notion that medical "advances" will cure not only diseases but also cure the ultimate disease of death.

Bibliography

Abramson, JD; Rosenberg, HG; Jewell, N; Wright, JM, 2013. Should people at low risk of cardiovascular disease take a statin? *BMJ* 347, 16123.

Abyzov, A; Mariani, J; Palejev, D; Zhang, Y; Haney, MS; *et al.*, 2012. Somatic copy number mosaicism in human skin revealed by induced pluripotent stem cells. *Nature* 492, 438–442.

Adams, PC; Reboussin, DM; Barton, JC; McLaren, CE; Eckfeldt, JH; *et al.*, 2005. Hemochromatosis and iron-overload screening in a racially diverse population. *N Engl J Med* 352, 1769–1778.

Alexandrov, LB; Nik-Zainal, S; Wedge, DC; Aparicio, SAJR; Behjati, S; *et al.*, 2013. Signatures of mutational processes in human cancer. *Nature* 500, 415–421.

Allam, AH; Thompson, RC; Wann, S; Miyamoto, MI; el-Din, A e-HN; *et al.*, 2011. Atherosclerosis in ancient Egyptian mummies. *J Am Coll Cardiol Img* 4, 315–327.

Amadio, JP; Walsh, CA, 2006. Brain evolution and uniqueness in the human genome. *Cell* 22, 1033–1035.

Anthes, E, 2014. Live faster, die younger. *Nature* 508, S16–S17.

Ardlie, KG; Kruglyak, L; Seielstad, M, 2002. Patterns of linkage disequilibrium in the human genome. Nat Rev Genet 3, 299–309.

Armanios, M; Blackburn, EH, 2012. The telomere syndromes. *Nat Rev Genet* 13, 693–704.

Asher, G; Schibler, U, 2013. Crosstalk between components of circadian and metabolic cycles in mammals. *Cell Metab* 13, 125–137.

Attanasio, C; Nord, AS; Zhu, Y; Blow, MJ; Li, Z; *et al.*, 2013. Fine tuning of craniofacial morphology by distant-acting enhancers. *Science* 342, 1241006.

Baek, J-H; McKenna, T; Eriksson, M. Hutchinson–Gilford Progeria syndrome. In Puiu, M (Ed.), 2013. *Genetic Disorders*, Ch. 3, InTech, doi: 10.5772/53794.

Bainer, R; Weaver, V, 2013. Strength under tension. *Science* 341, 965–966.
Ball, P, 2013. Celebrate the unknowns. *Nature* 496, 419–420.
Balter, M, 2005. Are human brains still evolving? Brain genes show signs of selection. *Science* 309, 1662–1663.
Banito, A; Lowe, SW, 2013. A new development in senescence. *Cell* 155, 977–978.
Barbosa-Morais, NL; Irimia, M; Pan, Q; Xiong, HY; Gueroussov, S; *et al.*, 2012. The evolutionary landscape of alternative splicing in vertebrate species. *Science* 338, 1587–1593.
Barnea, A; Pravosudov, V, 2011. Birds as a model to study adult neurogenesis: Bridging evolutionary, comparative and neuroethological approaches. *Eur J Neurosci* 34, 884–907.
Bateson, P; Barker, D; Clutton-Brock, T; Deb, D; D'Udine, B; *et al.*, 2004. Developmental plasticity and human health. *Nature* 430, 419–421.
Baudisch, A; Vaupel, JW, 2012. Getting to the root of aging. *Science* 338, 618–619.
Bayer, R; Fairchild, AL; Hopper, K; Nathanson, CA, 2013. Confronting the sorry state of U.S. health. *Science* 341, 962–963.
Beekman, M; Nederstigt, C; Suchiman, HED; Kremer, D; van der Breggen, R; *et al.*, 2010. Genome-wide association study (GWAS)-identified disease risk alleles do not compromise human longevity. *Proc Natl Acad Sci U S A* 107, 18046–18049.
Bekaert, S; Derradji, H; Baatout, S, 2004. Telomere biology in mammalian germ cells and during development. *Dev Biol* 274, 15–30.
Bell, JT; Spector, TD, 2011. A twin approach to unraveling epigenetics. *Trends Genet* 27, 116–125.
Benoit, SC; Tracy, AL; Davis, JF; Choi, D; Clegg, DJ, 2008. Novel functions of orexigenic hypothalamic peptides: From genes to behavior. *Nutrition* 24, 843–847.
Benson, EK; Lee, SW; Aaronson, SA, 2010. Role of progerin-induced telomere dysfunction in HGPS premature cellular senescence. *J Cell Sci* 12, 2605–2612.
Bergmann, O; Frisen, J, 2013. Why adults need new brain cells. *Science* 340, 695–696.
Berndt, JD; Moon, RT, 2013. Making a point with Wnt signals. *Science* 339, 1388–1389.
Bersaglieri, T; Sabeti, PC; Patterson, N; Vanderploeg, T; Schaffner, SF; *et al.*, 2004. Genetic signatures of strong recent positive selection at the lactase gene. *Am J Hum Genet* 74, 1111–1120.
Berthoud, H-R, 2011. Metabolic and hedonic drives in the neural control of appetite: Who is the boss? *Curr Opin Neurobiol* 21, 888–896.
Beutler, E, 2004. Iron absorption in carriers of the C282Y hemochromatosis mutation. *Am J Clin Nutr* 80, 799–800.
Bhattacharjee, Y, 2014. The vigilante. *Science* 343, 1306–1309.
Bintu, L; Ishibashi, T; Dangkulwanich, M; Wu, Y-Y; Lubkowska, L; *et al.*, 2012. Nucleosomal elements that control the topography of the barrier to transcription. *Cell* 151, 738–749.
Bird, A, 2013. Genome biology: Not drowning but waving. *Cell* 154, 951–952.
Biro, FM; Greenspan, LC; Galvez, MP; Pinney, SM; Teitelbaum, S; *et al.*, 2013. Onset of breast development in a longitudinal cohort. *Pediatrics* 132, 1–9.
Blackburn, EH; Epel, ES, 2012. Too toxic to ignore. *Nature* 490, 169–171.

Blaxter, M, 2014. Development: The maternal-zygotic transition revisited. *Curr Biol* 24, R72–R75.

Bloch, G; Barnes, BM; Gerkema, MP; Helm, B, 2013. Animal activity around the clock with no overt circadian rhythms: Patterns, mechanisms and adaptive value. *Proc R Soc B* 280, 20130019.

Blumenthal, DM; Gold, MS, 2010. Neurobiology of food addiction. *Curr Opin Clin Nutr Metab Care* 13, 359–365.

Boccardi, V; Herbig, U, 2012. Telomerase gene therapy: A novel approach to combat aging. *EMBO Mol Med* 4, 685–687.

Bocquet-Appel, J-P, 2011. When the world's population took off: The springboard of the Neolithic demographic transition. *Science* 333, 560–561.

Boehm, A-M; Khalturin, K; Anton-Erxleben, F; Hemmrich, G; Klostermeier, UC; *et al.*, 2012. *FoxO* is a critical regulator of stem cell maintenance in immortal Hydra. *Proc Natl Acad Sci USA* 109, 19697–19702.

Boks, MP; Derks, EM; Weisenberger, DJ; Strengman, E; Janson, E; *et al.*, 2009. The relationship of DNA methylation with age, gender and genotype in twins and healthy controls. *PLoS One* 4, e6767, doi: 10.1371.

Bonaguidi, MA; Wheeler, MA; Shapiro, JS; Stadel, RP; Sun, GJ; *et al.*, 2011. *In vivo* clonal analysis reveals self-renewing and multipotent adult neural stem cell characteristics. *Cell* 145, 1142–1155.

Bonnal, S; Valcarcel, J, 2008. Spliceosome meets telomerase. *Nature* 456, 879–880.

Boyko, AR, 2011. The domestic dog: Man's best friend in the genomic era. *Genome Biol* 12, 216.

Boyko, AR; Quignon, P; Li, L; Schoenebeck, JJ; Degenhardt, JD; Lohmueller, KE; *et al.*, 2010. A simple genetic architecture underlies morphological variation in dogs. *PLoS Biol* 8, e1000451, doi: 10.1371/journal.pbio.1000451.

Box, JA; Bunch, JT; Tang, W; Baumann, P, 2008. Spliceosome cleavage generates the 3′ end of telomerase RNA. *Nature* 456, 910–914.

Breen, MS; Kemena, C; Vlasov, PK; Notredame, C; Kondrashov, FA, 2012. Epistasis as the primary factor in molecular evolution. *Nature* 490, 535–538.

Brownell, KD; Warner, KE, 2009. The perils of ignoring history: Big tobacco played dirty and millions died. How similar is big food? *Milbank Q* 87, 259–294.

Buchen, L, 2010. In their nuture. *Nature* 467, 146–148.

Budovskaya, YV; Wu, K; Southworth, LK; Jiang, M; Tedesco, P; *et al.*, 2008. An elt-3/elt-5/elt-6 GATA transcription circuit guides aging in *C. elegans*. *Cell* 134, 291–303.

Buettner, D, 2008. *The Blue Zones: Lessons for Living Longer from the People who've Lived the Longest*. Washington D.C., National Geographic.

Buzsaki, G, 2013. Time, space and memory. *Nature* 497, 568–569.

Byars, SG; Ewbank, D; Govindaraju, DR; Stearns, SC, 2010. Natural selection in a contemporary human population. *Proc Natl Acad Sci USA* 107(Suppl. 1), doi: 10.1073/pnas.0906199106.

Callaway, E, 2014. Epigenomics starts to make its mark. *Nature* 508, 22.

Cameron HA; Gould, E, 1994. Adult neurogenesis is regulated by adrenal steroids in the dentate gyrus. *Neuroscience* 61, 203–209.

Campisi, J, 2005. Senescent cells, tumor suppression, and organismal aging: Good citizens, bad neighbors. *Cell* 120, 513–522.

Campisi, J, 2013. The beginning of the end. *Nature* 505, 35–36.

Canadian Cancer Society. *Canadian Cancer Statistics 2011*. Released 2011, http://www.cancer-asian.com/images/news/Canadian_Cancer%20Statistics_2011_English.pdf

Cancer Research UK. *Cancer Statistics for the UK*. Released 2008–2010, http://www.cancerresearchuk.org/cancer-info/cancerstats/

Cao, R; Obrietan, K, 2010. mTOR signaling and entrainment of the mammalian circadian clock. *Mol Cell Pharmacol* 2, 125–130.

Cao, K; Blair, CD; Faddah, DA; Kieckhaefer, JE; Olive, M; Erdos, MR; *et al.*, 2011. Progerin and telomere dysfunction collaborate to trigger cellular senescence in normal human fibroblasts. *J Clin Invest* 121, 2833–2844.

Carone, BR; Rando, OJ, .2012. Rewriting the epigenome. *Cell* 149, 1422–1423.

Caspari, R; Lee, S-H, 2004. Older age becomes common late in human evolution. *Proc Natl Acad Sci USA* 101, 10895–10900.

Caspi, A; Sugden, K; Moffitt, TE; Taylor, A; Craig, IW; *et al.*, 2003. Influence of life stress on depression: Moderation by a polymorphism in the *5-HTT* gene. *Science* 301, 386–389.

Catchpole, B; Adams, JP; Holder, AL; Short, AD; Ollier, WER; *et al.*, 2013. Genetics of canine diabetes: Are the diabetes susceptibility genes identified in humans involved in breed susceptibility to diabetes mellitus in dogs? *Vet J* 195, 139–147.

Chakravarti, A; Clark, AG; Mootha, VK, 2013. Distilling pathophysiology from complex disease genetics. *Cell* 155, 21–26.

Champagne, FA, 2010. Epigenetic influence of social experiences across the lifespan. *Dev Psychbiol* 52, 299–311.

Chen, M; Manley, JL, 2009. Mechanisms of alternative splicing regulation: Insights from molecular and genomic approaches. *Nat Rev Mol Cell Biol* 10, 741–754.

Chow, TT; Zhao, Y; Mak, SS; Shay, JW; Wright, WE, 2012. Early and late steps in telomere overhang processing in normal human cells: The position of the final RNA primer drives telomere shortening. *Genes Dev* 26, 1167–1178.

Church, G, 2013. Improving genome understanding. *Nature* 502, 143.

Clark, AG; Glanowski, S; Nielsen, R; Thomas, PD; Kejariwal, A; *et al.*, 2003. Inferring nonneutral evolution from human-chimp-mouse orthologous gene trios. *Science* 302, 1960–1963.

Clevers, H; Nusse, R, 2012. Wnt/B-Catenin signaling and disease. *Cell* 149, 1192–1205.

Coelho, M; Dereli, A; Haese, A; Kuhn, S; Malinovska, L; *et al.*, 2013. Fission yeast does not age under favorable conditions, but does so after stress. *Curr Biol* 23, 1844–1852.

Cohen, AA, 2004. Female post-reproductive lifespan: A general mammalian trait. *Biol Rev* 79, 733–750.

Cortijo, S; Wardenaar, R; Colome-Tatche, M; Gilly, A; Etcheverry, M; *et al.*, 2014 Mapping the epigenetic basis of complex traits. *Science*, 2014 343, 1145–1148.

Corwin, RL; Grigson, PS, 2009. Symposium overview-food addiction: Fact or fiction? *J Nutr* 139, 617–619.

Cotney, J; Leng, J; Yin, J; Reilly, SK; DeMare, LE; *et al.*, 2013. The evolution of lineage-specific regulatory activities in the human embryonic limb. *Cell* 154, 185–196.

Cournil, A; Kirkwood, TBL, 2001. If you would live long, choose your parents well. *Trends Genet* 17, 233–235.

Couzin-Frankel, J, 2013. How does fetal environment influence later health? *Science* 340, 1160–1161.

Crosio, C; Heitz, E; Allis, CD; Borrelli, E; Sassone-Corsi, P, 2003. Chromatin remodeling and neuronal response: multiple signaling pathways induce specific histone H3 modifications and early gene expression in hippocampal neurons. *J Cell Sci* 28, 4905–4914.

Cruz, F; Vila, C; Webster, MT, 2008. The legacy of domestication: Accumulation of deleterious mutations in the dog genome. *Mol Biol Evol* 25, 231–236.

Csoka, AB; English, SB; Simkevich, CP; Ginzinger, DG; Butte, AJ; *et al.*, 2004. Genome-scale expression profiling of Hutchinson–Gilford progeria syndrome reveals widespread transcriptional misregulation leading to mesodermal/ mesenchymal defects and accelerated atherosclerosis. *Aging Cell* 3, 235–243.

Culliton, BJ, 1968. 64 armadillos threaten a theory. *Sci News* 94, 555–557.

Curry, A, 2013. Archaeology: The milk revolution. *Nature* 500, 20–22.

Curtsinger, JW; Fukui, HH; Townsend, DR; Vaupel, JW, 1992. Demography of genotypes: Failure of the limited life-span paradigm in *Drosophila melanogaster*. *Science* 258, 461–463.

D'Apice, MR; Tenconi, R; Mammi, I; van den Ende, J; Novelli, G, 2004. Paternal origin of *LMNA* mutations in Hutchinson–Gilford progeria. *Clin Genet* 65, 52–54.

Danese, A; Moffitt, TE; Harrington, HL; Milne, B; Polanczyk, G; *et al.*, 2009. Adverse childhood experiences and adult risk factors for age-related disease. *Arch Pediatr Adolesc Med* 163, 1135–1143.

Davis, S; Bozon, B; Laroche, S. How necessary is the activation of the immediate early gene zif268 in synaptic plasticity and learning? *Behav Brain Res*, 2003 142, 17–30.

Dauer, WT; Worman, HJ, 2009. The nuclear envelope as a signaling node in development and disease. *Dev Cell* 17, 626–638.

Day, JJ; Sweatt, JD, 2011. Epigenetic mechanisms in cognition. *Neuron* 70, 813–829.

Dean, LG; Vale, GL; Laland, KN; Flynn, E; Kendal, RL, 2014. Human cumulative culture: A comparative perspective. *Biol Rev* 89, 284–301.

Decker, ML; Chavez, E; Vulto, I; Lansdorp, PM, 2009. Telomere length in Hutchinson–Gilford progeria syndrome. *Mech Ageing Develop* 130, 377–383.

Deelen, J; Beekman, M; Capri, M; Franceschi, C; Slagboom, PE, 2013. Identifying the genomic determinants of aging and longevity in human population studies: Progress and challenges. *Bioessays* 35, 386–396.

De Jesus, BB; Vera, E; Schneeberger, K; Tejera, AM; Ayuso, E; *et al.*, 2012. Telomerase gene therapy in adult and old mice delays aging and increases longevity without increasing cancer. *EMBO Mol Med* 4, 691–704.

De Laat, W; Duboule, D, 2013. Topology of mammalian developmental enhancers and their regulatory landscapes. *Nature* 502, 499–506.

De Magalhaes, J; Costa, J; Church, GM, 2007. An analysis of the relationship between metabolism, developmental schedules, and longevity using phylogenetic independent contrasts. *J Geront Biol Sci* 62A, 149–160.

De Mendoza, A; Sebe-Pedros, A; Sestak, MS; Matejcic, M; Torruella, G; *et al.*, 2013. Transcription factor evolution in eukaryocytes and the assembly of the regulatory toolkit in multicellular lineages. *Proc Natl Acad Sci USA* 110, E4858–E4866.

Deng, Q; Ramskold, D; Reinius, B; Sandberg, R, 2014. Single-cell RNA-seq reveals dynamic, random monoallelic gene expression in mammalian cells. *Science* 343, 193–196.

Deweerdt, S, 2012. Looking for a master switch. *Nature*, 492, S10–S11.

Dias, BG; Ressler, KJ, 2014. Parental olfactory experience influences behavior and neural structure in subsequent generations. *Nat Neurosci* 17, 89–96.

Dickins, TE; Rahman, Q, 2013. Epigenetic adaptations: A reply to Suter, Boffelli and Martin. *Proc R Soc B* 280, 20131820.

Dittmer, T; Misteli, T, 2011. The lamin protein family. *Genome Biol* 12, 222.

Dobson, JM, 2013. Breed-predispositions to cancer in Pedigree dogs. *ISRN Vet Sci* p. 23, doi: 10.1155/2013/941275.

Doherty, CJ; Kay, SA, 2012. Circadian surprise — it's not all about transcription. *Science* 338, 338–340.

Driskell, RR; Lichtenberger, BM; Hoste, E; Kretzschmar, K; Simons, BD; *et al.*, 2013. Distinct fibroblast lineages determine dermal architecture in skin development and repair. *Nature* 504, 277–281.

Dudley, KJ; Sloboda, DM; Connor, KL; Beltrand, J; Vickers, MH, 2011. Offspring of mothers fed a high fat diet display hepatic cell cycle inhibition and associated changes in gene expression and DNA methylation. *PLoS One* 6, e21662, doi: 10.1371/journal.pone.0021662.

Duffin, J, 2005. *Lovers and Livers: Disease Concepts in History*. Toronto, Canada: University of Toronto Press.

Dunlap, KD; McCarthy, EA; Jashari, D, 2008. Electrocommunication signals alone are sufficient to increase neurogenesis in the brain of adult electric fish, *Apteronotus leptorhynchus*. *Dev Neurobiol* 68, 1420–1428.

Duong, HA; Robles, MS; Knutti, D; Weitz, CJ, 2011. A molecular mechanism for circadian clock negative feedback. *Science* 332, 1436–1439.

Eaton, SB; Konner, M; Shostak, M, 1988. Stone agers in the fast lane: Chronic degenerative diseases in evolutionary perspective. *Am J Med* 84, 739–749.

Eccleston, A; Cesari, F; Skipper, M, 2013. Transcription and epigenetics. *Nature* 502, 461.

Ehrlich, PR; Ehrlich, AH, 2013a. Can a collapse of global civilization be avoided? *Proc R Soc B* 280, 20122845.

Ehrlich, PR; Ehrlich, AH, 2013b. Future collapse: How optimistic should we be? *Proc R Soc B* 280, 20131373.

Elowitz, MB; Levine, AJ; Siggia, ED; Swain, PS, 2002. Stochastic gene expression in a single cell. *Science* 297, 1183–1186.

Erber, A; Riemer, D; Hofemeister, H; Bovenschulte, M; Stick, R; *et al.*, 1999. Characterization of the Hydra lamin and its gene: A molecular phylogeney of metazoan lamins. *J Mol Evol* 49, 260–271.

Evans, PD; Gilbert, SL; Mekel-Bobrov, N; Vallender, EJ; Anderson, JR; *et al.*, 2005. Microcephalin, a gene regulating brain size, continues to evolve adaptively in humans. *Science* 309, 1717–1720.

Ewing, CM; Ray, AM; Lange, EM; Zuhlke, KA; Robbins, CM; *et al.*, 2012. Germline mutations in *HOXB13* and prostate-cancer risk. *N Engl J Med* 366, 141–149.

Falk, D, 2004. Hominin brain evolution-new century, new directions. *Coll Antropol* 28(Suppl. 2), 59–65.

Falk, D; Zollikofer, CPE; Morimoto, N; Ponce de Leon, MS, 2012. Metopic suture of Taung (*Australopithecus africanus*) and its implications for hominin brain evolution. *Proc Natl Acad Sci USA* 109, 8467–8470.

Farmer, J; Zhao, X; Van Pragg, H; Wodtke, K; Gage, FH; *et al.*, 2004. Effects of voluntary exercise on synaptic plasticity and gene expression in the dentate gyrus of adult male Sprague–Dawley rats *in vivo*. *Neuroscience* 124, 71–79.

Faulk, C; Dolinoy, DC, 2011. Timing is everything. The when and how of environmentally induced changes in the epigenome of animals. *Epigenetics* 6, 791–797.

Fay, JC; Wyckoff, GJ; Wu, C-I, 2001. Positive and negative selection on the human genome. *Genetics* 158, 1227–1234.

Filipowicz, W; Bhattacharyya, SN; Sonenberg, N, 2008. Mechanisms of post-transcriptional regulation by microRNAs: Are the answers in sight? *Nature* 9, 102–114.

Fisher, TS; Zakian, VA, 2005. Ku: A multifunctional protein involved in telomere maintenance. *DNA Repair (Amst)* 4, 1215–1226.

Fleming, JM; Creevy, KE; Promislow, DEL, 2011. Mortality in North American dogs from 1984 to 2004: An investigation into age-, size-, and breed-related causes of death. *J Vet Intern Med* 25, 187–198.

Flicek, P, 2013. The handiwork of tinkering. *Nature* 500, 158–159.

Foote, AD, 2008. Mortality rate acceleration and post-reproductive lifespan in matrilineal whale species. *Biol Lett* 4, 189–191.

Foote, CG; Daunt, F; Gonzalez-Solis, J; Nasir, L; Phillips, RA; *et al.*, 2011. Individual state and survival prospects: Age, sex, and telomere length in a long-lived seabird. *Behav Ecol* 22, 156–161.

Fowden, AL; Coan, PM; Angiolini, E; Burton, GJ; Constancia, M, 2011. Imprinted genes and the epigenetic regulation of placental phenotype. *Prog Biophys Mol Biol* 106, 281–288.

Fraga, MF; Ballestar, E; Paz, MF; Ropero, S; Setien, F; *et al.*, 2005. Epigenetic differences arise during the lifetime of monozygotic twins. *Proc Natl Acad Sci USA* 102, 10604–10609.

Freund, J; Brandmaier, AM; Lewejohann, L; Kirste, I; Ktitzler, M; *et al.*, 2013. Emergence of individuality in genetically identical mice. *Science* 340, 756–759.

Frost, JM; Moore, GE, 2010. The importance of imprinting in the human placenta. *PLoS Genet* 6, e1001015, doi: 10.1371/journal.pgen.1001015.

Furey, TS; Sethupathy, P, 2013. Genetics driving epigenetics. *Science* 342, 705–706.

Furusawa, C; Kaneko, K, 2012. A dynamical-systems view of stem cell biology. *Science* 338, 215–217.
Futreal, PA; Coin, L; Marshall, M; Down, T; Hubbard, T; *et al.*, 2004. A census of human cancer genes. *Nat Rev Cancer* 4, 177–183.
Galis, F; Van Der Sluijs, I; Van Dooren, TJM; Metz, JAJ; Nussbaumer, M, 2007. Do large dogs die young? *J Exp Zool (Mol Dev Evol)* 308B, 119–126.
Garcia-Cao, M; O'Sullivan, RO; Peters, AHFM; Jenuwein, T; Blasco, MA. 2004. Epigenetic regulation of telomere length in mammalian cells by the Suv39h1 and Suv39h2 histone methyltransferases. *Nat Genet* 36, 94–99.
Gendrel, A-V; Attia, M; Chen, C-J; Diabangouaya, P; Servant, N; *et al.*, 2014. Developmental dynamics and disease potential of random monoallelic gene expression. *Dev Cell* 28, 366–380.
GfK Public Affairs and Corporate Communications. *The AP-GfK Poll: A survey of the American general population (ages 18+)*. March 2014, http://surveys.ap.org/data/GfK/AP-GfK%20March%202014%20Poll%20Topline%20%20Final_OBAMA.pdf
Gibbons, A, 2010. Tracing evolution's recent fingerprints. *Science* 329, 740–742.
Gibbons, A; Pennisi, E, 2013. How a fickle climate made us human. *Science* 341, 474–479.
Gibbs, WW, 2014. The clock-watcher. *Nature* 508, 169–170.
Gilad, Y; Man, O; Paabo, S; Lancet, D, 2013. Human specific loss of olfactory receptor genes. *Proc Natl Acad Sci USA* 100, 3324–3327.
Gilad, Y; Segre, D; Skorecki, K; Nachman, MW; Lancet, D; *et al.* Dichotomy of single-nucleotide polymorphism haplotypes in olfactory receptor genes and pseudogenes. *Nat Genet* 26, 221–224.
Glaeser, E, 2011. Cities, productivity, and quality of life. *Science* 333, 592–594.
Gokhman, D; Lavi, E; Prufer, K; Fraga, MF; Riancho, JA; *et al.*, 2014. Reconstructing the DNA methylation maps of the Neanderthal and the Denisovan. *Sci Exp* 344, 523–527.
Gonzalex-Porta, M; Calvo, M; Sammeth, M; Guigo, R., 2012. Estimation of alternative splicing variability in human populations. *Genome Res* 22, 528–538.
Gonzalez-Suarez, I; Gonzalo, S. 2008. Crosstalk between chromatin structure, nuclear compartmentalization, and telomere biology. *Cytogenet Genome Res* 122, 202–210.
Gonzalez-Suarez, I; Redwood, AB; Perkins, SM; Vermolen, B; Lichtensztejin, D; *et al.*, 2009. Novel roles for A-type lamins in telomere biology and the DNA damage response pathway. *EMBO J* 28, 2414–2427.
Gorbunova, V; Bozzella, MJ; Seluanov, A, 2008. Rodents for comparative aging studies: From mice to beavers. *Age* 30, 111–119.
Gorbunova, V; Seluanov, A, 2009. Coevolution of telomerase activity and body mass in mammals: From mice to beavers. *Mech Ageing Dev* 130, 3–9.
Gordon, LB; Cao, K; Collins, FS, 2012. Progeria: Translational insights from cell biology. *J Cell Biol* 199, 9–13.
Gordon, LB; Rothman, FG; Lopez-Otin, C; Misteli, T, 2014. Progeria: A paradigm for translational medicine. *Cell* 156, 400–407.

Gorkin, DU; Ren, B, 2014. Closing the distance on obesity culprits. *Nature* 507, 309–310.

Gould, SJ, 1996. *Full House: The Spread of Excellence from Plato to Darwin*. New York, NY: Three Rivers Press.

Greaves, M, 2014. Was skin cancer a selective force for black pigmentation in early hominin evolution? *Proc R Soc B* 281, 20132955.

Gross, L, 2010. A dog's eye view of morphological diversity. *PLoS Biol* 8, e1000452, doi: 10.1371/journal.pbio.1000452.

Gross, M, 2013a. Will our civilization survive this century? *Curr Biol* 23, R1017–R1020.

Gross, M, 2013b. Elements of consciousness in animals. *Curr Biol* 23, R981–R983.

Guerrero-Bosagna, C; Settles, M; Lucker, B; Skinner, MK, 2010. Epigenetic transgenerational actions of vinclozolin on promoter regions of the sperm epigenome. *PLoS One* 5, e13100, doi: 10.1371/journal.pone.0013100.

Guerrero-Bosagna, C; Skinner, MK, 2012. Environmentally induced epigenetic transgenerational inheritance of phenotype and disease. *Mol Cell Endocrinol* 354, 3–8.

Guibert, S; Weber, M, 2012. Erase for a new start. *Nature* 492, 363–364.

Gunders, D, 2012. Wasted: How America is losing up to 40 percent of its food from farm to fork to landfill. *NRDC Issue Pap* Released August, 2012, IP: 12-06-B, http://www.nrdc.org/food/files/wasted-food-ip.pdf

Gurven, M; Kaplan, HK, 2007. Longevity among hunter-gatherers: A cross-cultural examination. *Popul Dev Rev* 33, 321–365.

Guzowski, JF, 2002. Insights into immediate-early gene function in hippocampal memory consolidation using antisense oligonucleotide and fluorescent imaging approaches. *Hippocampus* 12, 86–104.

Hackett, JA; Surani, MA, 2013a. DNA methylation dynamics during the mammalian life cycle. *Phil Trans R Soc B* 368, 20110328.

Hackett, JA; Surani, MA, 2013b. Beyond DNA: Programming and inheritance of parental methylomes. *Cell* 153, 737–738.

Hall, SS, 2014. The accidental epigeneticist. *Nature* 505, 14–17.

Hampsey, M, 2012. A new direction for gene loops. *Science* 338, 624–625.

Han, Y-H; Zhou, H: Kim, J-H; Yan, T-D; Lee, K-H; *et al.*, 2009. Mechanisms of signal transduction: A unique cytoplasmic localization of retinoic acid receptor-γ and its regulations. *J Biol Chem* 284, 18503–18514.

Hannum, G; Guinney, J; Zhao, L; Zhang, L; Hughes, G; *et al.*, 2013. Genome-wide methylation profiles reveal quantitative views of human aging rates. *Mol Cell* 48, 1–9.

Haque, FN; Gottesman, II; Wong, AHC, 2009. Not really identical: Epigenetic differences in monozygotic twins and implications for twin studies in psychiatry. *Am J Med Genet C* 151C, 136–141.

Hare, B; Brown, M; Williamson, C; Tomasello, M, 2002. The domestication of social cognition in dogs. *Science* 298, 164–166.

Haslam, M, 2014. Dating chimpanzees. *Nature* 508, 322–323.

Hathaway, NA; Bell, O; Hodges, C; Miller, EL; Neel, DS; *et al.*, 2012. Dynamics and memory of heterochromatin in living cells. *Cell* 149, 1447–1460.

Hawks, J; Wang, ET; Cochran, GM; Harpending, HC; Moyzis, RK, 2007. Recent acceleration of human adaptive evolution. *Proc Natl Acad Sci USA* 104, 2075–20758.

Hay, N; Sonenberg, N, 2004. Upstream and downstream of mTOR. *Genes Dev* 18, 1926–1945.

Heintzman, ND; Stuart, RK; Hon,G; Fu, Y; Ching, CW; *et al.*, 2007. Distinct and predictive chromatin signatures of transcriptional promoters and enhancers in the human genome. *Nat Genet* 39, 311–318.

Hejjas, K; Kubinyi, E; Ronai, Z; Szekely, A; Vas, J; *et al.*, 2009. Molecular and behavioral analysis of the intron 2 repeat polymorphism in the canine dopamine D4 receptor gene. *Genes Brain Behav* 8, 330–336.

Hejjas, K; Vas, J; Topal, J; Szantai, E; Ronai, Z; *et al.*, 2007. Association of polymorphisms in the dopamine D4 receptor gene and the activity-impulsivity endophenotype in dogs. *Anim Genet* 38, 629–633.

Herskind, AM; McGue, M; Holm, NV; Sorensen, TIA; Harvald, B; *et al.*, 1993. The heritability of human longevity: a population-based study of 2872 Danish twin pairs born 1870–1900. *Hum Gent* 97, 319–323.

Ho, CY; Lammerding, J, 2012. Lamins at a glance. *J Cell Sci* 125, 2087–2093.

Hollander, CF; Solleveld HA; Zurcher, C; Nooteboom, AL; Van Zwieten, MJ, 1984. Biological and clinical consequences of longitudinal studies in rodents: Their possibilities and limitations. An overview. *Mech Ageing Dev* 28, 249–260.

Hopkin, M, 2005. Ethiopia is top choice for cradle of *Homo sapiens*. *Nat News* doi: 10.1038/news 050214-10.

Hopkins, PN, 2013. Molecular biology of atherosclerosis. *Physiol Rev* 93, 1317–1542.

Horowitz, A, 2009. *Inside of a Dog: What Dogs See, Smell, and Know*. New York, NY: Scribner.

Horvath, S, 2013. DNA methylation age of human tissues and cell types. *Genome Biol* 14, R115.

Houle, D; Govindaraju, DR; Omholt, S, 2010. Phenomics: The next challenge. (Boehm 11, 855–866.

Howe, MW; Tierney, PL; Sandberg, SG; Phillips, PEM; Graybiel, AM, 2013. Prolonged dopamine signaling in striatum signals proximity and value of distant rewards. *Nature* 500, 575–579.

Hsiang, SM; Burke, M; Miguel, E, 2013. Quantifying the influence of climate on human conflict. *Science* 341, 1235367.

Hsieh, J; Eisch, AJ, 2010. Epigenetics, hippocampal neurogenesis, and neuropsychiatric disorders; unraveling the genome to understand the mind. *Neurobiol Dis* 39, 73–84.

Hsieh, J; Schneider, JW, 2013. Neural stem cells, excited. *Science* 339, 1534–1535.

Huang, S; Risques, RA; Martin, GM; Rabinovitch, PS; Oshima, J, 2008. Accelerated telomere shortening and replicative senescence in human fibroblasts overexpressing mutant and wild-type lamin A. *Exp Cell Res* 314, 82–91.

Hughes, V, 2014. The sins of the father. *Nature* 507, 22–24.

Hunemeier, T; Gomez-Valdes, J; Ballesteros-Romero, M; de Azevedo, S; Martinez-Abadias, N; *et al.*, 2012. Cultural diversification promotes phenotypic evolution in Xavante Indians. *Proc Natl Acad Sci USA* 109, 73–77.

Hunter, PJ; Borg, TK. Integration from proteins to organs: The Physiome Project. *Nat Rev Mol Cell Biol* 4, 237–243.

Ibrahim, MX; Sayin, VI; Akula, MK; Liu, M; Fong, LG; *et al.*, 2013. Targeting isoprenylcysteine methylation ameliorates disease in a mouse model of progeria. *Science* 340, 1330–1333.

Inoki, K; Kim, J; Guan, K-L, 2012. AMPK and mTOR in cellular energy homeostasis and drug targets. *Annu Rev Pharmacol Toxicol* 52, 381–400.

Itan, Y; Powell, A; Beaumont, MA; Burger, J; Thomas, MG, 2009. The origins of lactase persistence in Europe. *PLoS Comput Biol* 5, e1000491, doi: 10.1371/journal.pcbi.1000491.

Jacob, V; Bergamin, E; Donoghue, MTA; Mongeon, V; LeBlanc, C; *et al.*, 2014. Selective methylation of histone H3 variant H3.1 regulates heterochromatin replication. *Science* 343, 1249–1253.

Jacobs, KB; Yeager, M; Zhou, W; Wacholder, S; Wang, Z; *et al.*, 2012. Detectable clonal mosaicism and its relationship to aging and cancer. *Nat Genet* 44, 651–658.

Jaffe, AE; Irizany, RA, 2014. Accounting for cellular heterogeneity is critical in epigenome-wide association studies. *Genome Biol* 15, R31.

Jagatheesan, G; Thanumalayan, S; Muralikrishana, BH; Rangaraj, N; Karande, AA, 1999. Colocalization of intranuclear lamin foci with RNA splicing factors. *J Cell Sci* 112, 4651–4661.

Johansen, JP, 2013. Anxiety is the sum of its parts. *Nature* 496, 174–175.

Johnson, TE, 2013. Rapid aging rescue. *Science* 340, 1299–1300.

Johnson, SC; Rabinovitch, PS; Kaeberlein, M, 2009. mTOR is a key modulator of ageing and age-related disease. *Nature* 493, 338–345.

Johnstone, RA; Cant, MA, 2010. The evolution of menopause in cetaceans and humans: The role of demography. *Proc R Soc B* 277, 3765–3771.

Jones, EAV; le Noble, F; Eichmann, A, 2006. What determines blood vessel structure? Genetic prespecification *vs.* hemodynamics. *Physiology* 21, 388–395.

Jones, OR; Scheuerlein, A; Salguero-Gomez, R; Camarda, CG; Schaible, R; *et al.*, 2014. Diversity of ageing across the tree of life. *Nature* 505, 169–173.

Jung, H-J; Coffinier, C; Choe, Y; Beigneux, AP; Davies, BSJ; *et al.*, 2012. Regulation of prelamin A but not lamin C by miR-9, a brain-specific micro RNA. *Proc Natl Acad Sci USA* 109, E423–E431.

Junker, JP; van Oudenaarden, A, 2014. Every cell is special: Genome-wide studies add a new dimension to single-cell biology. *Cell* 157, 8–11.

Kaeberlein, M; Kennedy, BK, 2011. Hot topics in aging research: Protein translation and TOR signaling, 2010. *Aging Cell* 10, 185–190.

Kaiser, J, 2013. The downside of diversity. *Science* 339, 1543–1545.

Kaiser, J, 2014. The epigenetics heretic. *Science* 343, 362–363.

Kaminsky, ZA; Tang, T; Wang, S-C; Ptak, C; Oh, GHT; *et al.*, 2009. DNA methylation profiles in monozygotic and dizygotic twins. *Nat Genet* 2009 41, 240–245.

Kasowski, M; Kyriazopoulou-Panagiotopoulou, S; Grubert, F; Zaugg, JB; Kundaje, A; *et al.*, 2013. Extensive variation in chromatin states across humans. *Science* 342, 750–752.

Katsios, C; Roukos, DH, 2010. Individual genomes and personalized medicine: Life diversity and complexity. *Future Med* 7, 347–350, http://www.futuremedicine.com/doi/pdf/10.2217/pme.10.30

Katz, DL, 2014. Obesity is not a disease. *Nature* 508, S57.

Kazda, A; Zellinger, B; Rossler, M; Derboven, E; Kusenda, B; *et al.*, 2012. Chromosome end protection by blunt-ended telomeres. *Genes Dev* 26, 1703–1713.

Keller, A; Graefen, A; Ball, M; Matzas, M; Boisguerin, V; Maixner, F; *et al.*, 2012. New insights into the Tyrolean Iceman's origin and phenotype inferred by whole-genome sequencing. *Nat Commun* doi: 10.1038/ncomms1701.

Keller, R, 2012. Physical biology returns to morphogenesis. *Science* 338, 201–203.

Kempermann, G, 2013. What the bomb said about the brain. *Science* 340, 1180–1181.

Kendal, J; Tehrani, JJ; Odling-Smee, J, 2011. Human niche construction in interdisciplinary focus. *Phil Trans R Soc B* 366, 785–792.

Khan, Z; Ford, MJ; Cusanovich, DA; Mitrano, A; Pritchard, JK; *et al.*, 2013. Primate transcript and protein expression levels evolve under compensatory selection pressures. *Science* 342, 1100–1104.

Khazaeli, AA; Curtsinger, JW, 2010. Life history variation in an artificially selected population of *Drosophila melanogaster*: Pleiotropy, superflies, and age-specific adaptation. *Evolution* 64, 3409–3416.

Khurana, E; Fu, Y; Colonna, V; Mu, XJ; Kang, HM; *et al.*, 2013. Integrative annotation of variants from 1092 humans: application to cancer genomics. *Sci Mag* 342, 1235587, doi: 10.1126/science.1235587.

Kicheva, A; Cohen, M; Briscoe, J, 2012. Developmental pattern formation: Insights from physics and biology. *Science* 338, 210–212.

Kimmel, CB; Ballard, WW; Kimmel, SR; Ullmann, B; Schilling, TF, 1995. Stages of embryonic development of the zebrafish. *Dev Dynam* 203, 253–310.

Kirkpatrick, B; Messias, E; Harvey, PD; Fernandez-Egea, E; Bowie, CR, 2008. Is schizophrenia a syndrome of accelerated aging? *Schizophr Bull* 34, 1024–1032.

Kleckner, N; Zickler, D; Witz, G, 2013. Chromosome capture brings it all together. *Science* 342, 940–941.

Kocabas, AM; Crosby, J; Ross, PJ; Otu, HH; Beyhan, Z; *et al.*, 2006. The transcriptome of human oocytes. *Proc Natl Acad Sci USA* 103, 14027–14032.

Koike, N; Yoo, S-H; Huang, H-C; Kumar, V; Lee, C; *et al.*, 2012. Transcriptional architecture and chromatin landscape of the core circadian clock in mammals. *Science* 338, 349–354.

Kola, I; Bell, J, 2011. A call to reform the taxonomy of human disease. *Nat Rev Drug Discovery* 10, 641–642.

Korb, E; Finkbeiner, S. Arc in synaptic plasticity: From gene to behavior. *Trends Neurosci* 34, 591–598.

Kosik, KS, 2013. Circles reshape the RNA world. *Nature* 495, 322–324.

Koukoura, O; Sifakis, S; Spandidos, D, 2012. DNA methylation in the human placenta and fetal growth (review). *Mol Med Rep* 5, 883–889.

Kraus, C; Pavard, S; Promislow, DEL, 2013. The size-life span trade-off decomposed: Why large dogs die young. *Am Nat* 181, 492–505.

Kucharski, R; Maleszka, J; Foret, S; Maleszka, R, 2008. Nutritional control of reproductive status in honeybees via DNA methylation. *Science* 319, 1827–1830.

Kukekova, AV; Johnson, JL; Teiling, C; Li, L; Oskina, IN; *et al.*, 2011a. Sequence comparison of prefrontal cortical brain transcriptome from a tame and an aggressive silver fox (*Vulpes vulpes*). *BMC Genomics* 12, 482.

Kukekova, AV; Temnykh, SV; Johnson, JL; Trut, LN; Acland, GM, 2012. Genetics of behavior in the silver fox. *Mamm Genome* 23, 164–177.

Kukekova, AV; Trut, LN; Chase, K; Kharlamova, AV; Johnson, JL; *et al.*, 2011b. Mapping loci for fox domestication: Deconstruction/reconstruction of a behavioral phenotype. *Behav Genet* 41, 593–606.

Laland, KN, 2008. Exploring gene-culture interactions: Insights from handedness, sexual selection and niche-construction case studies. *Phil Trans R Soc B* 363, 3577–3589.

Laland, KN; Odling-Smee, J; Gilbert, SF, 2008. EvoDevo and niche construction: Building bridges. *J Exp Zool Mol Dev Evol* 310B, 549–566.

Lam, LL; Emberly, E; Fraser, HB; Neumann, SM; Chen, E; *et al.*, 2012. Factors underlying variable DNA methylation in a human community cohort. *Proc Natl Acad Sci USA* 109(Suppl. 2), 17253–17260.

Langergraber, KE; Rowney, C; Schubert, G; Crockford, C; Hobaiter, C; *et al.*, 2014. How old are chimpanzee communities? Time to the most recent common ancestor of the Y-chromosome in highly patrilocal societies. *J Hum Evol* 69, 1–7.

Lansdorp, PM; Verwoerd, NP; van de Rijke, FM; Dragowska, V; Little, M-T; *et al.*, 1996. Heterogeneity in telomere length of human chromosomes. *Hum Mol Genet* 5, 685–691.

Lappalainen, T; Sammeth, M; Friedlander, MR; 't Hoen, PAC; Monlong, J; *et al.*, 2013. Transcriptome and genome sequencing uncovers functional variation in humans. *Nature* 501, 506–511.

Laurie, CC; Laurie, CA; Rice, K; Doheny, KF; Zelnick, LR; *et al.*, 2012. Detectable clonal mosaicism from birth to old age and its relationship to cancer. *Nat Genet* 44, 642–649.

Laviola, G; Hannan, AJ; Macri, S; Solinas, M; Jaber, M, 2008. Effects of enriched environment on animal models of neurodegenerative diseases and psychiatric disorders. *Neurobiol Dis* 31, 159–168.

Lawler, A, 2012. Uncovering civilization's roots. *Science* 335, 790–793.

Leach, TJ; Chotkowski, HL; Wotring, MG; Dilwith, RL; Glaser, RL, 2000. Replication of heterochromatin and structure of polytene chromosomes. *Mol Cell Biol* 20, 6308–6316.

Ledford, H, 2013. Circular RNAs throw genetics for a loop. *Nature* 494, 415.

Lee, JT, 2012. Epigenetic regulation by long noncoding RNAs. *Science* 338, 1435–1439.

Lee, PC; Bussiere, LF; Webber, CE; Poole, JH; Moss, CJ, 2013. Enduring consequences of early experiences: 40 years effects on survival and success among African elephants (*Loxodonta africana*). *Biol Lett* 9, 20130011.

Lee, RD, 2003. Rethinking the evolutionary theory of aging: Transfers, not births, shape senescence in social species. *Proc Natl Acad Sci USA* 100, 9637–9642.

Leigh, SR, 2004. Brain growth, life history, and cognition in primate and human evolution. *Am J Primatol* 62, 139–164.

Lemos, JC; Wanat, MJ; Smith, JS; Reyes, BAS; Hollon, NG; *et al*., 2012. Severe stress switches CRF action in the nucleus accumbens from appetitive to aversive. *Nature* 490, 402–406.

Leonardi, M; Gerbault, P; Thomas, MG; Burger, J, 2012. The evolution of lactase persistence in Europe. A synthesis of archaeological and genetic evidence. *Int Dairy J* 22, 88–97.

Li, JZ; Absher, DM; Tang, H; Southwick, AM; Casto, AM; *et al*., 2008. Worldwide human relationships inferred from genome-wide patterns of variation. *Science* 319, 1100–1104.

Li, Y; Vonholdt, BM; Reynolds, A; Boyko, AR; Wayne, RK; *et al*., 2013. Artificial selection on brain-expressed genes during the domestication of dog. *Mol Biol Evol* 30, 1867–1876.

Liang, Q; Conte, N; Skarnes, WC; Bradley, A, 2008. Extensive genomic copy number variation in embryonic stem cells. *Proc Natl Acad Sci USA* 105, 17453–17456.

Liebermann, DA; Hoffman, B, 2008. Gadd45 in stress signaling. *J Mol Signal* 3, 15.

Lieberman, P, 2013. Synapses, language, and being human. *Science* 342, 944–945.

Lim, C; Allada, R, 2013. ATAXIN-2 activates PERIOD translation to sustain circadian rhythms in *Drosophila*. *Science* 340, 875–879.

Lindberg, J; Bjornerfeldt, S; Saetre, P; Svartberg, K; Seehus, B; *et al*., 2005. Selection for tameness has changed brain gene expression in silver foxes. *Curr Biol* 15, R915–R916.

Lindblad-Toh, K; Wade, CM; Mikkelsen, TS; Karlsson, EK; Jaffe, DB; *et al*., 2005. Genome sequence, comparative analysis and haplotype structure of the domestic dog. *Nature* 438, 803–819.

Linford, NJ; Kuo, T-H; Chan, TP; Pletcher, SD, 2011. Sensory perception and aging in model systems: From the outside in. *Annu Rev Cell Dev Biol* 27, 759–785.

Lopez-Mejia, IC; Vautrot, V; De Toledo, M; Behm-Ansmant, I; Bourgeois, CF; *et al*., 2011. A conserved splicing mechanism of the *LMNA* gene controls premature aging. *Hum Mol Genet* 20, 4540–4555.

Lopez-Otin, C; Blasco, MA; Partridge, L; Serrano, M; Kroemer, G, 2013. The hallmarks of aging. *Cell* 153, 1194–1217.

Lucitti, JL; Jones, EAV; Huang, C; Chen, J; Fraser, SE; *et al*., 2007. Vascular remodeling of the mouse yolk sac requires hemodynamic force. *Development* 134, 3317–3326.

Lukas, J; Lukas, C, 2013. Shielding broken DNA for a quick fix. *Science* 339, 652–653.

Lupski, JR, 2013. Genome Mosaicism — one human, multiple genomes. *Science* 341, 358–359.

Lutter, M; Nestler, EJ, 2009. Homeostatic and hedonic signals interact in the regulation of food intake. *J Nutr* 139, 629–632.

Lutz, CC; Robinson, GE, 2013. Activity-dependent gene expression in honey bee mushroom bodies in response to orientation flight. *J Exp Biol* 216, 2031–2038.

Ma, DK; Jang, M-H; Guo, JU; Kitabatake, Y; Chang, ML; *et al*., 2009. Neuronal activity-induced Gadd45b promotes epigenetic DNA demethylation and adult neurogenesis. *Science* 323, 1074–1077.

MacArthur, DG; Balasubramanian, S; Frankish, A; Huang, N; Morris, J; *et al*., 2012. A systematic survey of loss-of-function variants in human protein-coding genes. *Science* 335, 823–828.

Macosko, EZ; McCarroll, 2012. Exploring the variation within. *Nat Genet* 44, 614–616.

Macosco, EZ; McCarroll, 2013. Our fallen genomes. *Science* 342, 564–565.

Manolio, TA, 2010. Genome-wide association studies and assessment of the risk of disease. *N Engl J Med* 363, 166–176.

Marino, G; Ugalde, AP; Salvador-Montoliu, N; Varela, I; Quiros, PM; *et al.*, 2008. Premature aging in mice activates a systemic metabolic response involving autophagy induction. *Hum Mol Genet* 17, 2196–2211.

Mark, M; Ghyselinck, NB; Chambon, P, 2006. Function of retinoid nuclear receptors: Lessons from genetic and pharmacological dissections of the retinoic acid signaling pathway during mouse embryogenesis. *Annu Rev Phamacol Toxicol* 46, 451–480.

Martinez-Cadenas, C; Lopez, S; Ribas, G; Flores, C; Garcia, O; *et al.*, 2013. Simultaneous purifying selection on the ancestral *MC1R* allele and positive selection on the melanoma-risk allele *V60L* in South Europeans. *Mol Biol Evol* 30, 2654–2665.

Masri, S; Sassone-Corsi, P, 2013. The circadian clock: A framework linking metabolism, epigenetics and neuronal function. *Nat Rev Neurosci* 14, 69–75.

Mathavan, S; Lee, SGP; Mak, A; Miller, LD; Murthy, KRK; *et al.*, 2005. Transcriptome analysis of zebrafish embryogenesis using microarrays. *PLoS Genet* 1, e29, doi: 10.1371/journal.pgen.0010029.

Mattick, JS, 2013. Probing the phenomics of noncoding RNA. *eLife* 2, e01968, doi:10.7554/eLife.01968.

McClelland, S; Korosi, A; Cope, J; Ivy, A; Baram, TZ, 2011. Emerging roles of epigenetic mechanisms in the enduring effects of early-life stress and experience on learning and memory. *Neurobiol Learn Mem* 96, 79–88.

McConnell, MJ; Lindberg, MR; Brennand, KJ; Piper, JC; Voet, T; *et al.*, 2013. Mosaic copy number variation in human neurons. *Science* 342, 632–637.

McCord, RP; Nazario-Toole, A; Zhang, H; Chines, PS; Zhan, Y; *et al.*, 2013. Correlated alterations in genome organization, histone methylation, and DNA–lamin A/C interactions in Hutchinson–Gilford progeria syndrome. *Genome Res* 23, 260–269.

McGue, M; Vaupel, JW; Holm, N; Harvald, B, 1993. Longevity is moderately heritable in a sample of Danish twins born 1870–1880. *J Geront Biol Sci* 48, B237–B244.

McVean, GA and the 1000 Genomes Project Consortium, 2012. An integrated map of genetic variation from 1,092 human genomes. *Nature* 491, 56–65.

McVicker, G; van de Geijn, B; Degner, JF; Cain, CE; Banovich, NE; *et al.*, 2013. Identification of genetic variants that affect histone modifications in human cells. *Science* 342, 747–749.

Meader, S; Ponting, CP; Lunter, G, 2010. Massive turnover of functional sequence in human and other mammalian genomes. *Genome Res* 20, 1335–1343.

Mekel-Bobrov, N; Gilbert, SL; Evans, PD; Vallender, EJ; Anderson, JR; *et al.*, 2005. Ongoing adaptive evolution of *ASPM*, a brain size determinant in *Homo sapiens*. *Science* 309, 1720–1722.

Mekel-Bobrov, N; Lahn, BT, 2007. Response to comments by Timpson *et al.* and Yu *et al. Science* 317, 1036b.

Meshorer, E; Gruenbaum, Y, 2008. Gone with the Wnt/Notch: Stem cells in laminopathies, progeria, and aging. *J Cell Biol* 181, 9–13.

Meyer, M; Kircher, M; Gansauge, M-T; Li, H; Racimo, F; Mallick, S; *et al.*, 2012. A high-coverage genome sequence from an archaic *Denisovan* individual. *Science* 338, 222–226.

Michaelis, S; Hrycyna, CA, 2013. A protease for the ages. *Science* 339, 1529–1530.

Milani, L; Gupta, M; Andersen, M; Dhar, S; Fryknas, M; *et al.*, 2007. Allelic imbalance in gene expression as a guide to *cis*-acting regulatory single nucleotide polymorphisms in cancer cells. *Nucleic Acids Res* 35, e34, doi: 10.1093/nar/gkl1152.

Mitchell, BD; Hsueh, W-C; King, TM; Pollin, TI; Sorkin, J; *et al.*, 2001. Heritability of life span in the old order Amish. *Am J Med Genet* 102, 346–352.

Mitchell, C; Hobcraft, J; McLanahan, SS; Siegel, SR; Berg, A; *et al.*, 2014. Social disadvantage, genetic sensitivity, and children's telomere length. *Proc Natl Sci Acad USA* 111, 5944–5949.

Mizutani, Y; Tomita, N; Nizuma, Y; Yoda, K, 2013. Environmental perturbations influence telomere dynamics in long-lived birds in their natural habitat. *Biol Lett* 9, doi: 10.1098/rsbl.2013.0511.

Monaghan, P, 2014. Organismal stress, telomeres and life histories. *J Exp Biol* 217, 57–66.

Morata, G, 2014. Tethered wings. Non-essential spread. *Nature* 505, 162.

Moseley, JB, 2013. Cellular aging: Symmetry evades senescence. *Curr Biol* 23, R871–R873.

Moss, CJ, 2001. The demography of an African elephant (*Loxodonta africana*) population in Amboseli, Kenya. *J Zool Lond* 255, 145–156.

Moss, M. *The extraordinary science of addictive junk food*. Released February 24, 2013, http://www.nytimes.com/2013/02/24/magazine/the-extraordinary-science-of-junk-food.html?pagewanted=all&_r=0

Murphy Jr, WA; zur Nedden, D; Gostner, P; Knapp, R; Recheis, W; *et al.*, 2003. The iceman: Discovery and imaging. *Radiology* 226, 614–629.

Myers, S; Freeman, C; Auton, A; Donnelly, P; McVean, G, 2008. A common sequence motif associated with recombination hot spots and genome instability in humans. *Nat Genet* 40, 1124–1129.

Nelson, ADL; Shippen, DE, 2012. Blunt-ended telomeres: An alternative ending to the replication and end protection stories. *Genes Dev* 26, 1648–1652.

Nelson III, CA; Zeanah, CH; Fox, NA; Marshall, PJ; Smyke, AT; *et al.*, 2007. Cognitive recovery in socially deprived young children: The Bucharest early intervention project. *Science* 318, 1937–1940.

Nelson III, CA; Fox, NA; Zeanah, Jr, CH, 2013. Anguish of the abandoned child. *Sci Am* 308 (Apr.), 62–67.

Nemeth, Z; Bonier, F; McDougall-Shackleton, SA, 2013. Coping with uncertainty: Integrating physiology, behavior, and evolutionary ecology in a changing world. *Integr Compar Biol* 53, 960–964.

Neufeld, HN; Wagenvoort, CA; Edwards, JE, 1962. Coronary arteries in fetuses, infants, juveniles, and young adults. *Lab Invest* 11, 837–844.

Newman, SA, 2012. Physico-genetic determinants in the evolution of development. *Science* 338, 217–219.

Niculescu, MD, 2012. Challenges in nutrition-related DNA methylation studies. *Biomol Concepts* 3, 151–160.

Niehrs, C; Schafer, A, 2012. Active DNA demethylation by Gadd45 and DNA repair. *Trends Cell Biol* 22, 220–227.

Niv, Y, 2013. Dopamine ramps up. *Nature* 500, 533–534.

Nord, AS; Blow, MJ; Attanasio, C; Akiyama, JA; Holt, A; *et al.*, 2013. Rapid and pervasive changes in genome-wide enhancer usage during mammalian development. *Cell* 155, 1521–1531.

Norrgard, K; Schultz, J, 2008. Using SNP data to examine human phenotype differences. *Nat Educ* 2008 1, 85.

Nuyt, AM; Szyf, Mm 2007. Developmental programming through epigenetic changes. *Circ Res* 100, 452–455.

O'Donnell, A; Odrowaz, Z; Sharrocks, AD, 2012. Immediate-early gene activation by the MAPK pathways: What do and don't we know? *Biochem Soc Transactions* 40, 58–66.

Ogihara, T; Hata, T; Tanaka, K; Fukuchi, K; Tabuchi, Y; *et al.*, 1986. Hutchinson-Gilford Progeria syndrome in a 45-year-old man. *Am J Med* 81, 135–138.

O'Huallachain, M; Karczewski, KJ; Weissman, SM; Urban, AE; Snyder, MP, 2012. Extensive genetic variation in somatic human tissues. *Proc Natl Acad Sci USA* 109, 18018–18023.

Okamoto, K; Bartocci, C; Ouzounov, I; Diedrich, JK; Yates III, JR; *et al.*, 2013. A two-step mechanism for TRF2-mediated chromosome-end protection. *Nature* 494, 502–505.

Okuno, H, 2011. Regulation and function of immediate-early genes in the brain: Beyond neuronal activity markers. *Neurosci Res* 69, 175–186.

Olive, M; Harten, I; Mitchell, R; Beers, JK; Djabali, K; *et al.*, 2010. Cardivascular pathology in Hutchinson-Gilford Progeria: correlation with the vascular pathology of aging. *Arterioscler Thromb Vasc Biol* 30, 2301–2309.

Olson, AK; Eadie, BD; Ernst, C; Christie, BR, 2006. Environmental enrichment and voluntary exercise massively increase neurogenesis in the adult hippocampus via dissociable pathways. *Hippocampus* 16, 250–260.

Ong, C-T; Corces, VG, 2012. Enhancers: Emerging roles in a cell fate specification. *EMBO Rep* 13, 423–430.

Ooi, L; Wood, IC, 2007. Chromatin crosstalk in development and disease: Lessons from REST. *Nature* 8, 544–554.

Osborne, LR, 2008. Genomic rearrangements in the spotlight. *Nat Genet* 40, 6–7.

Ostrander, EA; Kruglyak, L, 2000. Unleashing the canine genome. *Genome Res* 10, 1271–1274.

Ouellet-Morin, I; Wong, CCY; Danese, A; Pariante, CM; Papadopoulos, AS; *et al.*, 2013. Increased serotonin transporter gene (*SERT*) DNA methylation is associated with bullying victimization and blunted cortisol response to stress in childhood: A longitudinal study of discordant monozygotic twins. *Psychol Med* 43, 1813–1823.

Owens, B, 2012. The single life. *Nature* 491, 27–29.

Oxfam Briefing Paper. *Working for the few*. Released January 20, 2014, http://www.oxfam.org/sites/www.oxfam.org/files/bp-working-for-few-political-capture-economic-inequality-200114-en.pdf

Palinski, W; Yamashita, T; Freigang, S; Napoli, C, 2007. Developmental programming: maternal hypercholesterolemia and immunity influence susceptibility to atherosclerosis. *Nutr Rev* 65, S182–S187.

Papasaikas, P; Valcarcel, J, 2012. Splicing in 4D. *Science* 338, 1547–1548.

Park, J; Venteicher, AS; Hong, JY; Choi, J; Jun, S; *et al.*, 2009. Telomerase modulates Wnt signaling by association with target gene chromatin. *Nature*, 460, 66–72.

Parker, HG; Kim, LV; Sutter, NB; Carlson, S; Lorentzen, TD; *et al.*, 2004. Genetic structure of the purebred domestic dog. *Science* 304, 1160–1164.

Parra, EJ, 2007. Human pigmentation variation: Evolution, genetic basis, and implications for public health. *Yearbook Phys Anthropol* 50, 85–105.

Pastalkova, E; Itskov, V; Amarasingham, A; Buzsaki, G, 2008. Internally generated cell assembly sequences in the rat hippocampus. *Science* 321, 1322–1327.

Patel, MS; Srinivasan, M. metabolic programming: causes and consequences. *J Biol Chem* 277, 1629–1632.

Patterson, DF, 2000. Companion animal medicine in the age of medical genetics. *J Vet Intern Med* 14, 1–9.

Pederson, N; Hessl, AE; Baatarbileg, N; Anchukaitis, KJ; Di Cosmo, N. Pluvials, 2014. Droughts, the Mongol Empire, and modern Mongolia. *Proc Natl Acad Sci USA* 111, 4375–4379.

Pennisi, E, 2013a. Evolution heresy? Epigenetics underlies heritable plant traits. *Science* 341, 1055.

Pennisi, E, 2013b. Old dogs teach a new lesson about canine origins. *Science* 342, 785–786.

Pennisi, E, 2014. Ancient DNA holds clues to gene activity in extinct humans. *Science* 344, 245–246.

Pereira, AC; Huddleston, DE; Brickman, AM; Sosunov, AA; Hen, R; *et al.*, 2007. An *in vivo* correlate of exercise-induced neurogenesis in the adult dentate gyrus. *Proc Natl Acad Sci USA* 104, 5638–5643.

Perry, GH; Dominy, NJ; Claw, KG; Lee, AS; Fiegler, H; *et al.*, 2007. Diet and the evolution of human amylase gene copy number variation. *Nat Genet* 39, 1256–1260.

Peter, A; Stick, R, 2012. Evolution of the lamin protein family. What introns can tell. *Nucleus* 3, 44–59.

Peter, M; Kitten, GT; Lehner, CF; Vorburger, K; Bailer, SM; *et al.*, 1989. Cloning and sequencing of cDNA clones encoding chicken lamins A and B1 and comparison of the primary structures of vertebrate A- and B-type lamins. *J Mol Biol* 208, 393–404.

Petrovich, GD; Holland, PC; Gallagher, M, 2005. Amygdalar and prefrontal pathways to the lateral hypothalamus are activated by a learned cue that stimulates eating. *J Neurosci* 25, 8295–8302.

Pettay, JE; Helle, S; Jokela, J; Lummaa, V, 2007. Natural selection on female life-history traits in relation to socio-economic class in pre-industrial human populations. *PLoS One* 2, e606, doi: 10.1371/journal.pone.0000606.

Pinhasi, R; von Cramon-Taubadel, N, 2009. Craniometric data supports demic diffusion model for the spread of agriculture into Europe. *PLoS One* 4, e6747, doi: 10.1371/journal.pone.0006747.

Plassmann, H; O'Doherty, J; Shiv, B; Rangel, A, 2008. Marketing actions can modulate neural representations of experienced pleasantness. *Proc Natl Acad Sci USA* 105, 1050–1054.

Poduri, A; Evrony, GD; Cai, X; Walsh, CA, 2013. Somatic mutation, genomic variation, and neurological disease. *Science* 341, 1237758.

Pollard, KS; Salama, SR; Lambert, N; Lambot, M-A; Coppens, S; *et al.*, 2006. An RNA gene expressed during cortical development evolved rapidly in humans. *Nature* 443, 167–172.

Pollex, RL; Hegele, RA, 2004. Hutchinson–Gilford Progeria syndrome. *Clin Genet* 66, 375–381.

Ponting, CP; Hardison, RC, 2011. What fraction of the human gene is functional? *Genome Res* 21, 1769–1776.

Potenza, MN. Obesity, food and addiction: Emerging neuroscience and clinical and public health implications. *Neuropsychopharamacol Rev* 39, 249–250.

Potts, R, 1996. Evolution and climate variability. *Science* 273, 922–923.

Poucet, B; Sargolini, F, 2013. A trace of your place. *Science* 340, 35–36.

Raices, M; Maruyama, H; Dillin, A; Karlseder, J, 2005. Uncoupling of longevity and telomere length in *C. elegans*. *PLoS Genet* 1, e30.

Ramos, FJ; Chen, SC; Garelick, MG; Dai, D-F; Liao, C-Y; *et al.*, 2012. Rapamycin reverses elevated mTORC1 signaling in lamin A/C deficient mice, rescues cardiac and skeletal muscle function, and extends survival. *Sci Transl Med* 4, 144ra103.

Rebke, M; Coulson, T; Becker, PH; Vaupel, JW, 2010. Reproductive improvement and senescence in a long-lived bird. *Proc Natl Acad Sci USA* 107, 7841–7846.

Redwood, AB; Perkins, SM; Vanderwaal, RP; Feng, Z; Biehl, KJ; *et al.*, 2011. A dual role for A-type lamins in DNA double-strand break repair. *Cell Cycle* 10, 2549–2560.

Reich, De; Schaffner, SF; Daly, MJ; McVean, G; Mullikin, JC; *et al.*, 2002. Human genome sequence variation and the influence of gene history, mutation and recombination. *Nat Genet* 32, 135–142.

Reik, W; Murrell, A, 2000. Silence across the border. *Nature* 405, 408–409.

Reul, JMHM; Hesketh, SA; Collins, A; Mecinas, MG, 2009. Epigenetic mechanisms in the dentate gyrus act as a molecular switch in hippocampus-associated memory function. *Epigenetics* 4, 434–439.

Richerson, P, 2013. Group size determines cultural complexity. *Nature* 503, 351–352.

Richerson, PJ; Boyd, R; Henrich, J, 2010. Gene-culture coevolution in the age of genomics. *Proc Natl Acad Sci USA* 107, 8985–8992.

Robison, AJ; Nestler, EJ, 2011. Transcriptional and epigenetic mechanisms of addiction. *Nat Rev Neurosci* 12, 623–637.

Robson, SL; Wood, B, 2008. Hominin life history: Reconstruction and evolution. *J Anat* 212, 394–425.

Rochette, J; Pointon, JJ; Fisher, CA; Perera, G; Arambepola, M; *et al.*, 1999. Multicentric origin of hemochromatosis gene (*HFE*) mutations. *Am J Hum Genet* 64, 1056–1062.

Roensch, K; Tazaki, A; Chara, O; Tanaka, EM, 2013. Progressive specification rather than intercalation of segments during limb regeneration. *Science* 342, 1375–1379.

Rothwell, PM, 2005. External validity of randomized controlled trials: "To whom do the results of this trial apply?" *Lancet* 365, 82–93.

Rusinol, AE; Sinensky, MS, 2006. Farnesylated lamins, progeroid syndromes and farnesyl transferase inhibitors. *J Cell Sci* 119, 3265–3272.

Rutledge, CE; Thakur, A; O'Neill, KM; Irwin, RE; Sato, S; *et al.*, 2014. Ontogeny, conservation and functional significance of maternally inherited DNA methylation at two classes of non-imprinted genes. *Development* 141, 1313–1323.

Rutter, M, 2012. Achievements and challenges in the biology of environmental effects. *Proc Natl Acad Sci USA* 109, 17149–17153.

Sabeti, PC; Varilly, P; Fry, B; Lohmueller, J; Hostetter, E; *et al.*, 2007. Genome-wide detection and characterization of positive selection in human populations. *Nature* 449, 913–918.

Sachser, N; Kaiser, S; Hennessy, MB, 2013. Behavioral profiles are shaped by social experience: When, how and why. *Phil Trans R Soc B* 368, 20120344.

Salomons, HM; Mulder, GA; van de Zande, L; Haussmann, MF; Linskens, MHK; *et al.*, 2009. Telomere shortening and survival in free-living corvids. *Proc R Soc B* 276, 3157–3165.

Sanchez, A; Golding, I, 2013. Genetic determinants and cellular constraints in noisy gene expression. *Science* 342, 1188–1193.

Sasaki, T; Kishi, S, 2013. Molecular and chemical genetic approaches to developmental origins of aging and disease in zebrafish. *Biochim Biophys Acta* 1832, 1362–1370.

Saunders, EJ; Dadaev, T; Leongamornlert, DA; Jugurnauth-Little, S; Tymrakiewicz, M; *et al.*, 2014. Fine-mapping the *HOXB* region detects common variants tagging a rare coding allele: Evidence for synthetic association in prostate cancer. *PLoS Genet* 10, e1004129, doi: 10.1371/journal.pgen.1004129.

Savage-Rumbaugh, S; Fields, WM; Taglialatela, J, 2000. Ape consciousness-human consciousness: a perspective informed by language and culture. *Am Zool* 40, 910–921.

Scaffidi, P; Misteli, T, 2006. Lamin A-dependent nuclear defects in human aging. *Science* 312, 1059–1063.

Scaffidi, P; Misteli, T, 2008. Lamin A-dependent misregulation of adult stem cells associated with accelerated ageing. *Nat Cell Biol* 10, 452–459.

Scheinfeldt, LB; Tishkoff, SA, 2013. Recent human adaptation: Genomic approaches, interpretation and insights. *Nature* 14, 692–702.

Schiltz, CA; Bremer, QZ; Landry, CF; Kelley, AE, 2004. Food-associated cues alter forebrain functional connectivity as assessed with immediate early gene and proenkephalin expression. *BMC Biol* 5, 16.

Schubeler, D, 2007. Enhancing genome annotation with chromatin. *Nat Genet* 39, 284–285.

Schubeler, D, 2012. Epigenetic islands in a genetic ocean. *Science* 338, 756–757.

Seluanov, A; Chen, Z; Hine, C; Sasahara, THC; Ribeiro, AACM; *et al.*, 2007. Telomerase activity coevolves with body mass, not lifespan. *Aging Cell* 6, 45–52.

Shapiro, B; Hofreiter, M, 2014. A paleogenomic perspective on evolution and gene function: New insights from ancient DNA. *Science* 343, 1236573.

Shearin, AL; Ostrander, EA, 2010. Leading the way: Canine models of genomics and disease. *Dis Model Mech* 3, 27–34.

Shennan, S, 2011. Property and wealth inequality as cultural niche construction. *Phil Trans R Soc B* 366, 918–926.

Shevelyov, YY; Nurminsky, DI, 2012. The nuclear lamina as a gene-silencing hub. *Curr Iss Mol Biol* 14, 27–38.

Shultz, S; Opie, C; Atkinson, QD, 2011. Stepwise evolution of stable sociality in primates. *Nature* 479, 219–222.

Silk, JB, 2011. The path to sociality. *Nature* 479, 182–183.

Silveira, AB; Trontin, C; Cortijo, S; Barau, J; Del Bem, LEV; *et al.*, 2013. Extensive natural epigenetic variation at a *de novo* originated gene. *PLoS Genet* 9, e1003437. doi: 10.1371/journal.pgen.1003437.

Simeonova, I; Jaber, S; Draskovic, I; Bardot, B; Fang, M; *et al.*, 2013. Mutant mice lacking the p53 C-terminal domain model telomere syndromes. *Cell Rep* 3, 2046–2058.

Singer, MA, 2011. Insights into biomedicine from animal adaptations. *Compr Physiol* 1, 2063–2081.

Singer, MA, 2013. *Are Chronic Degenerative Diseases Part of the Ageing Process? Insights from Comparative Biology*. New York, NY: Nova Science Publishers, Inc.

Siriaco, G; Tamkun, JW, 2013. A histone timer for zygotic genome activation. *Dev Cell* 26, 558–559.

Skinner, MK, 2011. Role of epigenetics in developmental biology and transgenerational inheritance. *Birth Defects Res C* 93, 51–55.

Skoglund, P; Malmstrom, H; Omrak, A; Raghavan, M; Valdiosera, C; *et al.*, 2014. Genomic diversity and admixture differs for stone-age Scandinavian foragers and farmers. *Sci Mag* 344, 747–750, doi: 10.1126/science.1253448.

Slobodchikoff, C, 2012. *Chasing Doctor Doolittle: Learning the Language of Animals*. New York, NY: St. Martins Press.

Smithsonian Institute. *Human Origins Program*. Last updated June 19, 2014, http://humanorigins.si.edu

Somel, M; Liu, X; Khaitovich, P, 2013a. Human brain evolution: Transcripts, metabolites and their regulators. *Nat Rev Neurosci* 14, 112–126.

Somel, M; Sayres, MAW; Jordan, G; Huerta-Sanchez, E; Fumagalli, M; *et al.*, 2013b. A scan for human-specific relaxation of negative selection reveals unexpected polymorphism in proteasome genes. *Mol Biol Evol* 30, 1808–1815.

Song, J; Zhong, C; Bonaguidi, MA; Sun, GJ; Hsu, D; *et al.*, 2012. Neuronal circuitry mechanism regulating adult quiescent neural stem-cell fate decision. *Nature* 489, 150–154.

Soon, CS; Brass, M; Heinze, H-J; Haynes, J-D, 2008. Unconscious determinants of free decisions in the human brain. *Nat Neurosci* 11, 543–545.

Souroullas, GP; Sharpless, NE, 2013. Down's syndrome link to ageing. *Nature* 501, 325–326.

Spalding, KL; Bergmann, O; Alkass, K; Bernard, S; Salehpour, M; *et al.*, 2013. Dynamics of hippocampal neurogenesis in adult humans. *Cell* 153, 1219–1227.

Statistics Canada. *Leading causes of death in Canada, 2009, CANSIM Table 102-0561*. Released July 25, 2012, http://www.statcan.gc.ca/pub/84-215-x/84-215-x2012001-eng.htm

Stearns, SC; Byars, SG; Govindaraju, DR; Ewbank, D, 2010a. Measuring selection in contemporary human populations. *Nat Rev Genet* 11, 611–622.

Stearns, SC; Nesse, RM; Govindaraju, DR; Ellison, PT, 2010b. Evolutionary perspectives on health and medicine. *Proc Natl Acad Sci USA* 107, 1691–1695.

Stessman, HA; Bernier, R; Eichler, EE, 2014. A genotype-first approach to defining the subtypes of a complex disease. *Cell* 156, 872–877.

Stewart, JR; Stringer, CB, 2012. Human evolution out of Africa: The role of refugia and climate change. *Sci Mag* 335, 1317–1321, doi: 10.1126/science.1215627.

Storer, M; Mas, A; Robert-Moreno, A; Pecoraro, M; Ortells, MC; *et al.*, 2013. Senescence is a developmental mechanism that contributes to embryonic growth and patterning. *Cell* 155, 1119–1130.

Storrs, EE; Williams, RJ, 1968. A study of monozygous quadruplet armadillos in relation to mammalian inheritance. *Biochemistry* 60, 910–914.

Stranahan, AM; Khalil, D; Gould, E, 2006. Social isolation delays the positive effects of running on adult neurogenesis. *Nat Neurosci* 9, 526–533.

Stringer, C, 2012. What makes a modern human. *Nature* 485, 33–35.

Strobel, SA, 2013. Metal ghosts in the splicing machine. *Nature* 503, 201–202.

Stromberg, J, 2013. Expiration dates: Can your genes predict when you will die? *Smithsonian Mag* 43 (Jan.), p. 14.

Struhl, G, 2014. Tehered wings. Long-range thinking. *Nature* 505, 162–163.

Suryanarayanan, S; Hermanson, JC; Jeanne, RL, 2011. A mechanical signal biases caste development in a social wasp. *Curr Biol* 21, 231–235.

Suter, CM; Boffelli, D; Martin, DIK, 2013. A role for epigenetic inheritance in modern evolutionary theory? A comment in response to Dickins and Rahman. *Proc R Soc B* 280, 20130903.

Sutter, NB; Eberle, MA; Parker, HG; Pullar, BJ; Kirkness, EW; *et al.*, 2004. Extensive and breed-specific linkage disequilibrium in *Canis familiaris*. *Genome Res* 14, 2388–2396.

Swift, J; Ivanovska, IL; Buxboim, A; Harada, T; Dingal, PCDP; *et al.*, 2013. Nuclear lamin-A scales with tissue stiffness and enhances matrix-directed differentiation. *Science* 341, 1240104.

Swirnoff, AH; Milbrandt, J, 1995. DNA-binding specificity of NGFI-A and related zinc finger transcription factors. *Mol Cell Biol* 15, 2275–2287.

Tadros, W; Lipshitz, HD, 2009. The maternal-to-zygotic transition: A play in two acts. *Development* 136, 3033–3042.

Takeuchi, Y; Hashizume, C; Ha Chon, EM; Momozawa, Y; Masuda, K; *et al.*, 2005. Canine tyrosine hydroxylase (*TH*) gene and dopamine B-hydroxylase (*DBH*) gene: Their sequences, genetic polymorphisms, and diversities among five different dog breeds. *J Vet Med Sci* 67, 861–867.

Takeuchi, Y; Kaneko, F; Hashizume, C; Masuda, K; Ogata, N; *et al.*, 2009. Association analysis between canine behavioural traits and genetic polymorphisms in the Shiba Inu breed. *Anim Genet* 40, 616622.

Takizawa, R; Maughan, B; Arseneault, L, 2014. Adult health outcomes of childhood bullying victimization: Evidence from a five-decade longitudinal British birth cohort. *Am J Psychiatry*, doi: 10.1176.

Talens, RP; Christensen, K; Putter, H; Willemsen, G; Christiansen, L; *et al.*, 2012. Epigenetic variation during the adult lifespan: Cross-sectional and longitudinal data on monozygotic twin pairs. *Aging Cell* 11, 694–703.

Tang, B; Chang, W-L; Lanigan, CM; Dean, B; Sutcliffe, JG; *et al.*, 2009. Normal human aging and early-stage schizophrenia share common molecular profiles. *Aging Cell* 8, 339–342.

Tang, K; Thorton, KR; Stoneking, M, 2007. A new approach for using genome scans to detect recent positive selection in the human genome. *PLoS Biol* 5, e171, doi: 10.1371/journal.pbio.0050171.

Tejedor, JR; Valcarcel, J, 2010. Breaking the second genetic code. *Nature* 465, 45–46.

Timpson, N; Heron, J; Smith, GD; Enard, W, 2007. Comment on papers by Evans *et al.* and Mekel-Bobrov *et al.* on evidence for positive selection of *MCPH1* and *ASPM*. *Science* 317, 1036.

Tindell, AJ; Smith, KS; Berridge, KC; Aldridge, JW, 2009. Dynamic computation of incentive salience: "Wanting" what was never "liked." *J Neurosci* 30, 12220–12228.

Toll-Riera, M; Laurie, S; Alba, MM, 2011. Lineage-specific variation in intensity of natural selection in mammals. *Mol Biol Evol* 28, 383–398.

Tomanek, RJ, 2005. Formation of the coronary vasculature during development. *Angiogenesis* 8, 273–284.

Toomajian, C; Ajioka, RS; Jorde, LB; Kushner, JP; Kreitman, M, 2003. A method for detecting recent selection in the human genome from allele age estimates. *Genetics* 165, 287–297.

Travers, A, 2014. The 30-nm fiber redux. *Science* 344, 370–372.

Trut, LN. Early canid domestication: The farm-fox experiment. *Am Sci* 87, 160–169, doi: 10.1511/1999.2.160.

Trut, L; Oskina, I; Khariamova, A, 2009. Animal evolution during domestication: The domesticated fox as a model. *BioEssays* 31, 349–360.

Turner, DJ; Miretti, M; Rajan, D; Fiegler, H; Carter, NP; *et al.*, 2008. Germline rates of *de novo* meiotic deletions and duplications causing several genomic disorders. *Nat Genet* 40, 90–95.

Tyree, MT, 2003. The ascent of water. *Nature* 423, 923.

Underwood, E, 2014. Can Down syndrome be treated? *Science* 343, 964–967.

Urban HEART @ Toronto. 2014. Centre for Research on Inner City Health, St. Michael's Hospital, Toronto, Canada, http://www.torontohealthprofiles.ca/urbanheartattoronto.php

U.S. Food and Drug Administration. *Paving the way for personalized medicine*. Released October, 2013, http://www.fda.gov/downloads/scienceresearch/specialtopics/personalizedmedicine/ucm372421.pdf

U.S. Geological Survey. *World's oldest-known wild bird hatches another chick.* Released February 4, 2013, http://www.usgs.gov/newsroom/article.asp?ID=3504&from=rss

Valdes, AM; Glass, D; Spector, TD, 2013. Omics technologies and the study of human ageing. *Nat Rev Genet* 14, 601–607.

Van Bortle, K; Corces, VG, 2013. Spinning the web of cell fate. *Cell* 152, 1213–1217.

Van Heesbeen, H; Mesman, S; Veenvliet, JV; Smidt, MP, 2013. Epigenetic mechanisms in the development and maintenance of dopaminergic neurons. *Development* 140, 1159–1169.

Van Praag, H, 2008. Neurogenesis and exercise: Past and future directions. *Neuromol Med* 10, 59–66.

Van Praag, H; Christie, BR; Sejnowski, TJ; Gage, FH, 1999. Running enhances neurogenesis, learning, and long-term potentiation in mice. *Proc Natl Acad Sci USA* 96, 13427–13431.

Varki, A; Geschwind, DH; Eichler, EE, 2008. Explaining human uniqueness: Genome interactions with environment, behaviour and culture. *Nat Rev Genet* 9, 749–763.

Vaupel, JW, 2010. Biodemography of human ageing. *Nature* 464, 536–542.

Vaupel, JW; Baudisch, A; Dolling, M; Roach, DA; Gampe, J, 2004. The case for negative senescence. *Theor Popul Biol* 65, 339–351.

Veith, I, 1969. Historical reflections on the changing concepts of disease. *Calif Med* 110, 501–506.

Velican, C; Velican, D, 1976. Intimal thickening in developing coronary arteries and its relevance to atherosclerotic involvement. *Atherosclerosis* 23, 345–355.

Velican, C; Velican, D, 1979a. Some particular aspects of the microarchitecture of human coronary arteries. *Atherosclerosis* 33, 191–200.

Velican, D; Velican, C, 1979b. Study of fibrous plaques occurring in the coronary arteries of children. *Atherosclerosis* 33, 201–215.

Vlcek, S; Foisner, R, 2007. Lamins and lamin-associated proteins in aging and disease. *Curr Opin Cell Biol* 19, 298–304.

Vogel, C, 2013. Protein expression under pressure. *Science* 342, 1052–1053.

Voight, BF; Kudaravalli, S; Wen, X; Pritchard, JK, 2006. A map of recent positive selection in the human genome. *PLoS Biol* 4, e72, doi: 10.1371/journal.pbio.0040072.

Vonholdt, BM; Pollinger, JP; Lohmueller, KE; Han, E; Parker, HG; *et al.*, 2010. Genome-wide SNP and haplotype analyses reveal a rich history underlying dog domestication. *Nature* 464, 898–902.

Wagner, GP, 2012. The inner life of proteins. *Nature* 490, 493–494.

Wang, B; Matsuoka, S; Carpenter, PB; Elledge, SJ, 2002. 53BP1, a mediator of the DNA damage checkpoint. *Science* 298, 1435–1438.

Wang, G-d; Zhai, W; Yang, H-c; Fan, R-x; Cao, X; *et al.*, 2013. The genomics of selection in dogs and the parallel evolution between dogs and humans. *Nat Commun* 4, 1860, doi: 10.1038/ncomms2814.

Wang, H; Wang, L; Erdjument-Bromage, H; Vidal, M; Tempst, P; *et al.*, 2004. Role of histone H2A ubiquitination in polycomb silencing. *Nature* 431, 873–878.

Ward, CV; Kimbel, WH; Johanson, DC, 2011. Complete fourth metatarsal and arches in the foot of *Australopithecus afarensis*. *Science*, 331, 750–753.

Ward, EJ; Parsons, K; Holmes, EE; Balcomb III, KC; Ford, JKB, 2009. The role of menopause and reproductive senescence in a long-lived social mammal. *Front Zool* 6, 4, doi: 10.1186/1742-9994-6-4.

Ward, LD; Kellis, M, 2012. Evidence of abundant purifying selection in humans for recently acquired regulatory functions. *Science* 337, 1675–1678.

Waszak, SM; Deplancke, B, 2013. Rounding up natural gene expression variation during development. *Dev Cell* 27, 601–603.

Wayne, RK; Ostrander, EA, 2007. Lessons learned from the dog genome. *Trends Genet* 23, 557–567.

Weaver, ICG; D'Alessio, AC; Brown, SE; Hellstrom, IC; Dymov, S; *et al.*, 2007. The transcription factor nerve growth factor-inducible protein A mediates epigenetic programming: Altering epigenetic marks by immediate-early genes. *J Neurosci* 14, 1756–1768.

Wei, W; Pelechano, V; Jarvelin, AI; Steinmetz, LM, 2011. Functional consequences of bidirectional promoters. *Trends Genet* 27, 267–276.

Wells, JCK, 2012. The evolution of human adiposity and obesity: Where did it all go wrong? *Dis Model Mech* 5, 595–607.

Weninger, WJ; Muller, GB; Reiter, C; Meng, S; Rabl, SU, 1999. Intimal hyperplasia of the infant parasellar carotid artery. A potential developmental factor in atherosclerosis and SIDS. *Circ Res* 85, 970–975.

Wiener, P; Wilkinson, S, 2011. Deciphering the genetic basis of animal domestication. *Proc R Soc B* 278, 3161–3170.

Williamson, SH; Hubisz, MJ; Clark, AG; Payseur, BA; Bustmante, CD; *et al.*, 2007. Localizing recent adaptive evolution in the human genome. *PLoS Genet* 3, e90, doi: 10.1371/journal.pgen.0030090.

Willox, AC; Harper, SL; Ford, JD; Edge, VL; Landman, K; *et al.*, 2013. Climate change and mental health: an exploratory case study from Rigolet, Nunatsiavut, Canada. *Clim Change* 121, 255–270.

Willyard, C, 2014. The family roots of obesity. *Nature* 508, S58–S60.

Wilson, EO, 2012. *The Social Conquest of Earth*. New York, NY: WW Norton and Company.

Wollstonecroft, MM, 2011. Investigating the role of food processing in human evolution: A niche construction approach. *Archaeol Anthropol Sci* 3, 141–150.

Wong, CCY; Caspi, A; Williams, B; Craig, IW; Houts, R; *et al.*, 2010. A longitudinal study of epigenetic variation in twins. *Epigenetics* 5, 516–526.

Wurbel, H, 2000. Behavior and the standardization fallacy. *Nat Genet* 26, 263.

Yamazawa, K; Kagami, M; Nagai, T; Kondoh, T; Onigata, K; *et al.*, 2008. Molecular and clinical findings and their correlations in Silver–Russell syndrome: Implications for a positive role of IGF2 in growth determination and differential imprinting regulation of the IGF2-H19 domain in bodies and placentas. *J Mol Med* 86, 1171–1181.

Yao, Z; Shafer, OT, 2014. The *Drosophila* circadian clock is a variably coupled network of multiple peptidergic units. *Science* 343, 1516–1520.

Yen, AW; Fancher, TL; Bowius, CL, 2006. Revisting hereditary hemochromatosis: Current concepts and progress. *Am J Med* 119, 391–399.

Yi, H; Xue, L; Guo, M-X; Ma, J; Zeng, Y; *et al.*, 2010. Gene expression atlas for human embryogenesis. *FASEB J* 24, 3341–3350.

Young, J; Morbois-Trabut, L; Couzinet, B; Lascols, O; Dion, E; *et al.*, 2005. Type A insulin resistance syndrome revealing a novel lamin A mutation. *Diabetes* 54, 1873–1878.

Young, SG; Jung, H-J; Coffinier, C; Fong, LG, 2012. Understanding the roles of nuclear A- and B- type lamins in brain development. *J Biol Chem* 287, 16103–16110.

Yu, F; Hill, S; Schaffner, SF; Sabeti, PC; Wang, ET; *et al.*, 2007. Comment on "Ongoing adaptive evolution of *ASPM*, a brain size determinant in *Homo sapiens*." *Science* 316, 370.

Yu, J; Ren, X; Lao, K; Xie, XS, 2006. Probing gene expression in live cells, one protein molecule at a time. *Science* 311, 1600–1603.

Zaidan, H; leshem, M; Gaisler-Salomon, I, 2013. Prereproductive stress to female rats alters corticotrophin-releasing factor type 1 expression in ova and behavior and brain corticotrophin-releasing factor type 1 expression in offspring. *Biol Psychiatry* 74, 680–687.

Zaret, KS, 2014. Genome reactivation after the silence of mitosis: Recapitulating mechanisms of development? *Dev Cell* 29, 132–134.

Zenobi, R, 2013. Single-cell metabolomics: Analytical and biological perspectives. *Science* 342, 1243259.

Zhang, J; Lian, Q; Zhu, G; Zhou, F; Sui, L; *et al.*, 2011. A human iPSC model of Hutchinson-Gilford Progeria reveals vascular smooth muscle and mesenchymal stem cell defects. *Cell Stem Cell* 8, 31–45.

Zhang, T-Y; Meaney, MJ, 2010. Epigenetics and the environmental regulation of the genome and its function. *Annu Rev Psychol* 61, 439–466.

Zhao, C; Deng, W; Gage, FH, 2008. Mechanisms and functional implications of adult neurogenesis. *Cell* 132, 645–660.

Zhao, X; Ueba, T; Christie, BR; Barkho, B; McConnell, MJ; *et al.*, 2003. Mice lacking methyl-CpG binding protein 1 have deficits in adult neurogenesis and hippocampal function. *Proc Natl Acad Sci USA* 100, 6777–6782.

Zhong, M; Zhang, Y; Lange, K; Fan, R, 2011. A cross population extended haplotype-based homozygosity score test to detect positive selection in genome-wide scans. *Stat Interface* 4, 51–63.

Ziller, MJ; Gu, H; Muller,F; Donaghey, J; Tsai, LT-Y; *et al.*, 2013. Charting a dynamic DNA methylation landscape of the human genome. *Nature* 500, 477–481.

Zlotogora, J, 1998. Germ line mosaicism. *Hum Genet* 102, 381–386.

Zong, C; Lu, S; Chapman, AR; Xie, XS, 2012. Genome-wide detection of single-nucleotide and copy-number variations of a single human cell. *Science* 338, 1622–1626.

Index